Microsoft Identity and Access Administrator SC-300 Exam Guide

Second Edition

Pass the SC-300 exam with confidence by using exam-focused resources

Aaron Guilmette

James Hardiman

Doug Haven

Dwayne Natwick

Microsoft Identity and Access Administrator SC-300 Exam Guide

Second Edition

Copyright © 2025 Packt Publishing

Authors: Aaron Guilmette, James Hardiman, Doug Haven, and Dwayne Natwick

Reviewer: Bart Van Vught

Relationship Lead: Anindya Sil

Content Engineer: David Sugarman

Production Designer: Salma Patel

Editorial Board: Vijin Boricha, Alex Mazonowicz, Aaron Nash, Gandhali Raut, and Ankita Thakur

First Published: March 2022

Second Edition: March 2025

Production Reference: 2040425

Published by Packt Publishing Ltd.
Grosvenor House
11 St Paul's Square
Birmingham
B3 1RB

ISBN 978-1-83620-039-0

www.packt.com

Contributors

About the Authors

Aaron Guilmette is a VP of technology focusing on dragging the defense industrial base into the modern technology area. Previously, he worked as a senior program manager for Microsoft 365 Customer Experience. His career spans 25 years across public and private sector technology consulting. As an author of over 15 other IT books you've probably seen recommended by Amazon, he specializes in identity, messaging, and automation technologies.

When he's not writing books or tools for his customers, trying to teach one of his kids to drive, or making tacos, Aaron can be found tinkering with cars and "investing" in *Star Wars* memorabilia. You can visit his blog at `https://aka.ms/aaronblog` or connect with him on LinkedIn at `https://www.linkedin.com/in/aaronguilmette`. Aaron resides in the metro Detroit area with his five children.

To Microsoft—thanks for constantly changing cloud technologies just enough to keep us on our toes and ensure we never run out of exams to take (or books to write).

To my kids—Liberty, Hudson, Glory, Anderson, and Victory—who remind me daily that troubleshooting technology problems is still easier than figuring out who left the empty pizza box in the fridge.

And to tacos, umbrella drinks, and fast cars—because if I ever do retire, I already know what paradise looks like.

– Aaron Guilmette

James Hardiman is a leading Identity and Access Management (IAM) expert with over two decades of experience in designing and implementing robust enterprise security solutions. His deep expertise encompasses identity governance, privileged access management, and cloud security, with a strong focus on ensuring compliance with evolving industry standards and regulations.

James holds a master of science degree in security studies with a concentration on cybersecurity from the University of Massachusetts and has earned numerous industry-recognized security certifications. Known for his ability to bridge the gap between technical complexity and business needs, James excels at aligning security architecture with strategic objectives. He prioritizes a user-centric approach, ensuring that security solutions enhance productivity while maintaining the highest levels of protection.

A resident of Massachusetts, James enjoys spending time with his wife and three children. He is passionate about driving innovation in the IAM field, particularly in safeguarding digital assets within highly regulated industries. Connect with James on LinkedIn: `https://www.linkedin.com/in/jameshardimanjr`.

Doug Haven is a security architect with over 20 years of experience in security engineering and architecture. His expertise spans policy development and security architecture across the manufacturing, government/DoD, transportation, financial services, and consulting industries. Throughout his career, Doug has worn many hats—including identity and access management architect, principal cloud security architect, network security architect, and enterprise security architect.

Doug earned a bachelor's degree in computer science and an AS in cybersecurity technology from Keiser University in Ft. Lauderdale, FL. Doug is also a Microsoft Certified Trainer (MCT) and holds the Microsoft Identity and Access Administrator certification. His industry-recognized certifications also include CISSP, CompTIA Security+, AWS Solution Architect Professional, and AWS Security Specialty.

Originally from New York, Doug now soaks up the sun in Florida with his wife and son. When he's not safeguarding digital realms, you might find him kayaking in the Everglades or playing guitar with his friends. To learn more about Doug's professional journey or to connect with him, visit his LinkedIn profile at `https://www.linkedin.com/in/doughaven/`.

I would like to extend my sincere appreciation to my family—my wife, Dawn, my son, Joshua, and my mother, Shirley—for their love and support.

I am also grateful to Aaron and my co-authors for their guidance throughout this project.

– Doug Haven

Dwayne Natwick is the CEO and principal architect of Captain Hyperscaler, a cloud and cybersecurity training company providing security and cloud certification training to IT professionals and organizations. He has been working in IT, security design, and architecture for over 30 years. His love of teaching led him to become a Microsoft Certified Trainer (MCT) Regional Lead and a Microsoft Most Valuable Professional (MVP). Dwayne has a master's degree in business IT from Walsh College, CISSP from ISC2, and 18 Microsoft certifications, including Identity and Access Administrator, Azure Security Engineer, and Microsoft 365 Security Administrator. Originally from Maryland, Dwayne currently resides in Michigan with his wife and three children.

Dwayne can be found providing and sharing information on social media, at industry conferences, on his blog site, and on his YouTube channel. You can follow Dwayne at `https://captainhyperscaler.com` or connect with him on LinkedIn at `https://www.linkedin.com/in/dnatwick`.

To my wife, Kristy, thank you for always being there and supporting me. You are the love of my life and my best friend. To my children, Austin, Jenna, and Aidan. Even with all my career accomplishments, you are what I am most proud of. You are all growing up to be such amazing people with kind hearts.

All four of you are my world and I could not make this journey without you.

All my love and support for everything that you do.

– Dwayne Natwick

About the Reviewer

Bart Van Vught has over 20 years of experience in helping organizations get the most out of their Microsoft products, both on and off the cloud. He is experienced in Microsoft 365 and Azure with a focus on modern security. Bart is a Microsoft Certified Trainer (MCT) and holds multiple certifications.

Table of Contents

2

Creating, Configuring, and Managing Microsoft Entra Identities   51

3

Implementing and Managing Identities for External Users and Tenants   87

4

Implementing and Managing Hybrid Identity   141

5

Planning, Implementing, and Managing
Microsoft Entra User Authentication   195

6

Planning, Implementing, and Managing
Microsoft Entra Conditional Access   249

7

Managing Risk Using Microsoft Entra ID Protection   271

8

Implementing Access Management for Azure Resources by Using Azure Roles   287

9

Implementing Global Secure Access   309

10

Planning and Implementing Identities for Applications and Azure Workloads   335

11

Planning, Implementing, and Monitoring the Integration of Enterprise Applications   355

12

Planning and Implementing App Registrations   387

13

Managing and Monitoring App Access Using
Microsoft Defender for Cloud Apps   407

14

Planning and Implementing Entitlement Management   441

15

Planning, Implementing, and Managing Access
Reviews in Microsoft Entra   461

16

Planning and Implementing Privileged Access   475

17

Monitoring Identity Activity Using Logs, Workbooks, and Reports   501

18

Planning and Implementing Microsoft Entra
Permissions Management   531

Preface

Identity and access management (**IAM**) is foundational for securing organizations. While it's long been an important operational concept, it was largely relegated to service desk operations for creating identities and assigning group memberships.

Several years ago, Microsoft started down the route of zero trust networking, with a foundational concept being identity as one of the key security boundaries.

Identity is increasingly used as a gateway to all of an organization's services—whether they're on-premises, in the Microsoft 365 cloud, or hosted by third-party cloud app providers. Being able to provision, manage, secure, and restrict the flow of data based on identity, device health or compliance, and access risk factors has become the preeminent way of protecting an organization's assets.

Throughout this book, we'll explore a broad range of IAM concepts, ranging from identity models such as cloud and hybrid identity to risk-based access and authorization policies. You'll discover cross-organization identity synchronization capabilities, design entitlement management strategies, and learn how to integrate third-party applications in a secure fashion—all of which will prepare you to successfully navigate the Microsoft Identity and Access Administrator exam.

Who This Book Is For

This book is intended for individuals who work in securing identity and workloads in a Microsoft 365 environment. This includes primarily identity and access administrators, but also security operations analysts and cybersecurity architects.

The content in this book assumes you have a basic understanding of Microsoft 365 concepts. You'll build on your foundational knowledge to gain the skills to be able to pass the SC-300 exam.

What This Book Covers

Chapter 1, Implementing and Configuring a Microsoft Entra Tenant, introduces the concepts of provisioning and administering a Microsoft 365 tenant.

Chapter 2, Creating, Configuring, and Managing Microsoft Entra Identities, walks through the steps necessary to create and modify users and groups, as well as assigning licenses.

Chapter 3, Implementing and Managing Identities for External Users and Tenants, introduces the concepts of external or guest users.

Chapter 4, Implementing and Managing Hybrid Identity, introduces hybrid identity models, managed through Entra Connect and Entra Connect Cloud Sync.

Chapter 5, Planning, Implementing, and Managing Microsoft Entra User Authentication, focuses on configuring authentication methods and capabilities such as multi-factor authentication and Windows Hello for Business.

Chapter 6, Planning, Implementing, and Managing Microsoft Entra Conditional Access, explores one of the core security technologies of Microsoft Entra—Conditional Access policies.

Chapter 7, Managing Risk by Using Microsoft Entra ID Protection, describes how to use risk-based policies with Microsoft Entra to protect user and workload identities.

Chapter 8, Implementing Access Management for Azure Resources by Using Azure Roles, demonstrates concepts such as role-based access controls and technologies such as Azure Key Vault.

Chapter 9, Implementing Global Secure Access, introduces the new Global Secure Access service for securing connectivity between endpoints and the Microsoft cloud.

Chapter 10, Planning and Implementing Identity for Applications and Azure Workloads, explains how to configure and secure identity for Azure services, applications, and workloads.

Chapter 11, Planning, Implementing, and Monitoring the Integration of Enterprise Applications, instructs on designing strategies for connecting and securing enterprise software-as-a-service apps, as well as using Microsoft Entra application proxy to publish on-premises applications.

Chapter 12, Planning and Implementing App Registrations, presents information on configuring app registrations, app roles, and API permissions.

Chapter 13, Managing and Monitoring App Access Using Microsoft Defender for Cloud Apps, explores using Defender for Cloud Apps to discover shadow IT and create policies to restrict access to apps and resources.

Chapter 14, Planning and Implementing Entitlement Management, leverages access packages for granting access to sites and applications, as well as managing the life cycle of external user accounts.

Chapter 15, Planning, Implementing, and Managing Access Reviews in Microsoft Entra, demonstrates how to use access reviews to manage the life cycle of access to resources in the Microsoft 365 tenant.

Chapter 16, Planning and Implementing Privileged Access, walks through deploying Privileged Identity Management to remove standing permissions to resources.

Chapter 17, Monitoring Identity Activity Using Logs, Workbooks, and Reports, outlines how to use the **Kusto Query Language** (KQL) to interpret logs and discover insights about identity and access in the Microsoft 365 environment.

Chapter 18, Planning and Implementing Microsoft Entra Permissions Management, introduces the new Entra Permissions Management product, allowing identity and access administrators to monitor permissions assignments and discover the over-granting of permissions in a Microsoft Entra environment.

How to Get the Most Out of This Book

This book is directly aligned with the Microsoft Certified: Identity and Access Administrator Associate exam and covers all the topics that an SC-300 aspirant needs to grasp in order to pass the exam.

It is advisable to stick to the following steps when preparing for the SC-300 exam:

Step 1: Read the complete book.

Step 2: Attempt the end-of-chapter practice questions in each chapter before moving on to the next one.

Step 3: Memorize key concepts using the flashcards on the website. (refer to the section *Online Practice Resources*)

Step 4: Attempt the online practice question sets. Make a note of the concepts you are weak in, revisit those in the book, and re-attempt the practice questions. (refer to the section *Online Practice Resources*)

Step 5: Review exam tips on the website. (refer to the section *Online Practice Resources*)

SC-300 aspirants will gain a lot of confidence if they approach their preparation as per the mentioned steps.

Online Practice Resources

With this book, you will unlock unlimited access to our online exam-prep platform (*Figure 0.1*). This is your place to practice everything you learn in the book.

> **How to access the resources**
>
> To learn how to access the online resources, refer to *Chapter 19, Accessing the Online Practice Resources* at the end of this book.

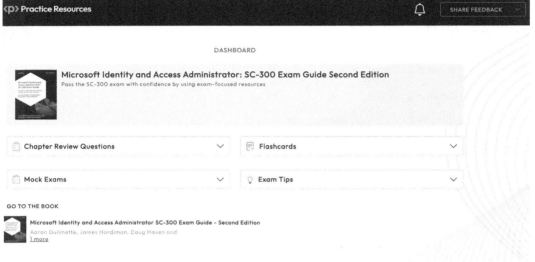

Figure 0.1 – Online exam-prep platform on a desktop device

Sharpen your knowledge of *Identity and Access Administrator* concepts with multiple sets of mock exams, interactive flashcards, and exam tips accessible from all modern web browsers.

To Make the Most Out of This Book

To get the most out of your studying experience, we recommend the following components:

- Azure tenant with a free-trial subscription (`https://azure.microsoft.com/en-us/free/ai-services/`)

- Microsoft 365 E5 trial subscription (`https://www.microsoft365.com`)

Download the Example Code Files

You can download the example code files for this book from GitHub at `https://github.com/PacktPublishing/Microsoft-Identity-and-Access-Administrator-SC-300-Exam-Guide`. If there's an update to the code, it will be updated in the GitHub repository.

We also have other code bundles from our rich catalog of books and videos available at `https://github.com/PacktPublishing/`. Check them out!

Download the Color Images

We also provide a PDF file that has color images of the screenshots and diagrams used in this book. You can download it here: `https://packt.link/L7Nt0`.

Conventions Used

There are a number of text conventions used throughout this book.

`Code in text`: Indicates code words in text, database table names, folder names, filenames, file extensions, pathnames, dummy URLs, user input, and X handles. Here is an example: "Microsoft will update your organization's SPF record with `v=spf1 include:spf.protection.outlook.com -all`."

A block of code is set as follows:

```
Get-MgUser -Filter "Department eq 'Project Management'"
-Top 10 -ConsistencyLevel Eventual -Property
DisplayName,UserPrincipalName,Department | Select
DisplayName,UserPrincipalName,Department
```

Any command-line input or output is written as follows:

```
Install-Module Microsoft.Graph
```

Bold: Indicates a new term, an important word, or words that you see onscreen. For instance, words in menus or dialog boxes appear in **bold**. Here is an example: "Roles can be easily managed within the Microsoft 365 admin center by expanding the navigation menu, expanding **Roles**, and then selecting **Role** assignments."

> **Tips or important notes**
> Appear like this.

Get in Touch

Feedback from our readers is always welcome.

General feedback: If you have questions about any aspect of this book, email us at customercare@packt.com and mention the book title in the subject of your message.

Errata: Although we have taken every care to ensure the accuracy of our content, mistakes do happen. If you have found a mistake in this book, we would be grateful if you would report this to us. Please visit www.packtpub.com/support/errata and fill in the form. We ensure that all valid errata are promptly updated in the GitHub repository at https://github.com/PacktPublishing/Microsoft-Identity-and-Access-Administrator-SC-300-Exam-Guide.

Piracy: If you come across any illegal copies of our works in any form on the internet, we would be grateful if you would provide us with the location address or website name. Please contact us at copyright@packt.com with a link to the material.

If you are interested in becoming an author: If there is a topic that you have expertise in and you are interested in either writing or contributing to a book, please visit authors.packtpub.com.

Share Your Thoughts

Once you've read *Microsoft Identity and Access Administrator SC-300 Exam Guide, Second Edition*, we'd love to hear your thoughts! Scan the QR code below to go straight to the Amazon review page for this book and share your feedback.

https://packt.link/r/1836200390

Your review is important to us and the tech community and will help us make sure we're delivering excellent quality content.

Download a Free PDF Copy of This Book

Thanks for purchasing this book!

Do you like to read on the go but are unable to carry your print books everywhere?

Is your eBook purchase not compatible with the device of your choice?

Don't worry, now with every Packt book you get a DRM-free PDF version of that book at no cost.

Read anywhere, any place, on any device. Search, copy, and paste code from your favorite technical books directly into your application.

The perks don't stop there, you can get exclusive access to discounts, newsletters, and great free content in your inbox daily.

Follow these simple steps to get the benefits:

1. Scan the QR code or visit the link below:

https://packt.link/free-ebook/9781836200390

2. Submit your proof of purchase.
3. That's it! We'll send your free PDF and other benefits to your email directly.

1

Implementing and Configuring a Microsoft Entra Tenant

The Microsoft 365 tenant serves as the primary boundary for security and content for your organization inside the Microsoft cloud, logically separating your organization's identities and data from that of other organizations also using the Microsoft 365 service. Although the initial setup of a tenant may seem straightforward—requiring just the input of contact information and payment details—the design and implementation of a tenant and its features involve multiple considerations to ensure secure access to an organization's data.

In this chapter, you will explore the essential elements of planning your Microsoft 365 experience, particularly as they map to the SC-300 exam. The objectives and skills covered in this chapter include the following:

- Configuring and managing built-in and custom Microsoft Entra roles
- Recommending when to use administrative units
- Configuring and managing administrative units
- Evaluating effective permissions for Microsoft Entra roles
- Configuring and managing custom domains
- Configuring company branding settings
- Configuring tenant-wide settings and properties

By the end of this chapter, you should be able to perform the initial configuration steps for a Microsoft 365 tenant and explain how to administer organization-wide settings.

Provisioning a Tenant

A **tenant**, from a Microsoft 365 perspective, is the top-level container that both identifies your organization and provides its security boundary. The tenant container object is a logical boundary that separates your organization's users, applications, and data from that of other organizations using the Microsoft 365 service. Creating a tenant is the prerequisite step to working with Microsoft 365.

While provisioning a tenant itself isn't on the SC-300 exam, you should be familiar with how the process works, as some of the choices you make up front may determine what features and capabilities your tenant will use.

Planning a Tenant

The first choice you need to make is which kind of tenant you'll acquire. Tenants are available for various different types of organizations. You'll choose a tenant based on a number of factors, including what size organization you have, as well as potentially what industry or vertical your organization is in.

Selecting a Tenant Type

Microsoft has made a variety of suites and packages available, targeting different types of organizations, as shown in *Figure 1.1*:

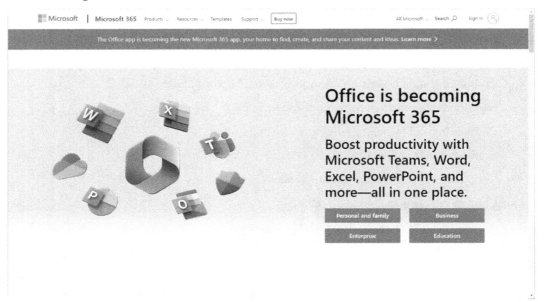

Figure 1.1: Types of tenants

Table 1.1 lists the types of tenants available and their target customers:

Tenant type	Target customer
Microsoft 365 Personal	Single person or home user
Microsoft 365 Family	Single person, up to 6 users
Microsoft 365 Business	Up to 300 users
Microsoft 365 Enterprise	Unlimited users
Microsoft 365 Government	Unlimited users
Microsoft 365 Education	Unlimited users

Table 1.1: Tenant types and target customers

The SC-300 exam tests you on the Microsoft 365 Enterprise plans and features available in the worldwide commercial cloud. The exam may question you about which tenant type is appropriate for your organization based on the organization's size.

> **Tenant type deep dive**
>
> The SC-300 exam focuses on the feature set and service bundles available in Microsoft 365 Enterprise plans, though the technologies available are largely the same across all plans. Microsoft 365 Government (also known as Government Community Cloud or GCC) is available only for local, state, and federal US government customers (and their partners or suppliers) and has a subset of the currently commercially available features. Microsoft 365 Education exists in the Worldwide Commercial cloud, has the same feature set as the commercial enterprise set, but also has a few added features targeted to educational institutions. Microsoft 365 for Education is only available to schools and universities.

Selecting a Managed Domain

After choosing what type of tenant you'll acquire, one of the next choices you'll need to make is selecting a tenant name. When you start a Microsoft 365 subscription, you are prompted to choose a name in Microsoft's `onmicrosoft.com` managed namespace. The tenant name must be unique across all other Microsoft 365 customers.

Tenant name considerations

After many (many!) years of customer requests, the **tenant-managed domain name** can be changed after it has been selected. Technically, you can't *change* the tenant domain, but you can add a *new* tenant **fallback domain**. As such, it's still important to choose something that is appropriate for your organization. The tenant name is visible in a handful of locations, so be sure to select a name that doesn't reveal any personally identifiable information or trade secrets and looks professionally appropriate for the type of organization you're representing. There is also a SharePoint tenant rename process in preview, but it's limited to organizations that have less than 10,000 sites provisioned. For more information, see `https://learn.microsoft.com/en-us/microsoft-365/admin/setup/add-or-replace-your-onmicrosoftcom-domain?view=o365-worldwide`.

Provisioning a Tenant

Provisioning a tenant is a relatively simple task requiring you to fill out a basic contact form and choose a tenant name. Microsoft offers a variety of trial subscriptions to help people understand the capabilities of the platform.

Trial information

Microsoft has updated its subscription plans by removing Teams from the included applications. The offers are now labeled **No Teams**, though Teams can be added through the Microsoft 365 admin center once the trial is activated. You can view available Microsoft 365 and Office 365 offers here: `https://www.microsoft.com/en-us/microsoft-365/enterprise/microsoft365-plans-and-pricing`.

Currently, available trial subscriptions require you to provide payment information. Trials will roll over as a fully paid subscription after the trial period ends. If you're standing up a trial tenant to study for the exam, you'll want to make sure you cancel it as soon as you're done using it. *Figure 1.2* shows the trial sign-up page:

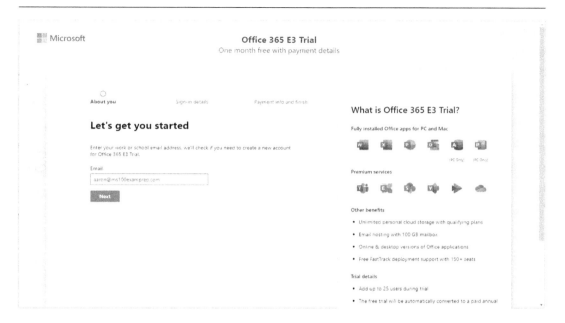

Figure 1.2: Starting a trial subscription

The sign-up process may prompt you for a phone number to be used during verification (either through a text/SMS or call) to help ensure that you're a valid potential customer and not an automated system.

After verifying your status as a human, you'll be prompted to select your managed domain, as shown in *Figure 1.3*:

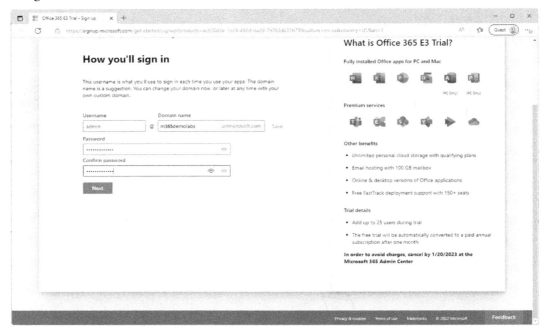

Figure 1.3: Choosing a managed domain

In the **Domain name** field, you'll be prompted to enter a domain name. If the domain name value you select is already taken, you'll receive an error and will be prompted to select a new name.

After you've finished, you can enter payment information for a trial subscription. Note the end date of the trial; if you fail to cancel by this date, you'll be automatically billed for the number of licenses you have configured during your trial!

Now that you've got a tenant activated, it's time to move on to the actual SC-300 objectives as discussed in the next section!

> **Advanced setup guides**
>
> While this book focuses primarily on the requirements for the SC-300 exam objectives, there is a lot to learn in a tenant. You can use the advanced deployment guides in the admin center to explore and set up other features of your Microsoft 365 environment. For more information, see `https://learn.microsoft.com/en-us/microsoft-365/enterprise/setup-guides-for-microsoft-365?view=o365-worldwide`.

Configuring and Managing Built-In and Custom Microsoft Entra Roles

Entra ID roles are used to delegate permissions to perform tasks in Entra ID, Microsoft 365, and Azure. Many people are familiar with the Global Administrator role, as it is the first role that's granted when you create a tenant. However, there are dozens of other roles available that can be used to provide a refined level of delegation throughout the environment. As the number of applications and services available in the Microsoft 365 ecosystem has grown, so has the number of security and administrative roles.

Roles for applications, services, and functions are intuitively named and generally split into two groups: *Administrator* and *Reader*. However, there are some roles that either don't follow that nomenclature or have additional levels of permission associated with them (such as *Printer Technician* or *Attack Simulator Payload Author*).

The Global Administrator role can administer all parts of the tenant organization, including creating and modifying users or groups and delegating other administrative roles. In most cases, users with the Global Administrator role can access and modify all parts of an individual Microsoft 365 service—for example, editing Exchange transport rules, creating SharePoint Online sites, or setting up directory synchronization. Some features, such as eDiscovery, require specific roles in order to use them. Even though the Global Administrator role doesn't have the ability to perform all tasks initially, the role does allow you to grant application- or workload-specific roles to enable their use.

> **Further reading**
>
> There are currently over 70 built-in administrative roles specific to Entra ID services and applications. For an up-to-date list of the roles available, see `https://learn.microsoft.com/en-us/azure/active-directory/roles/permissions-reference`.

For the SC-300 exam, you should be familiar with the core Microsoft 365 and Entra ID roles, as described in *Table 1.2*:

Role name	Role description
Global Administrator	Can manage all aspects of Entra ID and Microsoft 365 services.
Hybrid Identity Administrator	Can manage Entra Connect and Entra Cloud Sync configuration settings, including pass-through authentication (PTA), password hash synchronization (PHS), seamless single sign-on (SSO), and federation settings.
Billing Administrator	Can perform billing tasks such as updating payment information.

Role name	Role description
Compliance Administrator	Can read and manage the compliance configuration and reporting in Entra ID and Microsoft 365.
Exchange Administrator	Can manage all aspects of the Exchange Online service.
Guest Inviter	Can invite guest users regardless of the **Members can invite guests** setting.
Office Apps Administrator	Can manage Office apps, including policy and settings management.
Reports Reader	Can read sign-in and audit reports.
Security Reader	Can read security information and reports in Entra ID and Office 365.
SharePoint Administrator	Can manage all aspects of the SharePoint service.
Teams Administrator	Can manage all aspects of the Microsoft Teams service.
User Administrator	Can manage all aspects of users and groups, including resetting passwords for limited admins.

Table 1.2: Core Entra ID and Microsoft 365 roles

Planning for Role Assignments

One of the core tenets of security is the use of a least-privilege model. **Least privilege** means delegating the minimum level of permissions to accomplish a particular task, such as creating a user or resetting a password. In the context of Microsoft 365 and Entra ID, this translates to using the built-in roles for services, applications, and features where possible instead of granting the Global Administrator role. Limiting the administrative scope for services based on roles is commonly referred to as **role-based access control (RBAC)**.

In order to help organizations plan for a least-privileged deployment, Microsoft currently maintains a list of least-privileged roles necessary to accomplish certain tasks, grouped by application or content area: `https://learn.microsoft.com/en-us/azure/active-directory/roles/delegate-by-task`. Related tasks are grouped into roles. These roles can then be assigned to users based on their job duties.

When planning for role assignments in your organization, you can choose to assign roles directly to users or via a specially designated Entra ID group. If you have several users that need a variety of roles, you may want to create a group to ease the administrative burden of adding multiple users to multiple roles.

If you want to create and use groups for role assignment, you must enable the group for role assignment (the Entra ID **isAssignableToRole** property) during the group creation. For example, when using the Azure portal to create a group as shown in *Figure 1.4*, the **Azure AD roles can be assigned to the group** toggle needs to be set to **Yes** in order for the group to be provisioned with that capability.

> **Note**
>
> The role assignment property *cannot* be updated once the group has been created. If you create a group that you want to be used for role assignment and you fail to set this option during group creation, you'll need to delete the group and start over. This is to prevent privilege escalation attempts.

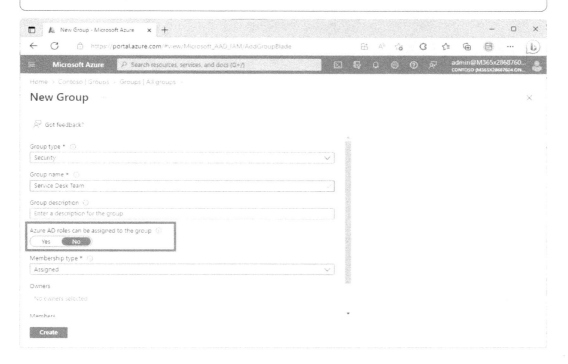

Figure 1.4: Configuring the isAssignableToRole property on a new group

If you want to create role-eligible groups in Entra ID, those groups must be configured to use assigned membership. As soon as you move the slider to enable a role-assignable group, the ability to change the membership type is grayed out to prevent accidentally elevating a user to a privileged role through a dynamic rule.

Managing Roles in the Microsoft 365 Admin Center

Roles can be easily managed within the Microsoft 365 admin center by expanding the navigation menu, expanding **Roles**, and then selecting **Role assignments**.

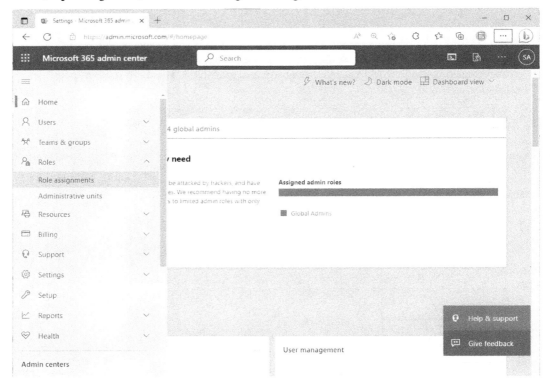

Figure 1.5: Role assignments

Roles are displayed across four tabs: **Azure AD**, **Exchange**, **Intune**, and **Billing**, as shown in *Figure 1.6*:

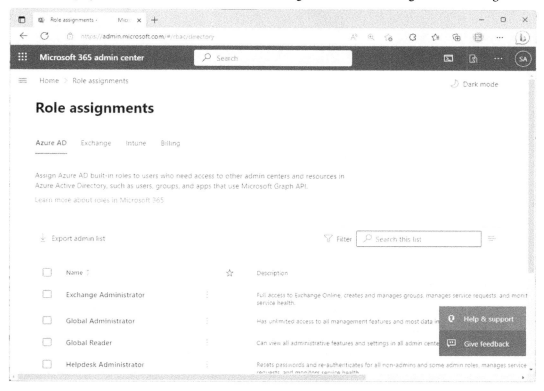

Figure 1.6: The Role assignments page

To add people to a role, simply select the role from the list, choose the **Assigned** tab, and then add either users or groups to the particular role.

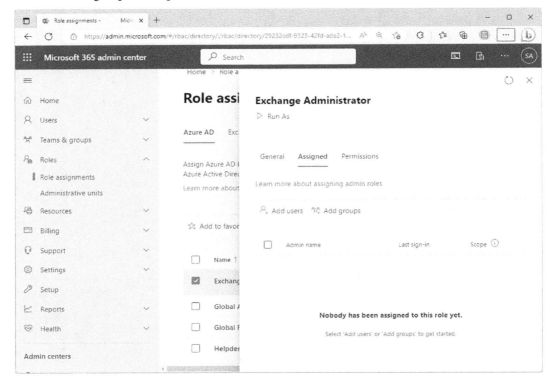

Figure 1.7: Making role assignments

Depending on the role being granted through this interface, you may be able to use Microsoft 365 groups, role-assignable security groups, or mail-enabled security groups.

Managing Role Groups for Microsoft Defender, Microsoft Purview, and Microsoft 365 Workloads

Now that you're familiar with role groups and concepts, you will learn how to manage roles for the following specific workload and feature areas of Microsoft 365:

- Microsoft Defender
- Microsoft Purview
- Microsoft 365 workloads

There are some nuances of managing each that are covered in the following sub-sections.

Microsoft Defender

Like other products in the Microsoft 365 suite, Defender uses roles to manage groups of permissions for tasks. All of the Microsoft Defender roles can be administered from either the Entra admin center (`https://entra.microsoft.com`) or the Azure portal (`https://portal.azure.com`). Both interfaces also provide the ability to define custom roles or role groups. Microsoft 365 Defender also has a new RBAC model available. The Microsoft 365 Defender RBAC model is in preview and is subject to change.

Microsoft 365 Defender users can be configured to use either the global Entra ID roles or custom roles from the Microsoft 365 Defender portal. When using Entra ID's global roles to assign permissions for Microsoft 365 Defender, it's important to note that the Entra ID roles will grant access to multiple workloads.

By default, Global Administrators and Security Administrators have access to Microsoft 365 Defender features. To delegate individual administrative duties where a broader Microsoft 365 Defender role might not be appropriate for your organization's needs, you can use custom roles, as shown in *Figure 1.8*:

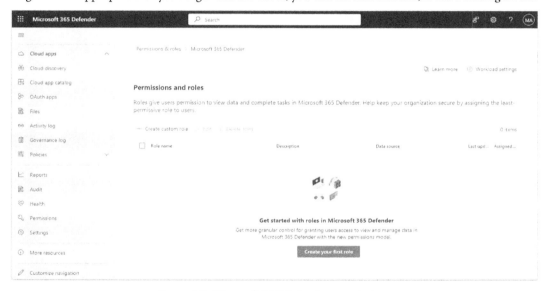

Figure 1.8: Microsoft 365 Defender permissions

To create a custom role, follow these steps:

1. Navigate to the Microsoft 365 Defender portal (`https://security.microsoft.com`) with an account that is either a member of Global Administrators or Security Administrators.

2. In the navigation menu, select **Permissions**.

3. Click **Create custom role**.

4. On the **Basics** page, enter a **Role name** value and click **Next**.

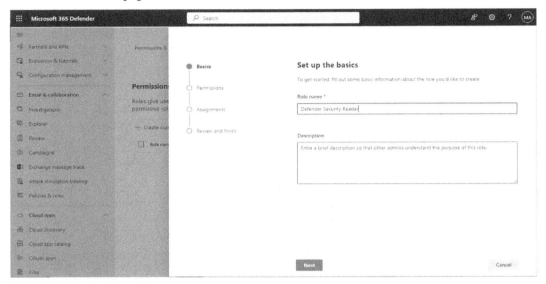

Figure 1.9: Creating a new custom role

5. Select permissions from the available permissions groups. For example, select **Security Operations**, then choose the **Select all read-only permissions** radio button as shown in *Figure 1.10*, and click **Apply**. Then, click **Next**.

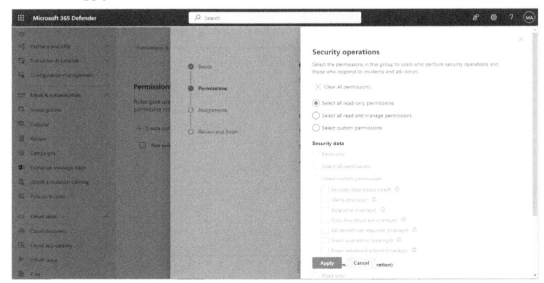

Figure 1.10: Selecting permissions

6. On the **Assignments** page, click **Add assignment**.

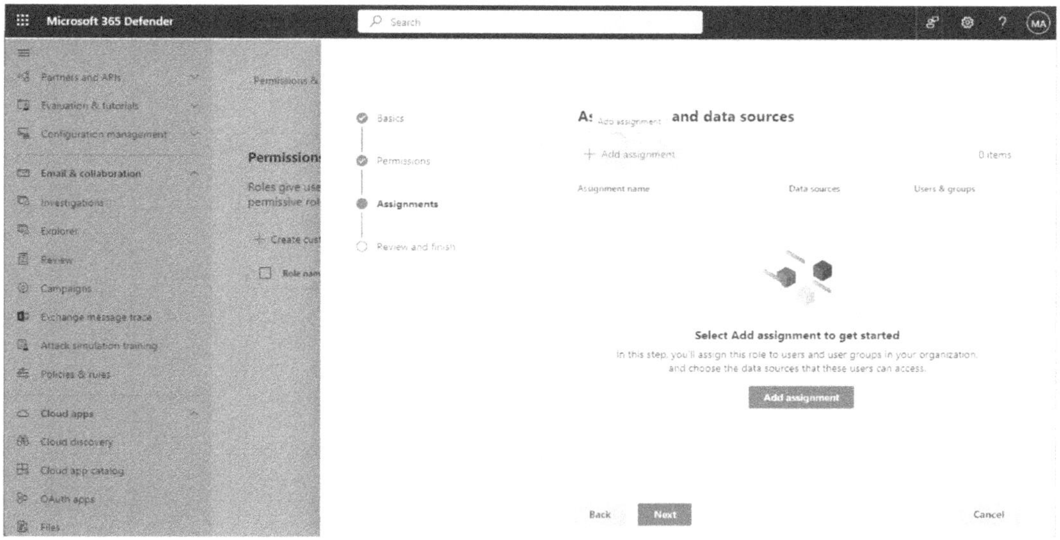

Figure 1.11: Adding user and data assignments

7. On the **Add assignment** page, enter an **Assignment name** value for this permissions assignment.

8. On the **Add assignment** page, select the data sources to which this assignment applies. You can select **Choose all data sources (including current and future supported data sources)** to make a broadly scoped role or select specific individual data sources.

9. On the **Add assignment** page, select which users or groups will be configured with this assignment, as shown in *Figure 1.12*. Click **Add** when finished.

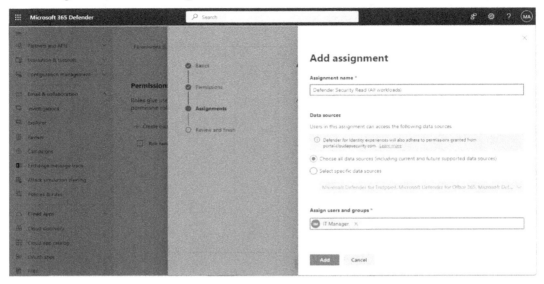

Figure 1.12: Selecting assignment options

10. Add any other assignments if necessary and then click **Next** to continue.

11. On the **Review and finish** page, confirm the selections and then click **Submit**. See *Figure 1.13*.

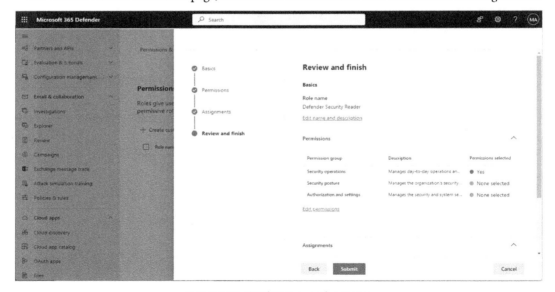

Figure 1.13: Confirming configuration

Once roles and assignments have been configured, users can log in and view or manage the features to which they've been granted permission.

> **Further reading**
>
> For more information on the nuances of the Microsoft 365 Defender custom roles and available permissions, see `https://learn.microsoft.com/en-us/microsoft-365/security/defender/custom-permissions-details`.

Next, you will explore the roles and permissions for Microsoft Purview.

Microsoft Purview

Like Microsoft 365 Defender, Microsoft Purview can leverage both Entra ID global roles (available throughout the Microsoft 365 platform) as well as roles and role groups specifically designed for Microsoft Purview that are only available in Microsoft Purview. Some features (such as eDiscovery) can only be configured using the Purview-specific roles.

You can view the global Entra ID roles by navigating to the Microsoft Purview compliance center, expanding **Roles & scopes**, selecting **Permissions**, and then selecting **Roles** under **Azure AD**. See *Figure 1.14*:

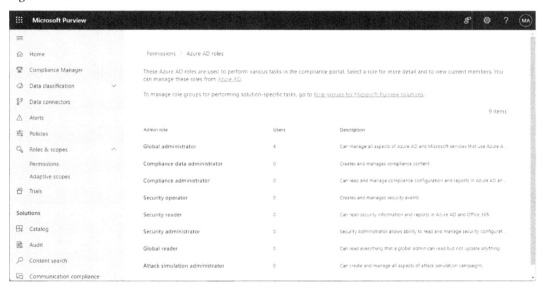

Figure 1.14: Azure AD roles in Microsoft Purview permissions

The Microsoft Purview-specific roles can be seen in the Microsoft Purview compliance center (`https://compliance.microsoft.com`) by expanding **Roles & scopes**, selecting **Permissions**, and then selecting **Roles** under **Microsoft Purview solutions**. See *Figure 1.15*:

Figure 1.15: Microsoft Purview solutions roles

Like Microsoft 365 Defender, you can also create custom role groups for Microsoft Purview solutions. Microsoft Purview roles also support scoping with administrative units. Currently, the following features support administrative units:

Solution or feature	Configuration areas
Data life cycle management	Retention policies, retention label policies, role groups
Data loss prevention (DLP)	DLP policies, role groups
Communications compliance	Adaptive scopes
Records management	Retention policies, retention label policies, adaptive scopes, role groups
Sensitivity labels	Sensitivity label policies, auto-labeling policies, role groups

Table 1.3: Microsoft Purview support for administrative units

Next, you will review role groups for Microsoft 365 workloads and how they can be managed.

Microsoft 365 Workloads

The core Microsoft 365 workloads, such as Exchange Online and SharePoint Online, have built-in support for a number of role groups. In the case of Exchange Online, there are additional management roles that can be assigned within the Exchange admin center's existing RBAC mechanisms. They're only visible inside the Exchange service and only apply to Exchange-specific features.

Figure 1.16: Microsoft 365 workload roles

While many workloads will have a single role group (such as Kaizala Administrator or SharePoint Administrator), some workloads such as Teams have multiple role groups that can be used to further delegate administration. You can review the current list of roles available in the Microsoft 365 admin center by navigating to the admin center (`https://admin.microsoft.com`), expanding **Roles**, and selecting **Role assignments**.

Next, we'll explore the role administrative units play in delegated administration.

Recommending When to Use Administrative Units

Administrative units are groups of users and devices that can be managed by specific administrators.

In an on-premises Active Directory setup, you can delegate administrative functions using the **Delegation of Control** wizard in Active Directory Users and Computers or Active Directory Administrative Center.

Unlike the hierarchical structure of on-premises Active Directory, Entra requires defining boundaries such as administrative units to delegate control. **Administrative units** are logical boundaries that can contain users, groups, and devices.

Administrative units in Entra can be role-scoped, allowing administrators to be granted specific roles (such as Helpdesk Administrator), thereby limiting their administrative capabilities to only the assigned administrative units.

Configuring and Managing Administrative Units

The easiest way to create and manage administrative units is through the Microsoft 365 admin center (though they can also be created and managed inside the Entra ID portal).

In this section, we'll explore how to create and manage administrative units.

Creating Administrative Units

In this example, you will create an administrative unit called **California** that will be used to manage users who live and work in that geographical region. During creation, you will configure administrators to be able to perform role-scoped activities inside that administrative unit:

1. Navigate to the Microsoft 365 admin center (`https://admin.microsoft.com`) and log in with a global administrator credential.

2. Expand **Roles** and click **Administrative units**. If you don't see **Roles** in the navigation menu, you may need to click **Show all** at the bottom of the menu to display all of the menu nodes (see *Figure 1.17*). Then, select +**Add unit**.

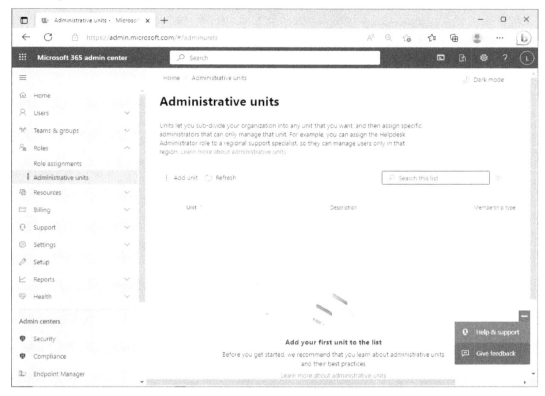

Figure 1.17: Administrative units page

3. On the **Basics** page, enter a **Name** value and a **Description** value for the administrative unit and click **Next**.

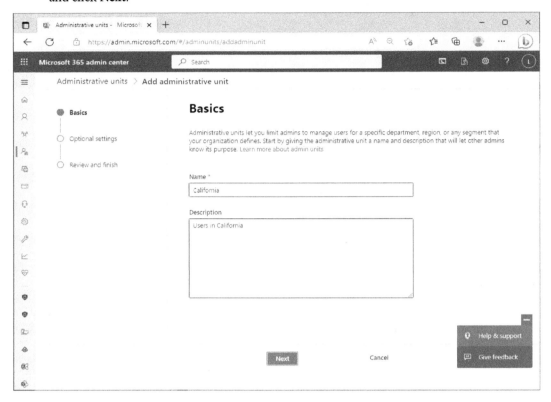

Figure 1.18: The Basics page

4. On the **Add members** sub-page, add any additional users to the administrative unit or click **Next** to proceed.

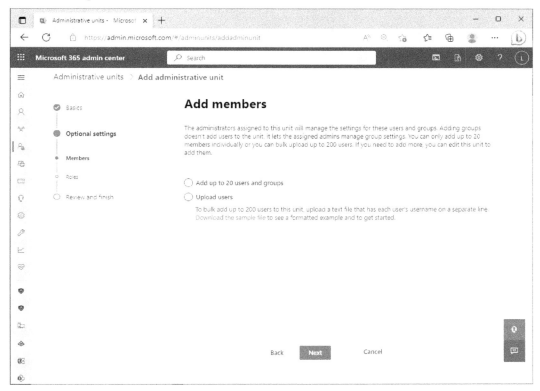

Figure 1.19: The Add members page

5. On the **Assign admins to scoped roles** page, review the roles listed. Not all roles can be scoped to administrative units. In this example, select the checkbox next to **User Administrator** and then click the role name itself to bring up its properties. See *Figure 1.20*.

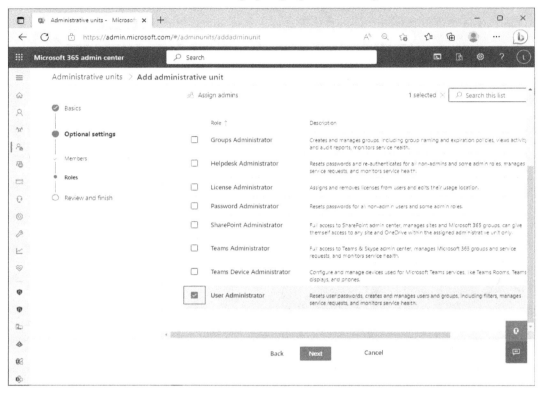

Figure 1.20: Adding roles

6. On the **User Administrator** flyout, click the **Assigned** tab.

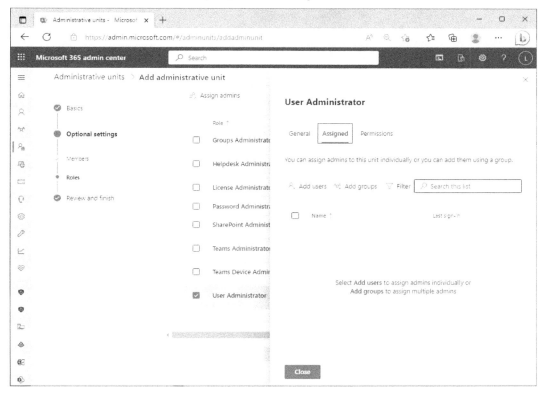

Figure 1.21: The User Administrator flyout

7.   Click **Add users** or **Add groups** to assign administrators to this role. Click **Close** when finished.

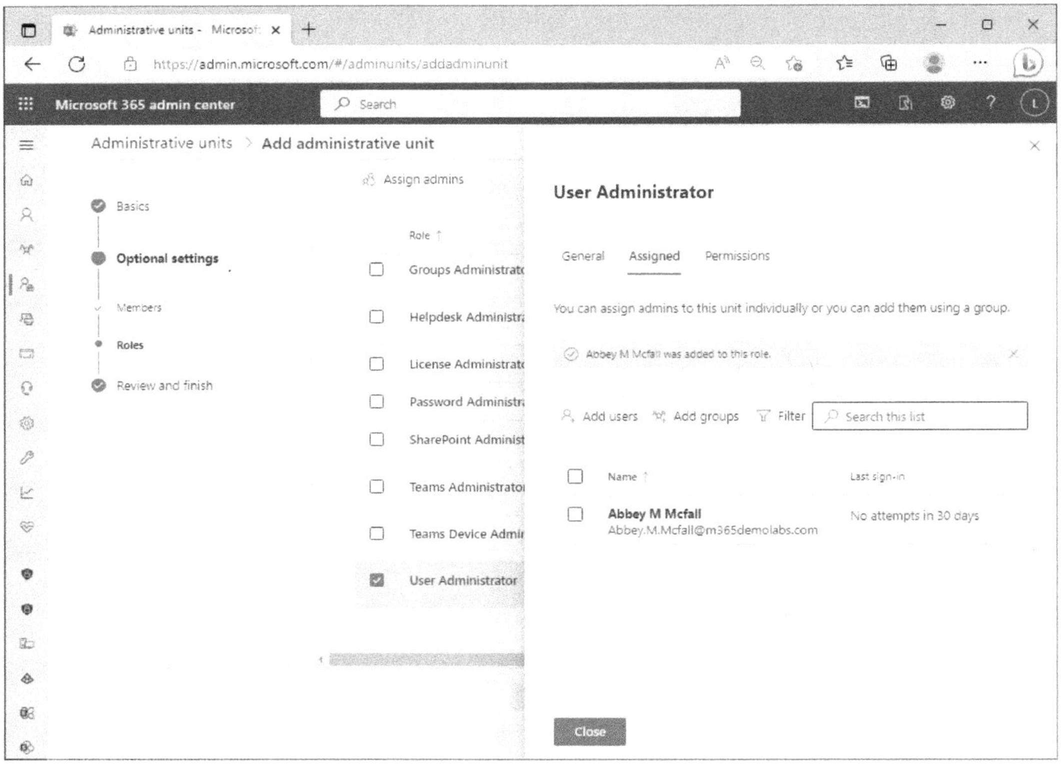

Figure 1.22: Adding users to a role

8.   On the **Assign admins to scoped roles** page, click **Next**.

9.   On the **Review and finish** page, review your selections and then click **Add**.

10. Click **Done** to return to the **Administrative units** page.

One of the features of role-scoped administration is the ability to limit what objects can be impacted by a particular administrator. As you noticed during the configuration, only a subset of the roles available in the tenant honor administrative unit scoping. You may want to periodically review your administrative unit configuration to see whether any additional scoped roles are available to be added to it. This will be discussed next.

Viewing and Updating Administrative Units

After you create administrative units, you can review them and modify their members and administrators from either the Azure AD portal or the Microsoft 365 admin center under **Roles | Administrative units**.

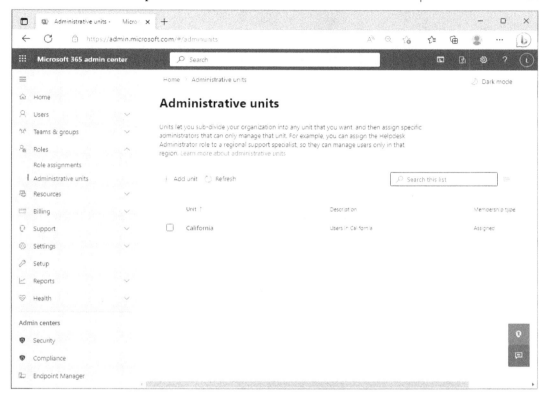

Figure 1.23: Viewing administrative units

By selecting a group, you can assign users and groups to the administrative unit.

While you can assign groups to administrative units, this does not automatically add the group member objects to the administrative scope—it only enables managing the properties of the group itself. You need to add the members of the group to the administrative unit separately in order for them to also be in scope.

> **Note**
>
> **Dynamic administrative units** are a preview feature that allows you to use filters and queries to automatically populate administrative units. Like dynamic groups, dynamic administrative units can only have one object type (either users or devices). Dynamic administrative units can only be configured in the Entra ID portal at this time. This feature is not available in GCC High currently.

When setting up administrative structures and delegation in your organization, ensure that you understand the limits of scoping controls. For example, if you assign an administrator to both an administrative unit and a role such as Exchange or SharePoint Administrator, they can modify users within their administrative unit. However, they might also be able to change application settings that impact all users across the entire tenant.

> **Note**
>
> Exchange Online features additional RBAC scoping controls to offer finer-grained administration delegation.

Next, we'll look at a few ways to retrieve permissions data for Microsoft Entra roles.

Evaluating Effective Permissions for Microsoft Entra Roles

When it comes to managing permissions and roles in Entra ID, it's important to understand that Entra role assignments are based on an additive model. This means that your effective permissions are the sum of all your role assignments.

You can explore the output of all role assignments (including privileged assignment escalations) in the Entra admin center (`https://entra.microsoft.com`) by expanding **Identity**, selecting **Roles & admins**, and then clicking **Download assignments**.

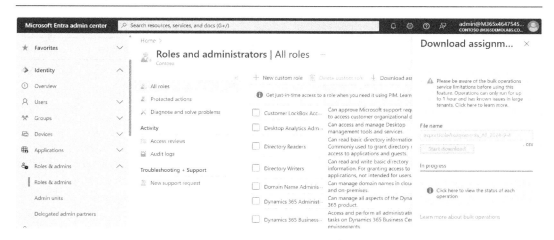

Figure 1.24: Downloading role assignment data

You can also explore the Entra admin center on a per-role basis and look for groups with memberships. The **Assignments** column only shows active roles, so it's recommended to periodically review them.

> **Further reading**
>
> The Microsoft 365 admin center and Entra admin center don't provide a great interface to be able to see all role assignments at a glance. To get this information, you'll have to resort to either PowerShell or the Microsoft Graph API. To make this task a little easier, you can use a tool such as Vasil Michev's role reporting script: `https://github.com/michevnew/PowerShell/blob/master/AADRolesInventory-Graph.ps1`.

Next, we'll shift gears to configuring a tenant to support custom (sometimes called *vanity*) domains.

Configuring and Managing Custom Domains

The managed domain you choose when provisioning a tenant remains integral to the Microsoft 365 tenant throughout its entire life cycle. It functions as a fully operational domain namespace, equipped with a Microsoft-managed publicly available domain name. However, most organizations prefer to use their own domain names for activities such as email communication and Microsoft Teams interactions.

> **Note**
>
> Custom **Domain Name System (DNS)** records *cannot* be added to the Microsoft-managed namespace.

Organizations can add any public domain name to their Microsoft 365 tenant. Microsoft supports the configuration of up to 5,000 domains within a single tenant. This includes both top-level domains (for example, `contoso.com`) and subdomains (for example, `businessunit1.contoso.com` or `businessunit2.contoso.com`).

Acquiring a Domain Name

Most organizations come to Microsoft 365 with existing domain names. Those domain names can easily be added to your tenant. In addition, you can purchase new domain names to be associated with your tenant.

Third-Party Registrar

Most large organizations have existing relationships with third-party domain registrars, such as Network Solutions or GoDaddy. You can use any ICANN-accredited registrar for your region to purchase domain names.

> **About ICANN**
>
> The **Internet Corporation for Assigned Names and Numbers** (**ICANN**) is a non-profit organization established in 1998 to provide guidance and policy for the internet's unique identifiers, including domain names. Before ICANN's formation, Network Solutions managed the global DNS registry under a subcontract from the United States Defense Information Systems Agency.

You can start your search for a domain with a registrar. A partial list of domain registrars is available here: `https://www.icann.org/en/accredited-registrars`.

Microsoft

Some organizations may wish to use Microsoft as the registrar. Depending on your subscription, you may be able to purchase domains from within the Microsoft 365 admin center, as shown in *Figure 1.25*:

Figure 1.25: Purchasing a domain through the Microsoft 365 admin center

When purchasing a domain through the Microsoft admin center, you may be able to purchase directly from Microsoft or may be redirected to a traditional domain registrar partner. Also, if you've purchased Microsoft 365 through a partner, you may be redirected to the partner's website, depending on their relationship with Microsoft. If purchasing directly from Microsoft, you can select from the following top-level domains:

- `.biz`
- `.com`
- `.info`
- `.me`
- `.mobi`
- `.net`
- `.tv`
- `.co.uk`
- `.org.uk`

Domain purchases are billed separately from your Microsoft 365 subscription services. When purchasing a domain from Microsoft, you'll have very limited ability to manage DNS records. If you require custom DNS record configuration (such as configuring a **mail exchanger** (**MX**) record to point to a third-party mail gateway), you'll want to purchase your domains separately.

Configuring a Domain Name

Configuring a domain for your tenant is straightforward and requires access to your organization's public DNS service provider. Some large organizations host and manage their own DNS, while others opt to use external service providers, such as domain registrars, to provide these services.

> **Tip**
> If you're unsure of where the DNS for your domain is hosted, you can use a service such as `https://www.whois.com`.

In order to be compatible with Microsoft 365, a DNS service must support configuring the following types of records:

- **Canonical Name (CNAME)**: CNAME records are alias records for a domain, allowing a name to point to another name as a reference. For example, let's say you build a site named www. contoso.com on a web server. That site resolves to an IP address of 1.2.3.4. Later, your organization decides to develop sites for each region and you build websites for na.contoso. com, eu.contoso.com, and ap.contoso.com on that same server. You might then implement a CNAME record for www.contoso.com to point to na.contoso.com.

- **Text (TXT)**: A TXT record is a DNS record used to store unstructured information. **Request for Comments (RFC)** 1035 (https://tools.ietf.org/html/rfc1035) specifies that the value must be text strings but gives no specific format for the data. Over the years, **Sender Policy Framework (SPF)**, **DomainKeys Identified Mail (DKIM)**, **Domain-Based Message Authentication, Reporting, and Conformance (DMARC)**, and other authentication and verification data have used specially crafted TXT records to hold data. The Microsoft 365 domain verification process requires the administrator to place a certain value in a TXT record to confirm ownership of the domain.

- **Service Location (SRV)**: An SRV record is used to specify a combination of a hostname in addition to a port for a particular internet protocol or service.

- **MX**: The MX record is used to identify which hosts (servers or other appliances, services, or endpoints) are responsible for processing mail for a domain.

In order to use a **custom domain** (sometimes referred to as a **vanity domain**) with Microsoft 365, you'll need to add it to your tenant.

To add a custom domain, follow these steps:

1. Navigate to the Microsoft 365 admin center (https://admin.microsoft.com) and log in.

2. Expand **Settings** and select **Domains**, as shown in *Figure 1.26*:

Figure 1.26: The Domains page of the Microsoft 365 admin center

3. Click **Add domain**.

4. On the **Add domain** page, enter the custom domain name you wish to add to your tenant. Select **Use this domain** to continue. See *Figure 1.27*.

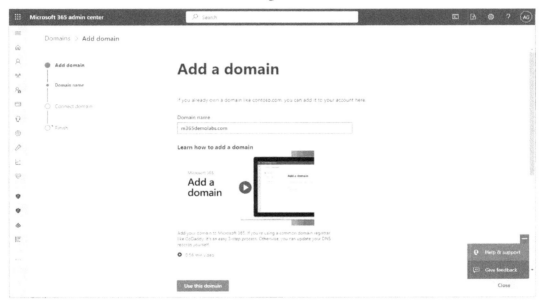

Figure 1.27: The Add a domain page

If your domain is registered at a host that supports **Domain Connect**, you can click **Verify** and then enter your registrar's credentials, as shown in *Figure 1.28*. Microsoft will automatically configure the necessary domain records on your behalf.

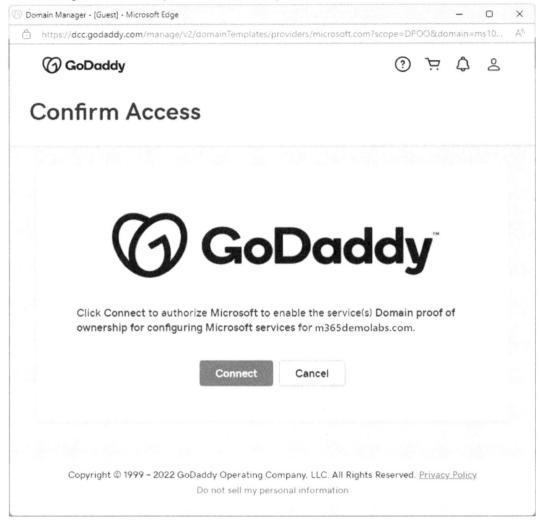

Figure 1.28: Authorizing Domain Connect with GoDaddy to update DNS records

You can also select **More options** to see all the potential verification methods available:

- If you are using a registrar that supports Domain Connect, you can enter the credentials for your registrar. When ready, click **Connect**.

- If you select **More options**, you will be presented with manual configuration choices. The default option (if your domain supports Domain Connect) will be to have the Microsoft 365 wizard update your organization's DNS records at the registrar. If you are going to be configuring advanced scenarios (such as Exchange Hybrid for mail coexistence and migration) or have other complex requirements, you may want to consider managing the DNS records manually or opting out of select services. If you choose to add your own domain records, you'll be presented with the values you need to configure.

- If you choose any of the additional verification options (such as **Add a TXT record to the domain's DNS records**), you'll need to manually add DNS records through your DNS service provider. Microsoft provides the values necessary for you to configure records with your own service provider. After configuring the entries with your service provider, you can come back to the wizard and select **Verify**, as shown in *Figure 1.29*:

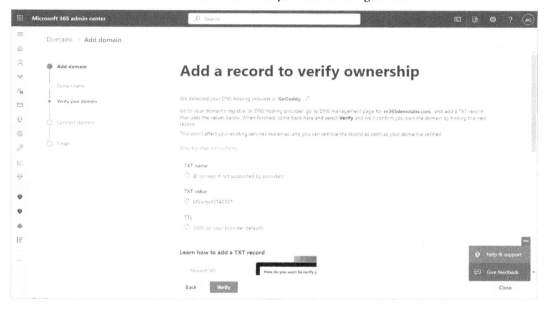

Figure 1.29: Completing verification records manually

If you are creating records manually, it may take anywhere from 10 minutes to 48 hours for the wizard to be able to detect the records.

5. After the domain has been verified, proceed to the **Connect domain** page. Depending on your choices and whether you're using a Domain Connect provider, you may have the option to apply the **Let Microsoft add your DNS records** setting to support your organization's services. If you choose the default option for letting Microsoft handle the records, you'll be presented with a series of choices. Each choice represents a service that Microsoft can configure. Click **Advanced options** (*Figure 1.30*) to expand the choices. The different options are described here:

- The first checkbox, **Exchange and Exchange Online Protection**, manages DNS settings for Outlook and email delivery. If you have an existing on-premises Exchange Server deployment (or another mail service solution), you should clear this checkbox before continuing as you'll need custom DNS settings. The default selected option means that Microsoft will make the following updates to your organization's DNS:

  - Your organization's MX record will be updated to point to Exchange Online Protection. If you have an existing mail service, this will break delivery to that service.

  - The Exchange Autodiscover record will be updated to point to `autodiscover.outlook.com`.

  - Microsoft will update your organization's SPF record with `v=spf1 include:spf.protection.outlook.com -all`.

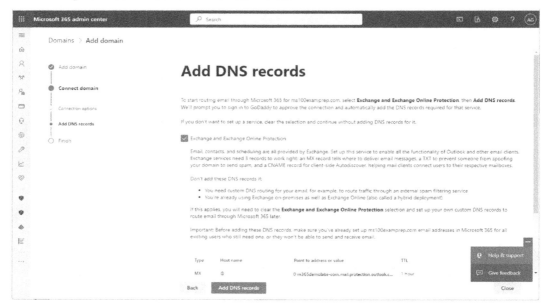

Figure 1.30: Adding DNS records

- The second setting, **Skype for Business**, will configure DNS settings for Skype for Business. If you have an existing Skype for Business Online deployment or you're using Skype for Business on-premises, you may need to clear this box until you verify your configuration to prevent external communication errors:

  - Microsoft will add two SRV records: `_sip._tls.@<domain>` and `_sipfederationtls._tcp@<domain>`.

  - Microsoft will also add two CNAMEs for Lync: `sip.<domain>` to point to `sipdir.online.lync.com` and `lyncdiscover.<domain>` to point to `webdir.online.lync.com`.

- The third checkbox, **Intune and Mobile Device Management for Microsoft 365**, configures applicable DNS settings for device registration. It is recommended to leave this enabled:

  - Microsoft will add the following CNAME entries to support mobile device registration and management: `enterpriseenrollment.<domain>` to `enterpriseenrollment.manage.microsoft.com` and `enterpriseregistration.<domain>` to `enterpriseregistration.windows.net`.

- The fourth option, DKIM, is not selected by default. If selected, Microsoft will add the following CNAME entries to support DKIM: `selector1._domainkey` to `selector1-<domain>._domainkey.<tenant.onmicrosoft.com>` and `selector2._domainkey` to `selector2-<domain>._domainkey.<tenant.onmicrosoft.com>`.

6. Click **Add DNS records**.

7. If prompted, select **Connect** to authorize Microsoft to update your registrar's DNS records. Click **Done** to exit the wizard.

You can continue adding as many domains as you need (up to the tenant maximum of 5,000 domains).

If you selected the DKIM option, you'll be presented with a notification that you'll need to go confirm the settings on the DKIM configuration page of the Microsoft 365 Defender portal (`https://security.microsoft.com/dkimv2`) once the DNS and service-side configuration changes have been completed.

If you attempt to enable the DKIM toggle before the configuration has been completed, you'll receive a dialog box instructing you to make the necessary changes (see *Figure 1.31*).

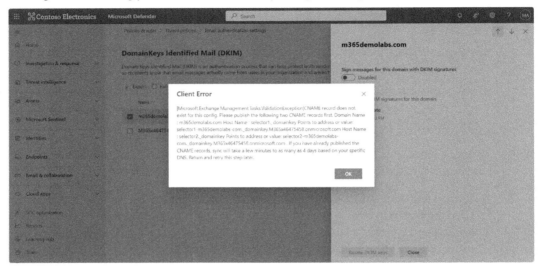

Figure 1.31: DKIM configuration error message

If you receive this message, verify that the DNS records have been added to your DNS host.

Adding a domain deep dive

To review alternative steps (such as configuration through PowerShell) or learn more information about the overall domain configuration process, see `https://learn.microsoft.com/en-us/microsoft-365/admin/setup/add-domain`.

Managing DNS Records Manually

If you've opted to add DNS records manually, you may need to go back to the Microsoft 365 admin center and view the settings. To do this, you can navigate to the **Domains** page in the Microsoft 365 admin center, select your domain, and then select **Manage DNS**:

Figure 1.32: Managing DNS settings for a domain

On the **Connect domain** page, click **More options** to expand the options, and then select **Add your own DNS records**. From here, you can view the specific DNS settings necessary for each service. You can also download a file that can be uploaded to your own DNS server.

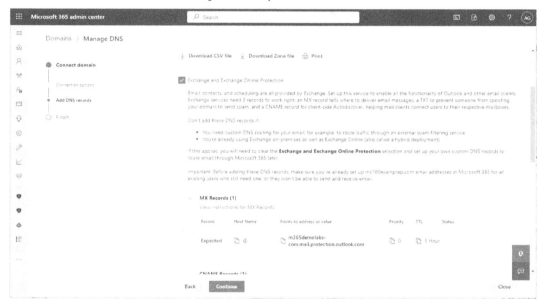

Figure 1.33: Viewing DNS settings

The CSV output is formatted as columns, while the zone file output is formatted for use with standard DNS services and can be imported into BIND or Microsoft DNS servers.

Configuring a Default Domain

After adding a domain, Microsoft 365 automatically sets the first custom domain as the default domain, which will be selected when creating new users. However, if you have additional domains, you may choose to select a different domain to be used as the default domain when creating objects.

To manage which domain will be set as your primary domain, select the domain from the **Domains** page and then click **Set as default** to make the change:

Figure 1.34: Setting the default domain

The default domain will be selected by default when creating cloud-based users and groups. You cannot set a federated domain (for example, one that is used with Active Directory Federation Service) as the default domain.

> **Custom domains and synchronization**
>
> When creating new cloud-based objects, you can select from any of the domains available in your tenant. However, when synchronizing users from an on-premises directory, objects will be configured with the domain that matches the on-premises object. If the corresponding domain hasn't been verified in the tenant, synchronized objects will be configured to use the tenant-managed domain.

Next, you will explore the core branding settings of a tenant.

Configuring Company Branding Settings

You can customize the Microsoft 365 experience for your users by setting custom company branding settings.

Microsoft 365 Admin Center

The Microsoft 365 admin center provides settings to customize the portal to your organization's color scheme. Themes can be set for the entire organization, as well as scoped to groups of users.

To access the theme settings, navigate to the Microsoft 365 admin center, expand **Settings**, select **Org settings**, and then click the **Organization profile** tab. You can select the **Custom themes** option to display the available choices, as shown in *Figure 1.35*:

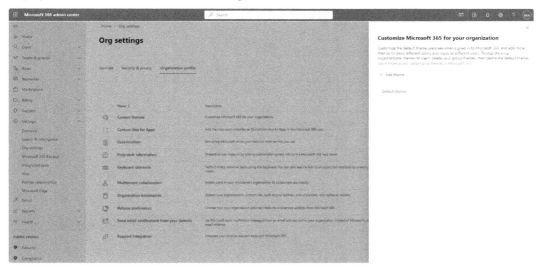

Figure 1.35: Viewing the available themes

You can click **Add theme** to create your own theme settings or edit the default theme. If you create a new theme, you can assign it to up to five Microsoft 365 groups.

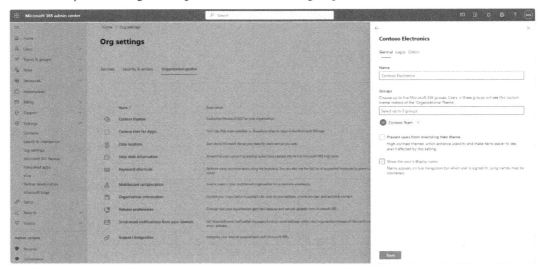

Figure 1.36: Creating a new theme

On the **Logos** tab, you can add a link to an image file that will be displayed next to the app launcher icon. The URL must be prefixed by https:// and will be scaled to 200 x 48 pixels.

On the **Colors** tab, you can select colors for the screen elements.

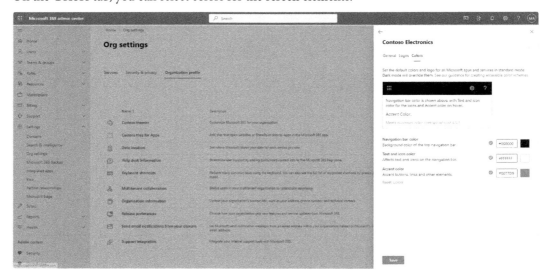

Figure 1.37: Configuring theme colors

Click **Save** when you are done creating or editing theme settings.

Microsoft Entra Admin Center

In addition to the branding settings available for the Microsoft 365 portal experience, you can also customize branding for the log-on experience. These settings are administered through the Microsoft Entra admin center (`https://entra.microsoft.com`) under **Identity | Company branding**. See *Figure 1.38*.

Figure 1.38: Configuring Microsoft Entra company branding settings

These settings will appear whenever a user is prompted to sign in to Microsoft 365. On this page, you can customize colors and background logos for the sign-in page, as well as specialized header and footer content, your own **Cascading Style Sheets** (**CSS**), and custom text for various notices. Many organizations use this area to provide information for service assistance, privacy terms, and terms of use.

Configuring Tenant-Wide Settings and Properties

In addition to branding, there are several other common settings that can be configured and impact the availability of features and services across all users.

Services

The **Services** tab of the **Org settings** window is used to configure broad tenant-wide features. Some of these may include actual services (such as Bookings or Teams), while others may impact how features are used (such as Microsoft 365 groups).

There are currently 40 different options on this page. Most of the settings are very high-level or coarse controls; fine-tuning configurations will require navigating to the appropriate service or workload's administrative interfaces.

Table 1.4 lists the features that can be configured on the **Services** page.

Option	Description
Account Linking	Allow users to connect Microsoft Entra ID to their Microsoft account to earn Bing rewards.
Adoption Score	Manage the selection criteria for whose data will be included in the adoption score as well as high-level filtering options.
Azure Speech Services	Allow Azure to process emails and documents to improve speech recognition accuracy.
Bookings	Enable or disable bookings and select **Bookings** features organization-wide.
Brand center	Enable Microsoft 365 to set up a SharePoint site used to store branding assets for your tenant customization.
Calendar	Manage high-level calendar sharing options.
Copilot for Sales	Enable Copilot for Sales features.
Cortana	Allow Cortana features to access Microsoft 365 data.
Directory synchronization	Link to download the Entra Connect synchronization tool.
Dynamics 365 Applications	Allow Dynamics 365 to generate insights data from Microsoft 365 users. Disabled by default.
Dynamics 365 Customer Voice	Configure Dynamics 365 Customer Voice to gather survey information.
Dynamics CRM	This links to the **Dynamics CRM** settings (Power Platform admin center).
Mail	This links to a page to manage various Exchange mail tasks, such as audit reports, message trace reports, and spam filtering policies.
Microsoft 365 Groups	Manage features of Microsoft 365 groups, such as allowing guests.
Microsoft 365 installation options	Select the channel for Microsoft 365 app updates.

Option	Description
Microsoft 365 Lighthouse	Enable management of tenants by a partner through Microsoft 365 Lighthouse.
Microsoft 365 on the web	Allow integration of Microsoft 365 files with third-party cloud services. Enabled by default.
Microsoft Azure Information Protection	Link to **Azure Information Protection (AIP)** configuration settings (Azure Rights Management).
Microsoft communication to users	Choose whether or not users receive emails from Microsoft about the services they're licensed for.
Microsoft Edge site lists	Manage lists of sites that should be opened in Edge versus legacy Internet Explorer mode.
Microsoft Forms	Manage sharing and external content options for Microsoft Forms.
Microsoft Graph Data Connect	Manage the use of Graph Data Connect with Azure Data Factory and Azure Synapse Analytics. If enabled, you can choose to separately allow dataset access for Viva Insights and SharePoint Online/OneDrive for Business.
Microsoft Loop	Enable access to Loop workspaces.
Microsoft Planner	Enable iCalendar access for Planner.
Microsoft Search on Bing homepage	Manage curated items to show up on Bing's home page for users.
Microsoft Teams	Enable automatic enablement of Teams, and configure guest access enablement.
Microsoft To Do	Enable push notifications for To Do.
Microsoft Viva Insights	Enable Insights features for users.
Modern authentication	Enable modern authentication and SMTP authentication for Exchange Online.
Multi-factor authentication	Link to administer legacy per-user multi-factor authentication settings.
News	Configure **News** settings for the organization. **News** is a feature of the Bing/Edge product that displays content related to the organization's industry and selected keywords.
Office Scripts	Enable Office Scripts within the Microsoft 365 tenant.

Option	Description
Reports	Manage privacy settings for reports displayed in the admin center and make usage analytics report data available to Power BI.
Search & intelligence usage analytics	Configure Microsoft search settings and digest email.
SharePoint	Manage high-level external sharing settings with a link to the SharePoint admin center for advanced sharing settings.
Sway	Enable external sharing and content integration options for Sway.
User owned apps and services	Grant users access to the Office Store as well as the ability to start trials and auto-claim licenses.
Viva Learning	Configure privacy and diagnostic data settings for Viva Learning.
What's new in Microsoft 365	Temporarily deprecated; redirects to Microsoft Learn.
Whiteboard	Enable the Whiteboard app in Microsoft Teams.

Table 1.4: Organization-wide services and features in Microsoft 365

While you won't be required to memorize this table for the exam, you should at least be broadly familiar with the items. As a rule, services and features are enabled through licenses—the only exception at this time is Bookings (controlled through this **Services** page toggle).

Security and Privacy

The security and privacy options are geared toward high-level administration or global settings of your tenant, not specifically toward content or data security. For example, Customer Lockbox doesn't have configurable features outside of enabling it based on licensing.

Table 1.5 depicts the items currently configurable on the **Security & privacy** page.

Option	Description
Customer Lockbox	Enable Customer Lockbox (if licensed). Custom Lockbox enables a workflow for authorizing support personnel's access to the tenant during incidents.
Help & support query collection	Choose whether Microsoft can collect information on the support requests made in the admin center.
Idle session timeout	Sign users out of Microsoft 365 web apps after a period of inactivity. This does not impact mobile or desktop client applications. This setting overrides session timeout policies configured in the Outlook web app and SharePoint.
Microsoft Graph Data Connect applications	Manage apps that use Microsoft Graph Data Connect.
Password expiration policy	Enable the **Password Never Expire** option for accounts. Does not affect synchronized accounts.
Privacy profile	Configure a privacy link and contact for your organization. This link is shown during the Teams meeting join experience.
Privileged access	Enable privileged access at the task level (separate from Entra Privileged Identity Management).
Pronouns	Enable users to display personal pronoun information in Microsoft 365 apps such as Teams and Outlook.
Self-service password reset	Link to the Azure portal to configure self-service password reset.
Sharing	Enable users to invite guests.

Table 1.5: Security and privacy settings

While you won't be required to memorize this table, you should at least be broadly familiar with the items available—primarily enabling Customer Lockbox and setting the tenant-wide idle session timeout and the tenant-wide setting for non-expiring passwords.

Organization Profile

The **Organization profile** settings (**Microsoft 365 admin center** | **Settings** | **Org settings**) are largely informational or used to manage certain tenant-wide aspects of the user experience. On this tab, you'll find the following settings:

Setting	Description
Custom themes	Create and apply themes to the Microsoft 365 portal for end users. You can also mandate specific themes, organization logos, and colors.
Custom tiles for apps	Configure additional tiles to display on the Microsoft 365 app launcher (sometimes referred to as the Launcher or the Waffle).
Data location	View the regional information where your tenants' data is stored.
Help desk information	Choose whether custom help desk support information for end users needs to be added to the Office 365 help pane.
Keyboard shortcuts	View the shortcuts available for use in the Microsoft 365 admin center.
Multitenant collaboration	These settings enable you to link multiple tenants together. Multitenant organizations support cross-tenant user synchronization and provisioning (through Microsoft Entra B2B collaboration).
Organization information	Update your organization's name and other contact information.
Release preferences	Choose the release settings for Office 365 features (excluding Microsoft 365 Apps). The available options are **Standard release for everyone**, **Targeted release for everyone**, and **Targeted release for select users**. The default setting is **Standard release for everyone**.
Send email notifications from your domain	Send system notification messages from an address linked to one of your verified domains instead of a Microsoft external address.
Support integration	Configure integration with third-party support tools, such as ServiceNow.

Table 1.6: Organization profile settings

Like the other **Organization settings** tabs, the settings on this page will be used infrequently—typically when just setting up your tenant and customizing the experience. As with the other **Organization profile** setting areas, you should spend some time in a test environment navigating the tenant to view these settings and updating them to see their effects.

Summary

In this chapter, you took your first steps on the journey to achieving the SC-300 certification! You learned about the core requirements and considerations for establishing a Microsoft 365 tenant, such as selecting a tenant name, adding a domain, and configuring basic role delegation features such as administrative units and roles.

In *Chapter 2*, you will begin exploring the types of identity in an Entra environment.

Exam Readiness Drill – Chapter Review Questions

Apart from mastering key concepts, strong test-taking skills under time pressure are essential for acing your certification exam. That's why developing these abilities early in your learning journey is critical.

Exam readiness drills, using the free online practice resources provided with this book, help you progressively improve your time management and test-taking skills while reinforcing the key concepts you've learned.

HOW TO GET STARTED

- Open the link or scan the QR code at the bottom of this page
- If you have unlocked the practice resources already, log in to your registered account. If you haven't, follow the instructions in *Chapter 19* and come back to this page.
- Once you log in, click the START button to start a quiz
- We recommend attempting a quiz multiple times till you're able to answer most of the questions correctly and well within the time limit.
- You can use the following practice template to help you plan your attempts:

Working On Accuracy		
Attempt	Target	Time Limit
Attempt 1	40% or more	Till the timer runs out
Attempt 2	60% or more	Till the timer runs out
Attempt 3	75% or more	Till the timer runs out
Working On Timing		
Attempt 4	75% or more	1 minute before time limit
Attempt 5	75% or more	2 minutes before time limit
Attempt 6	75% or more	3 minutes before time limit

The above drill is just an example. Design your drills based on your own goals and make the most out of the online quizzes accompanying this book.

First time accessing the online resources? 🔒

You'll need to unlock them through a one-time process. **Head to** *Chapter 19* **for instructions.**

Open Quiz

https://packt.link/sc300ch1

OR scan this QR code →

2

Creating, Configuring, and Managing Microsoft Entra Identities

Identity is one of the core pillars of the **Zero Trust** model. While you might not explicitly see Zero Trust mentioned by name on the SC-300 exam, rest assured that its principles such as **assume breach** and **least privilege** are used throughout. The Zero Trust model is critical to ensuring your environment is ready to face the latest security challenges.

> **Further Reading**
>
> If you need to brush up on Microsoft's Zero Trust concepts and terminology, see https://learn.microsoft.com/en-us/security/zero-trust/zero-trust-overview.

Entra identity encompasses many types of objects, including cloud-only user accounts, synchronized user accounts, and external guest accounts, along with groups, devices, and contacts. Entra identity also includes various types of service principals (such as managed identities). Even app registrations are a part of the identity framework. Each type of identity serves a specific purpose, making some more suitable for certain business cases than others. In this chapter, we will cover the following topics in alignment with the SC-300 exam objectives:

- Creating, configuring, and managing users
- Creating, configuring, and managing groups
- Managing custom security attributes
- Automating the management of users and groups by using PowerShell
- Managing licenses
- Assigning, modifying, and reporting on licenses

By the end of this chapter, you should be able to create and manage users, groups, and attributes (along with the requisite licensing needed to enable features).

Creating, Configuring, and Managing Users

Creating and managing users is a key part of running any information system, whether it's a small network application, a large enterprise directory, or a cloud service from a SaaS provider. Identities—used by people, apps, and devices—help authenticate and enable activities.

In Entra ID, there are three main types of user identity:

- Cloud-based users
- Synchronized users
- Guest users

When planning identity scenarios, it's crucial to know the benefits, features, drawbacks, and capabilities of each type, along with their authentication schemes. This includes factors such as ease of setup, integration with existing directories or security products, on-premises infrastructure requirements, and network availability.

In this section, we'll explore how to manage each type of user.

Creating and Managing Cloud Users

From an Entra ID standpoint, cloud users are the simplest to understand and manage. When you set up an Entra ID or Microsoft 365 tenant, one of the first steps is creating the initial Global Administrator user identity (`user@tenant.onmicrosoft.com`). This identity is stored in the Entra ID directory for your tenant. Cloud users are those whose primary identity source is Entra ID.

The initial `onmicrosoft.com` domain (or tenant domain) is always cloud-only since Entra ID is the authority for it. When adding new domains to a tenant, they are initially managed domains, meaning Entra ID manages the identity store.

One advantage of cloud-only users is the lack of dependency on other infrastructure or identity services. For many small organizations, cloud-only identity is ideal as it requires no hardware or software investment beyond the Microsoft 365 subscription. However, a downside is the lack of integration with on-premises directory solutions and applications.

> **Tip**
> Microsoft recommends keeping at least one cloud-only account as a best practice in case access to any on-premises environment is lost.

The easiest way to create cloud users is through the Microsoft 365 admin center (`https://admin.microsoft.com`). To set up a user, go to **Users**, select **Active users**, and select **Add a user**. The wizard will guide you through configuring the account, as shown in *Figure 2.1*:

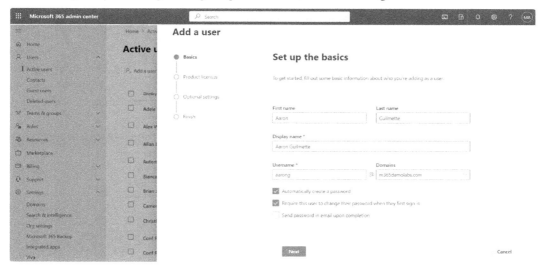

Figure 2.1: Adding a new user

By following the **Add a user** wizard, you can configure basic properties for a user as well as location and license assignment, as shown in *Figure 2.2*:

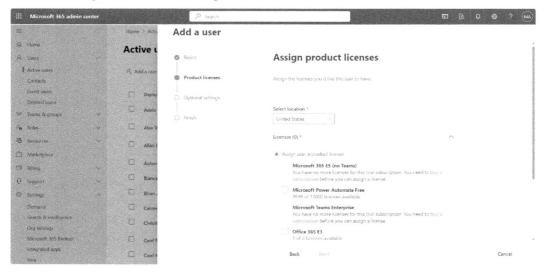

Figure 2.2: Configuring user options

You can configure optional settings as well, such as security roles, job titles and departments, addresses, and phone numbers.

You can also add users through the Entra admin center (`https://entra.microsoft.com`). The Entra admin center is structured quite a bit differently from the Microsoft 365 admin center, mainly because it's based on the Azure portal and exposes admin interfaces for a variety of resources and services across the Azure landscape, such as virtual machines and compute services. There are several differences in managing users and objects between the two interfaces. The Microsoft 365 admin center offers a simpler menu-driven experience, guiding administrators to configure common options and features during the provisioning process, while the Entra admin center provides many more configurable selections.

Figure 2.3: Microsoft Entra admin center

Most organizations that choose a cloud-only identity perspective will likely provision objects in the Microsoft 365 admin center due to its simplicity.

Creating and Managing Synchronized Users

As discussed in *Chapter 4, Implementing and Managing Hybrid Identity*, identity synchronization replicates your on-premises identity to Entra ID. Whether you're using Entra Connect Sync, Entra Cloud Sync, or a third-party tool, the process is similar: an on-premises agent or service connects to both Active Directory and Entra ID, reads the objects from Active Directory, and creates corresponding objects in Entra ID.

During this process, the on-premises and cloud objects are linked by a unique, immutable attribute that remains constant throughout the object's lifecycle.

Deep Dive

When Microsoft first debuted its directory synchronization software appliance, DirSync, an on-premises object was linked to its corresponding cloud object by converting the on-premises object's `objectGUID` attribute value to a Base64 string, which was then stored in the cloud object's `ImmutableID` attribute. In modern versions of Entra Connect, this process has evolved to also use the `ms-DS-ConsistencyGuid` attribute. By default, the `ms-DS-ConsistencyGuid` attribute in Active Directory is blank. When Entra Connect is set up to use `ms-DS-ConsistencyGuid` as the source anchor, an object's `objectGUID` value is copied to its corresponding `ms-DS-ConsistencyGuid` attribute. Since a new `objectGUID` is generated each time an object is created, using a static value such as `ms-DS-ConsistencyGuid` helps organizations maintain the relationship between identities during Active Directory domain migrations, which often occur during business mergers, acquisitions, and divestitures.

Guest User Accounts

Guest users are special accounts with limited rights to create and view content in the Entra environment, typically referred to as **Entra Business-to-Business** (**B2B**) identities (formerly known as Azure Business-to-Business identities).

Entra B2B guest accounts are usually created through an invitation process. This might involve inviting someone from an external organization to join a Microsoft SharePoint site, collaborate on a OneDrive document, or access files in a Teams channel. When an invitation is sent, an Entra identity object is created in the inviting organization's tenant, and an invitation email is sent to the external recipient.

Upon receiving the invitation, the recipient clicks the link in the email and is directed to an Azure sign-in flow. This prompts them to enter credentials corresponding to their own identity source. This could be another Entra ID or Microsoft 365 tenant, a consumer account (such as Microsoft, Google, or Facebook), or another third-party issuer using a SAML/WS-Fed-based identity provider. The process of the recipient accepting the invitation is known as redemption.

Further Reading

Although guests are typically added through an invitation process, the new Entra ID **cross-tenant synchronization** feature allows for the automated provisioning of guest objects between trusted tenants. This feature operates similarly to internal directory synchronization. Microsoft recommends using this feature only for Entra ID tenants that belong to the same organization. For more information on cross-tenant synchronization, see `https://learn.microsoft.com/en-us/azure/active-directory/multi-tenant-organizations/cross-tenant-synchronization-overview`.

While guest users can be viewed and manipulated in the Microsoft 365 admin center, they can only be provisioned through either the Microsoft Azure portal (`https://portal.azure.com`) or the Microsoft Entra admin center (`https://entra.microsoft.com`). Selecting **Add a guest user** in the Microsoft 365 admin center transfers you over to the Azure portal to complete the invitation process.

After logging in to either the Entra admin center or the Azure portal, you can begin the process of inviting guests. The process is a little different based on which portal you use, but for the most part, it is the same. When creating a guest user from the Entra admin center, for example, you'll be able to select whether the user is a **Member** (user) type or **Guest** (external user) type using a dropdown. See *Figure 2.4*:

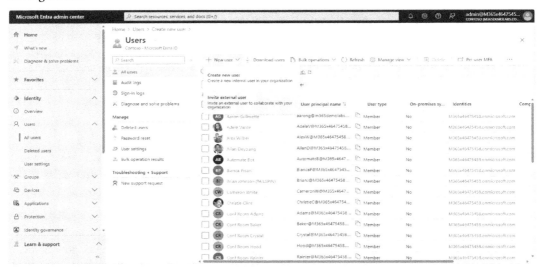

Figure 2.4: Choosing which type of user to create

The user interface elements for inviting a guest user are similar to those for creating a new cloud user. The main differences are in the selection of the external user or guest versus user or member template and, in the case of a guest user, you'll have the opportunity to supply message content (which will be included as part of the email invitation sent). See *Figure 2.5*:

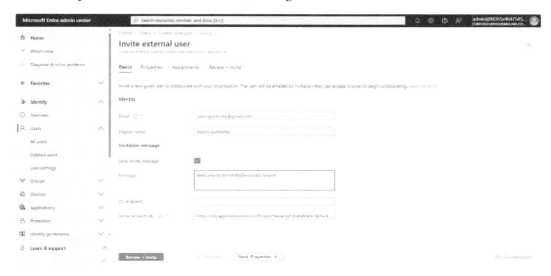

Figure 2.5: The Invite external user or guest wizard

Once a guest has been invited, take note of the properties:

- The guest identity's **User principal name** value is formatted as **emailalias_domain.com#EXT#@{tenant}.onmicrosoft.com**

- The **Identities** property is set to **{tenant}.onmicrosoft.com**

- The user type is set to **Guest**

- The invitation state is set to **Pending acceptance**

See *Figure 2.6* for reference:

Figure 2.6: Guest user invitation details

Upon receiving and accepting the invitation, the recipient is prompted to accept the terms of the invitation and grant permissions:

- Receive profile data including name, email address, and photo

- Collect and log activity including logins, data that has been accessed, and content associated with apps and resources in the inviting tenant

- Use profile and activity data by making it available to other apps inside the organization

- Administer the guest user account

After consenting, the invitation state in the Entra admin center or Azure portal is updated from **Pending acceptance** to **Accepted**. Additionally, depending on what identity source the guest user is authenticated against, the **Identities** property could be updated to one of several possible values:

- **External Entra ID**: Entra ID identity from another organization

- **Microsoft Account**: Microsoft account ID associated with Hotmail, Outlook.com, Xbox, LiveID, or other Microsoft consumer property

- **Google.com**: A user identity associated with either Google's consumer products (such as Gmail) or a Google Workspace offering

- **Facebook.com**: A user identity associated with Meta's Facebook service

- **{issuer URI}**: Another third-party SAML/Ws-Fed-based identity provider

Guest users can be assigned licenses, granted access to apps, and delegated administrative roles inside the inviter's tenant.

Creating, Configuring, and Managing Groups

Groups are directory objects used to perform operations, grant rights or permissions, or communicate with multiple users collectively. In Entra ID, there are several types of groups:

- **Security groups**: Typically used for granting permissions to resources, either on-premises or in Entra ID. Security groups can be mail-enabled so that they can function like a distribution list as well.

- **Distribution lists or distribution groups**: Usually used for sending emails to multiple recipients. They can also be used to restrict the scope of rules or for filtering purposes in Entra ID, SharePoint Online, and Exchange Online.

- **Microsoft 365 groups**: Formerly called modern groups (and sometimes referred to as unified groups), these are versatile groups that can function as a security group for assigning permissions or as a distribution group for handling email. Microsoft 365 groups are linked to SharePoint Online sites and form the basis for teams in Microsoft Teams. Each group is connected to an Exchange group mailbox, allowing it to store persistent messages such as emails or channel conversations in Teams. Microsoft 365 groups are only available in Entra ID and do not have an on-premises equivalent.

Each group type has specific capabilities and benefits, and one or more types may be suitable for a particular task. In Entra ID, security groups can be mail-enabled or not, while distribution groups and Microsoft 365 groups are always mail-enabled.

In Entra ID, cloud-based groups can have either assigned or dynamic membership. With assigned membership, an administrator updates group membership periodically. Dynamic groups are built using object queries that periodically add or remove members. For example, a dynamic group called *Sales* might automatically include users whose job title or department is set to *Sales*. Groups in Entra ID can contain users, contacts, devices, and other groups, and they can be converted between assigned and dynamic membership.

Key points to remember about working with groups are as follows:

- An Entra ID tenant can have both synchronized groups from on-premises environments and cloud-only groups.

- Both security and distribution groups can be synchronized from on-premises environments, except for on-premises dynamic distribution groups. These cannot be synchronized because they are based on queries not supported in Entra ID. To address this, you must recreate the dynamic groups in Entra ID using supported query parameters or modify the on-premises group to use the assigned membership.

- Microsoft 365 groups cannot be members of other groups, nor can they have other groups nested within them due to their unique construction.

- Microsoft 365 groups are the only type of object with a cloud source of authority that can be written back to on-premises environments.

Now, let's look at administering groups in Entra ID!

Microsoft 365 Admin Center

For most Entra ID group administration use cases, you'll probably use the Microsoft 365 admin center. To configure groups in the Microsoft 365 admin center, follow these steps:

1. Navigate to the Microsoft 365 admin center (`https://admin.microsoft.com`). Expand **Teams & groups** and then select **Active teams & groups**:

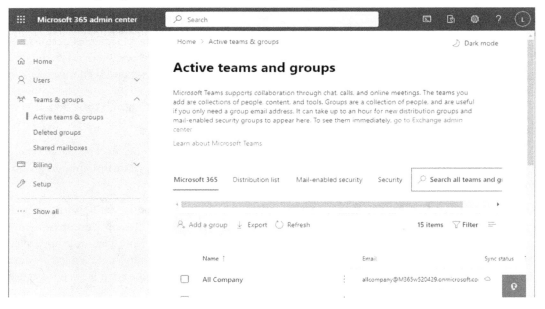

Figure 2.7: Active teams and groups

2.   Click **Add a group**.

3.   On the **Group type** page, select the type of group to create. With the exception of **Security** groups, all group types will require essentially the same information. As a note, non-mail-enabled security groups do not allow you to add owners or members to the workflow.

If you select **Microsoft 365** as your group type, you'll also have the option at the end of the wizard to create a team from the group.

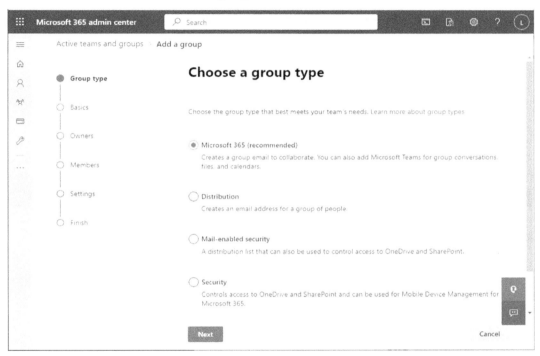

Figure 2.8: Choose a group type

4.   On the **Basics** page, enter a name and description for the group. Click **Next**.

5.   On the **Owners** page, click **Assign owners** to assign at least one owner. Microsoft recommends having at least two owners (in case one leaves the organization or is absent for a period of time). The owner *cannot* be an external guest user. Click **Next** when finished.

6.   On the **Members** page, click **Add members** and select members for the group. Click **Next** to proceed when finished with this optional step.

7. On the **Settings** page, configure settings for the group and then click **Next**:

    •   For distribution groups and mail-enabled security groups, specify an email address.

    •   For Microsoft 365 and security groups, assign any necessary Entra ID roles.

    •   For distribution groups, you can enable the ability for users outside the organization to email the groups (Microsoft 365 groups must have this setting configured manually in the Exchange properties for the group object).

    •   For Microsoft 365 groups, you can also configure privacy settings (either **Public** or **Private**). Public groups can be browsed and joined by anyone while private groups require an owner to add members.

    •   For Microsoft 365 groups, you can choose to enable a team for the group, though users must have a Teams license assigned to access the group.

8. On the **Finish** page, review the settings and click **Create group**.

After the group has been created, you can modify its settings in either the Azure portal, Entra admin center, or Microsoft 365 admin center, as shown in *Figure 2.9*:

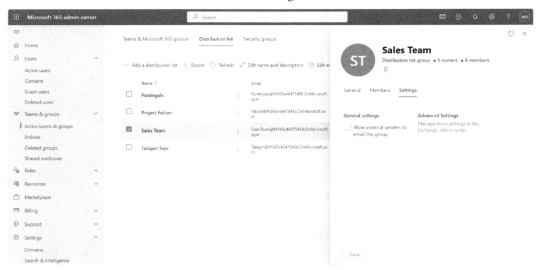

Figure 2.9: Modifying settings of a group in the Microsoft 365 admin center

Depending on the type of group being modified, different options may be available (such as the ability to associate with a **sensitivity label** or changing the privacy settings).

Entra Admin Center

The Entra admin center (and Azure portal) can also be used to create and manage groups. As with the user creation options, the Entra admin center provides a slimmed-down feel without the full wizard experience of the Microsoft 365 admin center.

To create and manage groups in the Entra admin center, follow these steps:

1. Navigate to the Entra admin center (`https://entra.microsoft.com`) or Azure portal (`https://aad.portal.azure.com`), expand **Groups**, and select **All groups**.

2. In the content pane, click **New group**.

3. On the **New Group** page, specify the group type (**Security** or **Microsoft 365**), the group name, and a group description. If you've selected **Microsoft 365** as the group type, you will also be required to enter a **Group email address** value:

Figure 2.10: Creating a new group

Security groups created in the Entra ID or Azure portals are not mail-enabled.

4. Choose whether Entra security roles can be assigned to the group. If you select **Yes**, then the group *must have* assigned membership.

5. Under **Membership type**, you can select from **Assigned**, **Dynamic User**, or **Dynamic Device** (if it is a security group) or from **Assigned** or **Dynamic User** if it is a Microsoft 365 group. Security groups with assigned membership can have all supported object types, but dynamic groups are constrained to a single object type.

Figure 2.11: Selecting the membership type

If you select a group with an **Assigned** membership type, you can add owners and members. If you select a group with either of the dynamic membership types, you must add a dynamic query, as shown in *Figure 2.12*:

Figure 2.12: Dynamic query required for dynamic groups

6. To configure a dynamic query, click **Add dynamic query**.

7. On the **Configure Rules** tab of the **Dynamic membership rules** page, configure an expression that represents the users or devices you want to have included in the group. For example, to create a membership rule that looks for either the value of **Consultant** in the **jobTitle** attribute or **Consulting** in the **department** attribute, select the appropriate **Property** from the list, select **Equals** in the **Operator** column, and then enter the appropriate data in the **Value** field, as shown in *Figure 2.13*:

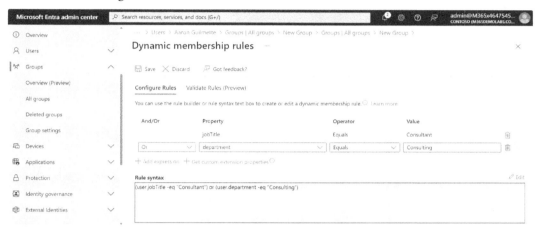

Figure 2.13: Configuring a dynamic query

You can add multiple expressions with the **And** or **Or** evaluation criteria. You can also manually edit the filter to build more complex rules.

8. You can select the **Validate Rules (Preview)** tab and add users you think should be in-scope or out-of-scope to verify that the rule is working correctly. Click **Add users** and then select users from the picker. In this example, none of the users selected met the criteria for the rule, so none of them were included in the group:

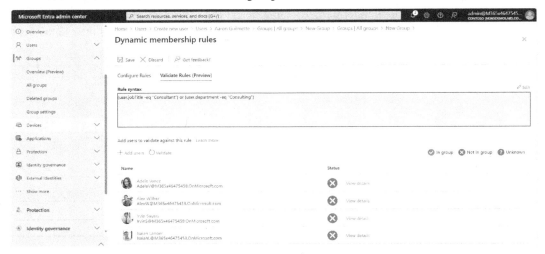

Figure 2.14: Testing a dynamic rule

9. When finished editing the rule, click **Save**.

10. Click **Create** to create the new group.

Using the Entra admin center, you can also update the membership rules for existing groups or change a group's membership from assigned to dynamic by selecting the group and then editing the details on its **Properties** menu, as shown in *Figure 2.15*:

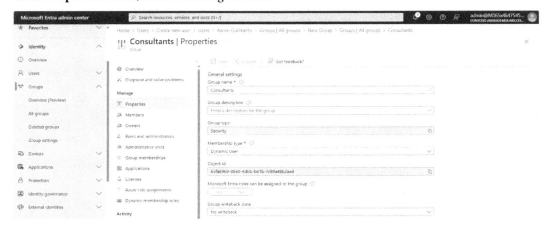

Figure 2.15: Editing a group

If you change a group's membership from *assigned* to *dynamic*, you'll need to create a query. It's important to note, though, that you cannot change a group's type (for example, from **Security** to **Microsoft 365**) or whether a group is eligible for Entra ID role assignment—those options can only be selected when creating a group.

Next, we'll look at managing custom security attributes.

Managing Custom Security Attributes

Custom security attributes in Microsoft Entra ID are business-specific attributes (key-value pairs) that you can define and assign to Microsoft Entra objects. These attributes can be used to store information, categorize objects, or enforce fine-grained access control over specific Azure resources. Custom security attributes are compatible with Azure **attribute-based access control** (**ABAC**).

Custom attributes might be useful in several scenarios:

- Extending user profiles, such as adding an *Hourly Rate* attribute to all employee profiles and ensuring only HR administrators can view the *Hourly Rate* attribute in employee profiles.

- Categorizing hundreds or thousands of applications to create a filterable inventory for auditing purposes.

> **Important**
>
> Entra ID custom security attributes should not be confused with extensions. They may appear similar but are used for different purposes. Extensions are a Microsoft Graph API functionality that allows you to store additional custom data for objects, but should never be used to store private, sensitive, or confidential information (since they're visible to anyone with read permissions for the object). Custom security attribute access can be managed through role-based access control.

Attributes must be part of an attribute set. Your tenant can have multiple attribute sets that you assign to different users or applications. You may choose to use one attribute set for all users and apps, or you can define multiple attribute sets and use them for departments, teams, projects, or business units. Your attribute set design choices will inform how you delegate access to those attribute sets.

For example, you can grant access to attribute sets at either the tenant or attribute set level, and then you can use filtering in the Entra admin center to identify users or applications that have that particular attribute configured.

Creating Custom Attribute Sets and Custom Security Attributes

In order to create and manage custom attributes, you must be explicitly granted the **Attribute Definition Administrator** role. Once you have that role, you can create and manage attribute sets and custom security attributes from the Entra admin center.

To create an attribute set and an attribute, follow this process:

1. Navigate to the Entra admin center (`https://entra.microsoft.com`).

2. Expand **Protection** and select **Custom security attributes**.

3. Select **Add attribute set**.

Figure 2.16: Adding an attribute set

4. Provide an attribute set name, description, and the maximum number of attributes that this set will contain (can be up to 500):

Figure 2.17: Configuring an attribute set

5. When finished, click **Add**.

6. Select the new attribute set name. Click **Add attribute**.

7. On the **New attribute** page, populate the **Attribute name** and **Data type** fields, as they're required. You can also populate the **Description** attribute (though it's not required) as well as whether the attribute will be a single value or multi-valued. Finally, you can choose whether to only allow predefined attribute values to be assigned (instead of allowing the attribute to be input free-form).

Figure 2.18: Creating a custom attribute

If you set **Only allow predefined attribute values to be assigned** to **Yes**, you'll need to click **Add value** and populate the choice of attribute values that you want others to be able to choose from.

8. Click **Save** when finished.

After the attribute has been saved, you can add more attributes or move on to managing access to the attributes and their data.

Managing Access to Attributes

Your organization may choose to store confidential information in custom security attributes since access to those can be restricted to certain individuals. To manage access at the scope level, you can add members to the appropriate **Attribute Definition** or **Attribute Assignment** roles inside the attribute set, as shown in *Figure 2.19*:

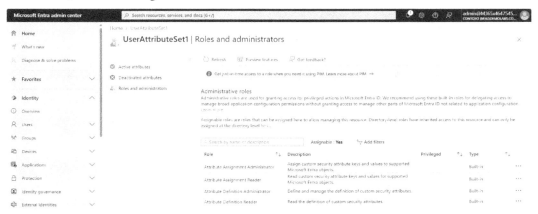

Figure 2.19: Granting scoped attribute access

If you want to grant access to manage or read attributes at the tenant level, simply administer the roles at the tenant level, as shown in *Figure 2.20*:

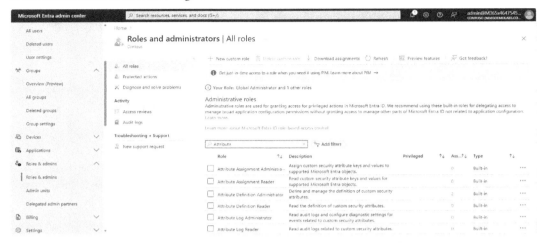

Figure 2.20: Granting attribute access at the tenant level

Next, we'll look at tying attributes to users and applications.

Assigning Attributes to Applications or Users

Once you've created an attribute set and defined custom security attributes, you can begin assigning those attributes to users and applications. The process is very similar for both, as you'll see.

Assigning Attributes to Users

To assign an attribute to a user, select the **Custom security attributes** navigation item on a user's property page in the Entra admin center and then choose **Add assignment**. See *Figure 2.21*:

Figure 2.21: User property sheet

You can select attributes from any set that you have access to, as depicted in *Figure 2.22*:

Figure 2.22: Assigning an attribute to a user

Next, we'll quickly look at assigning attributes to applications.

Assigning Attributes to Applications

Applications also support custom security attributes. You assign them the same way you assign them to users, using the Entra admin center. Simply navigate to **Applications | Enterprise application**, select an application, and then add a custom attribute assignment. See *Figure 2.23*:

Figure 2.23: Assigning an attribute to an application

That's all there is to it!

Deactivating Attribute Definitions

If you get to the point where you no longer need to use a certain attribute, you can deactivate it.

To deactivate an attribute, simply navigate to the attribute set containing the attribute, select the attribute to deactivate, and choose the **Deactivate attribute** button:

Figure 2.24: Deactivating an attribute

You cannot delete attribute sets or attributes at this time. You can only deactivate them.

Next, we'll explore using the PowerShell interface to help scale your effectiveness when administering Microsoft 365 environments.

Automating Bulk Operations by Using the Microsoft Entra Admin Center and PowerShell

While many midsized or large organizations will deploy a hybrid identity solution and manage accounts on-premises, you may face scenarios where you need to manage cloud identities or guests in bulk (such as creating bulk guest user invitations or during a tenant-to-tenant migration procedure), or running operations to query attributes of several objects at once.

These operations can be performed in several ways, including through the Microsoft 365 admin center, the Entra admin center, and various PowerShell commands (known as cmdlets).

For the purposes of the SC-300 exam, we'll just be looking at the Entra admin center and the new Microsoft Graph PowerShell cmdlets.

Entra Admin Center

Bulk operations can be performed through the Microsoft Entra admin center (as well as the Azure portal—the steps are nearly identical, with the core difference being the addition of additional menu navigation elements in the Entra admin center). The Entra admin center supports **Bulk create**, **Bulk invite** (for guest users), and **Bulk delete** operations. See *Figure 2.25*:

Figure 2.25: The Bulk operations menu in the Entra admin center

To get started, you'll need to use one of the templates provided in the Entra admin center. On the **Users** blade, select **Bulk operations** and then choose the appropriate operation.

> **Note**
>
> The templates in the Microsoft 365 admin center and Entra admin center are *not* interchangeable. You'll need to use the correct template for the interface that you're working with. If you use the wrong template, the import will fail.

On the corresponding flyout, you'll have the option to download a template. Once downloaded, you can edit it in any app that supports **comma-separated values (CSV)** files. See *Figure 2.26*:

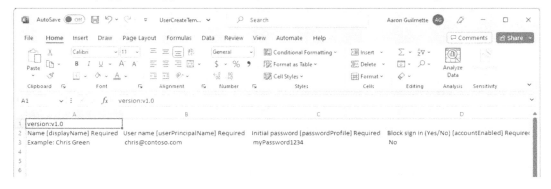

Figure 2.26: Entra admin center bulk user create template

It's important to note that the first two rows of any of the templates must be preserved and not modified in any way. Each identity to be modified is included in a separate row, starting at row 3. The first four fields (**displayName**, **userPrincipalName**, **passwordProfile**, and **accountEnabled**) are required. All other fields are optional.

When finished, you can upload the CSV back to the same flyout in the Entra admin center to process the request.

PowerShell

By far, the most flexible option for managing bulk users is through Windows PowerShell. There are currently three different PowerShell modules that can be used:

- **MSOnline**
- **Azure AD**
- **Microsoft Graph**

Both the MSOnline and Azure AD modules have been deprecated and replaced by the Microsoft Graph module (though the MSOnline and Azure AD modules still have functions in them not replicated by the Graph cmdlets). The modules work similarly, though the cmdlet names, parameters, syntax, and overall capabilities are different.

For the purposes of the updated SC-300 exam, we're only going to look at the new Microsoft Graph modules.

> **Note**
> As of July 2024, there is also a *new* Microsoft Entra PowerShell module that is intended to provide a little more compatibility and interchangeability with the outgoing Entra ID module. It is not a production module yet and is *not* included in the SC-300 exam.

Installing Modules

The module can be installed running the `Install-Module` cmdlet from an elevated PowerShell prompt on your system:

```
Install-Module Microsoft.Graph
```

Once the module has been installed, you can begin connecting to Entra ID and performing operations.

Connecting to Entra ID

To connect to Entra ID, use the following syntax from a PowerShell prompt:

```
Connect-MgGraph -Scopes "User.ReadWrite.All"
```

You'll need to provide credentials with the appropriate role permissions to create users (such as Global Admin or User Admin). In the case of the Microsoft Graph cmdlets, you'll also need to consent to the permission scopes.

About Scopes

The Microsoft Graph cmdlets rely on permissions scopes to allow the cmdlets to act on your behalf. When looking at a permission scope, you'll notice it has three parts, separated by periods, such as `User.ReadWrite.All`. The first segment indicates what type of object the permission scope applies to—in this case, `User` objects. The second segment details the type of permissions being requested, such as `Read` or `ReadWrite`. The third segment indicates what parameters, attributes, values, or scope of the object will be accessed through this permissions grant. In this case, the permission scope includes all properties. Putting it all together, the permissions scope being requested includes the ability to read or write all properties of the `User` object type. You can learn more about Microsoft Graph scopes at `https://learn.microsoft.com/en-us/graph/permissions-reference`.

Working with PowerShell

Once you're connected to Entra ID, you can begin working with objects such as users and groups. You're free to collect, organize, and manipulate the data in whatever way works best for you using any available cmdlets that you have access to. For example, if you need to gather a list of `User` objects and their properties, you can use one of the installed modules' `Get-*` cmdlets. You can choose to store, view, or manipulate the data in a myriad of ways—for example, storing it in a variable, displaying it to the console (screen), exporting it to a file, or passing the data through to another command.

PowerShell also supports a processing concept called piping. Piping can be used to redirect the output of one command into another command. It can be used to process intermediary computations or steps without writing data to disk.

Let's look at some common examples of how you might interact with one or more objects in bulk.

Retrieving User Data

Let's say you need to retrieve a list of the first 10 users in your organization that meet certain criteria (such as members of the *Project Management* department).

When working with the Microsoft Graph module, you'll need to use the following syntax:

```
Get-MgUser -Filter "Department eq 'Project Management'"
-Top 10 -ConsistencyLevel Eventual -Property
DisplayName,UserPrincipalName,Department | Select
DisplayName,UserPrincipalName,Department
```

You can see how this looks in Microsoft Graph in *Figure 2.27*:

```
PS C:\> Get-MgUser -Filter "Department eq 'Project Management'" -Top 10 -ConsistencyLevel Eventual -Property DisplayName
,UserPrincipalName,Department | Select DisplayName,UserPrincipalName,Department

DisplayName              UserPrincipalName                       Department
-----------              -----------------                       ----------
Terrianne E Briscoe      Terrianne.E.Briscoe@m365demolabs.com    Project Management
Micky L Gillette         Micky.L.Gillette@m365demolabs.com       Project Management
Gordie E Laughlin        Gordie.E.Laughlin@m365demolabs.com      Project Management
Nealson D Christianson   Nealson.D.Christianson@m365demolabs.com Project Management
Jude W Chavez            Jude.W.Chavez@m365demolabs.com          Project Management
Florie N Church          Florie.N.Church@m365demolabs.com        Project Management
Darelle A Hite           Darelle.A.Hite@m365demolabs.com         Project Management
Liliane N Bourgeois      Liliane.N.Bourgeois@m365demolabs.com    Project Management
Jean-Pierre Y Sawyers    Jean-Pierre.Y.Sawyers@m365demolabs.com  Project Management
Monica R Conyers         Monica.R.Conyers@m365demolabs.com       Project Management
```

Figure 2.27: The Get-MgUser cmdlet

Updating Users

One of the most common bulk administration tasks, after reporting, is updating objects. Let's say we needed to assign licenses to this group of users, but they didn't have their usage location set (which is required in order to assign users a license in the Microsoft 365 environment). In order to assign the proper licenses and service plans, we'll need to configure the correct location.

You can use the `Set-MgUser` cmdlet to update user objects with Microsoft Graph PowerShell. In this example, the piped data is processed using a `Foreach` command, instructing PowerShell to loop through the list of users returned, substituting the actual individual object (represented by `$_`) and the property of the object to be retrieved (represented by the `.id` placeholder):

```
Get-MgUser -Filter "Department eq 'Project Management'" -Top 5
-ConsistencyLevel Eventual -Property * | Foreach { Update-MgUser
-UserId $_.id -UsageLocation US }
```

Or, you can do it in Graph:

```
PS C:\> Get-MgUser -Filter "Department eq 'Project Management'" -Top 5 -ConsistencyLevel Eventual -Property * | Foreach
{ Update-MgUser -UserId $_.id -UsageLocation US }
```

Figure 2.28: Updating UsageLocation with Update-MgUser

Updating Licenses

License management is also a task that is often performed via scripting.

In the final license example, we'll use the Microsoft Graph PowerShell cmdlet. It works similarly to the Entra ID cmdlet, but the syntax requires a hash table to hold the license `SkuId` property:

```
$user = get-mguser -UserId karenb@M365w520429.OnMicrosoft.com
-Property *
$TeamsSku = Get-MgSubscribedSku -all | Where SkuPartNumber -eq "TEAMS_
EXPLORATORY"
Set-MgUserLicense -UserId $user.Id -AddLicenses @{SkuId = $TeamsSku.
SkuId} -RemoveLicenses @()
```

It is also possible to do it in Graph:

Figure 2.29: Adding a license with the Set-MgUserLicense cmdlet

Depending on your scenario, managing licensing through one of the PowerShell interfaces may be the most efficient way to craft custom license configurations.

Further Reading

Managing licenses via PowerShell can be a complex topic, especially when considering options for enabling or disabling individual service plans within a license or replacing licensing options for users. You can see more in-depth information on various techniques for managing licensing via PowerShell at `https://learn.microsoft.com/en-us/microsoft-365/enterprise/view-licenses-and-services-with-microsoft-365-powershell`.

Creating Users

There are several scenarios where you may need to bulk create users or contacts or bulk invite users to your tenant, such as hiring hundreds of seasonal workers or starting a collaboration with a partner organization. While the Entra admin center or Microsoft 365 admin center has limitations around the number of objects that can be imported, PowerShell is generally free of those limits.

Frequently, when these operations are required, you will be working with source data stored in a CSV text file.

Earlier in this section, you used a specially formatted CSV file to import objects into the Microsoft 365 admin center. You can use a similarly formatted CSV file to perform the action with PowerShell.

In this set of examples, we've entered a few names into a CSV file (as shown in *Figure 2.30*) to demonstrate bulk user processing. While some of the administrative interfaces (such as the Microsoft 365 admin center) limit you to a maximum of 249 objects, you can process thousands or tens of thousands of objects with PowerShell—the only real limitation is the memory on your computer. See *Figure 2.30*:

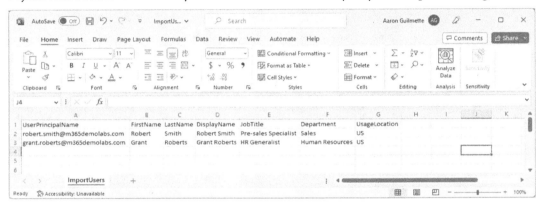

Figure 2.30: Bulk user template

You will use the Microsoft Graph-based New-MgUser cmdlet to script the creation of users from an input file. The sample script stores the input CSV in a variable, creates a password object that will be used for all of the user accounts, and then uses a Foreach loop to iterate over the objects in the input CSV:

```
$Users = Import-Csv C:\Temp\ImportUsers.csv
$PasswordProfile = @{ Password = «P@ssw0rd123» }
Foreach ($User in $Users) { New-MgUser -UserPrincipalName $User.
UserPrincipalName -GivenName $User.FirstName -Surname $User.LastName
-DisplayName $User.DisplayName -JobTitle $User.JobTitle -Department
$User.Department -UsageLocation $User.UsageLocation -Country $User.
UsageLocation -AccountEnabled -MailNickname $User.UserPrincipalName.
Split("@")[0] -PasswordProfile $PasswordProfile }
```

The output is shown in *Figure 2.31*:

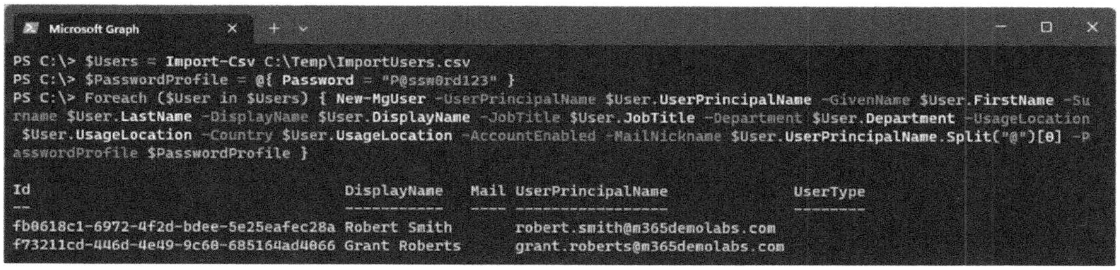

```
Microsoft Graph                  ×    +   ∨                                                   —   □   ×
PS C:\> $Users = Import-Csv C:\Temp\ImportUsers.csv
PS C:\> $PasswordProfile = @{ Password = "P@ssw0rd123" }
PS C:\> Foreach ($User in $Users) { New-MgUser -UserPrincipalName $User.UserPrincipalName -GivenName $User.FirstName -Su
rname $User.LastName -DisplayName $User.DisplayName -JobTitle $User.JobTitle -Department $User.Department -UsageLocation
 $User.UsageLocation -Country $User.UsageLocation -AccountEnabled -MailNickname $User.UserPrincipalName.Split("@")[0] -P
asswordProfile $PasswordProfile }


Id                                     DisplayName      Mail  UserPrincipalName              UserType
--                                     -----------      ----  -----------------              --------
fb0618c1-6972-4f2d-bdee-5e25eafec28a   Robert Smith           robert.smith@m365demolabs.com
f73211cd-446d-4e49-9c68-685164ad4066   Grant Roberts          grant.roberts@m365demolabs.com
```

Figure 2.31: Creating new users with the New-MgUser cmdlet

As you can see, the flexibility and capability of the PowerShell interface allow you to do far more than what's available in the graphical administration centers—with the trade-off that the parameters and syntax for the various modules can vary greatly.

Further Reading

We have only touched the surface describing the capabilities of the various PowerShell modules. You can learn more about all of the available modules and their associated cmdlets and best practices at `https://learn.microsoft.com/en-us/powershell/`.

In the next section, we'll shift the focus from managing users to managing devices.

Managing Device Join and Device Registration in Microsoft Entra ID

Devices are used to interact with Microsoft 365 services. Microsoft 365 supports three different kinds of device identities, as follows:

- **Microsoft Entra registration**: Microsoft Entra registered (sometimes referred to as Workplace joined) devices provide users with support for **bring your own device** (**BYOD**) or mobile device scenarios. In these scenarios, a user can access your organization's resources using a personal device.

- **Microsoft Entra join**: Entra joined devices are typically corporate-owned devices that are solely joined to the Microsoft 365 tenant. This might be equivalent to an Active Directory domain-joined device.

- **Microsoft Entra hybrid join**: Where Entra-joined devices are joined solely to a Microsoft 365 tenant, hybrid-joined devices are joined to both an Active Directory domain and a Microsoft 365 tenant.

You can view device details and manage device features such as BitLocker through the Microsoft Entra admin center, as shown in *Figure 2.32*:

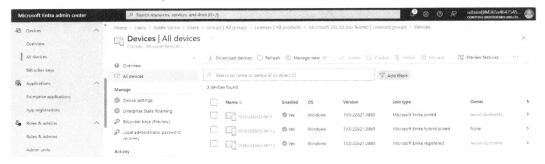

Figure 2.32: Viewing devices in Microsoft Entra

You can select devices to enable or disable from the main page.

If your devices are Intune-enrolled, you can configure additional settings, such as BitLocker recovery options or Microsoft's **Local Administrator Password Solution (LAPS)**. For example, after configuring a BitLocker policy and storing the recovery keys in Entra ID, you can view a device's keys through the Entra admin center:

Figure 2.33: Viewing a device's BitLocker recovery keys

Next, we'll look at managing and monitoring Microsoft 365 licenses.

Assigning, Modifying, and Reporting on Licenses

Every Microsoft 365 workload or service is tied to a license—whether that's individual product licenses for Exchange Online or SharePoint Online or bundled offerings such as Microsoft 365 G3 that include multiple services.

In Microsoft terminology, there are a number of key terms to be aware of:

- **Licensing plans**: A licensing plan is any purchased licensing item. For example, standalone Exchange Online P2, Entra ID P1, and Microsoft 365 E3 are all examples of licensing plans. Sometimes, these might be referred to as **SKUs**.

- **Services**: Also known as **service plans**, these are the individual workloads or services that exist inside of a licensing plan. For example, Exchange Online P2 has a single Exchange Online P2 service plan, while Microsoft 365 E3 has an Exchange Online service plan, a Microsoft 365 Apps service plan, an Entra ID P1 service plan, a OneDrive for Business service plan, a SharePoint Online service plan, and so on.

- **Licenses**: This is the actual number of individual licensing plans of a particular type that you have purchased. For example, if you have five subscriptions to Exchange Online P2 and five subscriptions to Microsoft 365 E3, you have 10 licenses (or five each of Exchange Online P2 and Microsoft 365 E3). Licenses are frequently mapped 1:1 with users or service principals, though some users may have more than one license plan associated with them.

- **SkuPartNumber**: When reviewing licensing objects in PowerShell, `SkuPartNumber` is the object that maps to a particular licensing plan. For example, *ENTERPRISEPACK* is the `SkuPartNumber` value that equates to the Office 365 licensing plan.

- **AccountSkuId**: This is the combination of your tenant name (such as *fabrikam*) and the SkuPartNumber or licensing plan. For example, the Office 365 E3 licensing plan belonging to the *fabrikam.onmicrosoft.com* tenant has an *AccountSkuId* of *fabrikam:ENTERPRISEPACK*.

- **ConsumedUnits**: These represent the number of items in a licensing plan that you have assigned to users. Let's say you have assigned a Microsoft 365 E3 licensing plan to three users. You now have three consumed units of the Microsoft 365 E3 licensing plan. If reviewing licensing from the Entra admin center, this field is sometimes displayed as **Assigned**.

- **ActiveUnits**: The number of units that you have purchased (or have in an active trial) for a particular licensing plan. If you are reviewing licensing from the Entra admin center, this field is sometimes displayed as **Total**.

- **WarningUnits**: The number of units that you haven't renewed of a particular licensing plan. These units will expire and be deprovisioned after the 30-day grace period. If reviewing licensing in the Entra admin center, this field is also sometimes displayed as **Expiring soon**.

You can easily view purchased licensing plan details in the Microsoft 365 admin center under **Billing | Licenses**:

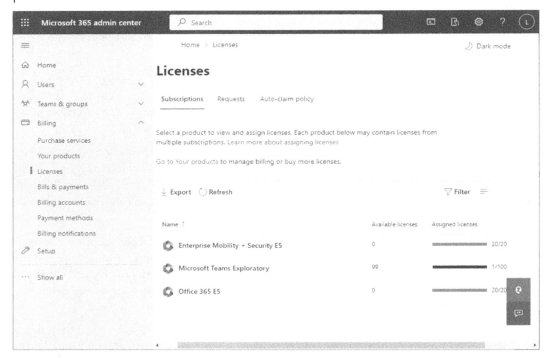

Figure 2.34: License details in the Microsoft 365 admin center

You can assign licenses in many ways:

- Through the **Licenses** page in the Microsoft 365 admin center (**Microsoft 365 admin center | Billing | Licenses**)

- On the properties of a user on the **Active users** page in the Microsoft 365 admin center (**Microsoft 365 admin center | Users | Active users | User properties**)

- To users through a user's properties page in the Entra admin center (**Entra admin center | Identity | Users | All users | User properties**)

- To groups through group-based licensing (**Entra admin center | Identity | Groups | Licenses | Licensed groups**)

- Through PowerShell cmdlets such as `Set-MgUserLicense`

Each licensing method allows you similar options for assigning license plans to users, including assigning multiple license plans or selectively enabling service plans inside an individual license plan.

For example, in the Microsoft 365 admin center, you can view and modify a user's licenses on the **Licenses and apps** tab of their profile.

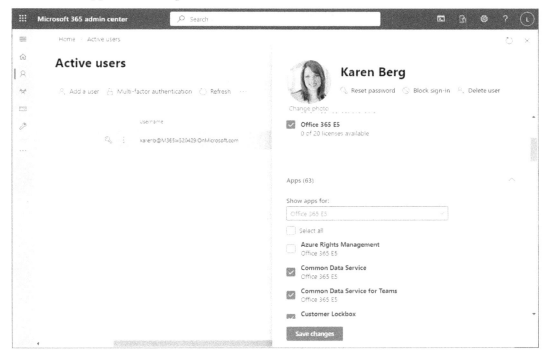

Figure 2.35: User license management

As you can see in *Figure 2.35*, the user has the **Office 365 E5** licensing plan enabled as well as individual services such as **Common Data Service**, **Common Data Service for Teams**, and **Customer Lockbox**, while the **Azure Rights Management** service plan for this licensing plan is disabled.

> **Note**
>
> In order to assign licenses, a **usage location** is required. The usage location is used to determine what service plans and features are available for a given user. Any user that does not have a usage location set will inherit the location of the Entra ID tenant.

Many organizations may choose to automate some or all of the licensing assignments. Group-based licensing allows you to specify one or more licenses to be assigned to one or more users or security groups.

To configure group-based licensing, you can follow these steps:

1. Navigate to the Microsoft Entra admin center (`https://entra.microsoft.com.com`).
2. Select **Identity** | **Billing** | **Licenses**.

3. Under **Manage**, select **All products**.

4. Select one or more licenses that you want to assign as a unit to a group and then click **Assign**.

Figure 2.36: Assigning selected licenses to a group

5. On the **Users and groups** tab, click **Add users and groups** and select one or more security groups from the list. You can only select security groups or mail-enabled security groups. The security groups can be cloud-only or synchronized.

6. Click the **Assignment options** tab.

7. Select which services you want to enable for each licensing plan by sliding the toggle to either **Off** or **On**.

Figure 2.37: Configuring assignment options

8. When finished, click **Review + assign**.

9. Confirm the configuration. When ready, click **Assign**.

> **Further Reading**
>
> For more information on configuring group-based licensing, see `https://learn.`
> `microsoft.com/en-us/azure/active-directory/fundamentals/active-`
> `directory-licensing-whatis-azure-portal`.

Summary

In this chapter, you learned how to administer users and groups from both the Entra admin center as well as PowerShell. From the day-to-day management aspect, mastering PowerShell will be one of the best returns on time investment, allowing you to manage your tenant at scale.

You also learned about managing attribute data. One of the newer features of Entra ID is the concept of custom security attributes, which can be used to store sensitive information for users and applications. This new feature can be especially useful if your organization begins developing custom applications that need to store private information as part of the user identity.

In the next chapter, we'll focus on managing external identities.

Exam Readiness Drill – Chapter Review Questions

Apart from mastering key concepts, strong test-taking skills under time pressure are essential for acing your certification exam. That's why developing these abilities early in your learning journey is critical.

Exam readiness drills, using the free online practice resources provided with this book, help you progressively improve your time management and test-taking skills while reinforcing the key concepts you've learned.

HOW TO GET STARTED

- Open the link or scan the QR code at the bottom of this page

- If you have unlocked the practice resources already, log in to your registered account. If you haven't, follow the instructions in *Chapter 19* and come back to this page.

- Once you log in, click the START button to start a quiz

- We recommend attempting a quiz multiple times till you're able to answer most of the questions correctly and well within the time limit.

- You can use the following practice template to help you plan your attempts:

Working On Accuracy		
Attempt	Target	Time Limit
Attempt 1	40% or more	Till the timer runs out
Attempt 2	60% or more	Till the timer runs out
Attempt 3	75% or more	Till the timer runs out
Working On Timing		
Attempt 4	75% or more	1 minute before time limit
Attempt 5	75% or more	2 minutes before time limit
Attempt 6	75% or more	3 minutes before time limit

The above drill is just an example. Design your drills based on your own goals and make the most out of the online quizzes accompanying this book.

First time accessing the online resources? 🔒

You'll need to unlock them through a one-time process. **Head to** *Chapter 19* **for instructions**.

Open Quiz

https://packt.link/sc300ch2

OR scan this QR code →

Implementing and Managing Identities for External Users and Tenants

While much of an organization's work happens within the virtual walls of its tenant, there are also several scenarios where users within the organization need to collaborate with others. This can happen when an external entity (vendor, customer, business partner, or any other type of organization) invites members from your organization to participate in their tenant or when one of your users invites an external user to participate in your tenant.

In order for your organization to maintain its productivity edge, you'll need to be able to deploy, administer, and secure external collaboration. To that end, we'll cover the following SC-300 objectives in this chapter:

- Managing external collaboration settings in Microsoft Entra ID
- Inviting external users, individually or in bulk
- Managing external user accounts in Microsoft Entra ID
- Implementing cross-tenant access settings
- Implementing and managing cross-tenant synchronization
- Configuring external identity providers

Mastering these objectives will not only prepare you for the SC-300 exam but also ensure your organization is ready to work collaboratively in a digital-first world.

Managing External Collaboration Settings in Microsoft Entra ID

External collaboration settings allow you to determine which roles in your organization can invite external users for B2B collaboration. These settings also provide the options to allow or block specific domains and even control what external guest users can view in your Microsoft Entra directory. The available options include the following:

- **Manage guest user access**: Microsoft Entra External ID lets you control what external guest users can see in your Microsoft Entra directory. For example, you can restrict guest users' access to group memberships or only allow them to see their own profile information. Depending on the security requirements of your organization, you may wish to limit that visibility.

- **Control who can invite guests**: By default, all users in your organization, (even external users), can invite external users for B2B collaboration. If you want to limit this capability, you can disable invitations for everyone or restrict them to specific roles.

- **Enable guest self-service sign-up with user flows**: For applications you create and manage inside your tenant, you can design user flows that allow a user to sign up for an app and create a new guest account. This feature can be enabled in your external collaboration settings, and you can then add a self-service sign-up user flow to your app.

- **Allow or block specific domains**: You can use collaboration restrictions to permit or deny invitations to specified domains.

In scenarios involving B2B collaboration with other Microsoft Entra organizations, you can control the flow of users and data between tenants, limiting access to specific users, groups, or applications from both the inbound and outbound perspectives.

To access these guest settings, navigate to the Microsoft Entra admin center (`https://entra.microsoft.com`). All of the settings are administered under the **Identity** | **External Identities** | **External collaboration settings** menu option, as shown in *Figure 3.1*:

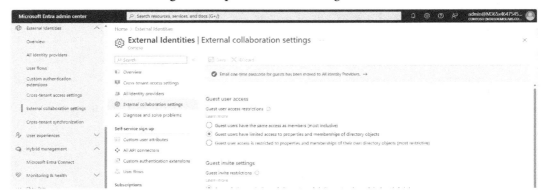

Figure 3.1: Navigating to External collaboration settings

Let's look at each of these areas in a bit more detail.

Guest Access Settings

The first section is **Guest user access**, which controls what level of access guests will have when accessing objects in your tenant.

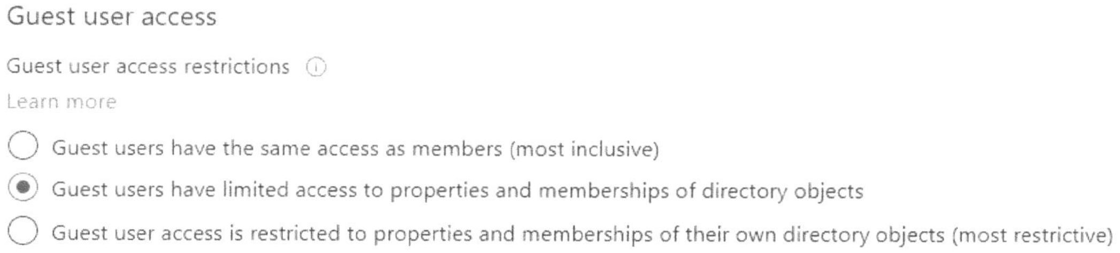

Guest user access

Guest user access restrictions ⓘ
Learn more

○ Guest users have the same access as members (most inclusive)

◉ Guest users have limited access to properties and memberships of directory objects

○ Guest user access is restricted to properties and memberships of their own directory objects (most restrictive)

Figure 3.2: Viewing guest user access settings

The choices are ordered from least restrictive, where guests can be granted essentially the same permissions as members, to most restrictive, where guests can only see information about their own objects. By default, the **Guest users have limited access to properties and memberships of directory objects** option is selected. This option allows guests to see the public properties of users and non-hidden groups. Your organization may have specific requirements governing how much data is visible to non-employees. As a best practice, you should consult with your organization's security leadership to determine the best option for your requirements. See *Figure 3.3*.

Guest Invite Settings

These restrictions control who is allowed to invite new guests into your tenant.

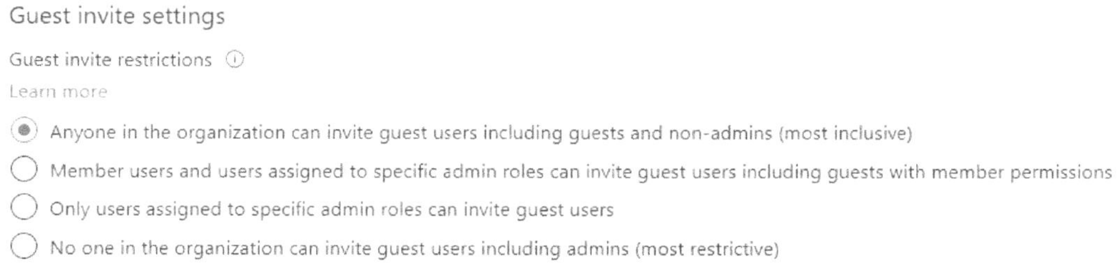

Guest invite settings

Guest invite restrictions ⓘ
Learn more

◉ Anyone in the organization can invite guest users including guests and non-admins (most inclusive)

○ Member users and users assigned to specific admin roles can invite guest users including guests with member permissions

○ Only users assigned to specific admin roles can invite guest users

○ No one in the organization can invite guest users including admins (most restrictive)

Figure 3.3: Viewing guest invite restrictions

By default, this feature is configured to be the most inclusive, meaning that anyone (including guests) can invite new guests into the tenant. While that is the default, many security-conscious organizations typically change this setting to something more restrictive.

Self-Service Sign-Up Flows

The **Enable guest self-service sign up via user flows** option is disabled by default, as shown in *Figure 3.4*:

Figure 3.4: Viewing the self-service sign-up option

This setting is typically only used when you have either third-party or custom in-house applications that your organization develops that are focused on external user integration.

Guest Leave Settings

The final core setting here is regarding how you handle guest users who want to leave your tenant. The default setting is enabled, which allows guests to self-select when they want to leave your environment.

External user leave settings

Allow external users to remove themselves from your organization (recommended) ⓘ

Learn more

Figure 3.5: Viewing external leave settings

That being said, most external users don't proactively remove themselves when they're done working with your organization, teams, groups, or data—as a best practice, you'll want to implement some other process (such as **Access Reviews**) to ensure you're governing the life cycle of your guests and closing down any unused accounts.

> **Further reading**
>
> For more information on Microsoft Entra access reviews (part of the entitlement management and identity governance product offering), see `https://learn.microsoft.com/en-us/entra/id-governance/access-reviews-overview`.

Now that we have adjusted our settings in preparation, let's start inviting our external collaborators in.

Inviting External Users, Individually or in Bulk

You can invite guest users individually in a variety of ways—from an end user application interface, such as Teams, SharePoint, OneDrive for Business, the Entra admin center, the Microsoft 365 admin center, or PowerShell.

For the purposes of the SC-300 exam, we'll focus on guest invitations from the administrative side.

Admin Center

From an administrative perspective, the easiest way to invite a guest user is through the Microsoft Entra admin center, as shown in *Figure 3.6*:

Figure 3.6: Inviting an individual guest user

When using this method, you're prompted to populate the **Email**, **Display name**, and **Message** fields for the external user. See *Figure 3.7*.

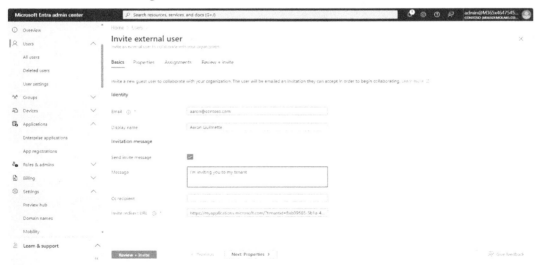

Figure 3.7: Filling out the user invitation form

Invite redirect URL is automatically populated with **https://myapplications.microsoft.com?/tenantid=<tenantguid>**, so you don't need to enter anything there.

Also, if you remember from *Chapter 2, Creating, Configuring, and Managing Microsoft Entra Identities*, you can also use the bulk user invitation option in the Entra admin center to upload a CSV.

Figure 3.8: Importing bulk users through the Entra admin center

Remember, you'll need to download the CSV template from this interface and populate it, as the input file needs to be formatted in a particular way.

You can also invite users from the Microsoft 365 admin center, but be warned—it actually redirects you to the legacy Azure.

Figure 3.9: Inviting guest users from the Microsoft 365 admin center

Next, we'll look at using PowerShell to invite guest users.

PowerShell

You can use the Microsoft Graph PowerShell module (introduced in *Chapter 2, Creating, Configuring, and Managing Microsoft Entra Identities*) to invite one or more guest users.

The `New-MgInvitation` cmdlet is used to invite users. The syntax is pretty straightforward, requiring a display name, email address, and the My Applications URL:

```
New-MgInvitation -InvitedUserDisplayName "John Doe"
-InvitedUserEmailAddress John@externaldomain.com -InviteRedirectUrl
"https://myapplications.microsoft.com" -SendInvitationMessage:$true
```

You can invite them one by one or use an input file (such as a CSV) and a scripting loop to cycle through all of the users:

```
$GuestUsers = Import-Csv .\Guests.csv
$GuestUsers | % { New-MgInvitation -InvitedUserDisplayName
$_.DisplayName -InvitedUserEmailAddress $_.EmailAddress
-InviteRedirectUrl "https://myapplications.microsoft.com"
-SendInvitiationMesage:$true }
```

The minimum permission necessary to generate an invitation is the `User.Invite.All` permission.

> **Further reading**
>
> For more information on the `New-MgInvitation` cmdlet, see `https://learn.microsoft.com/en-us/powershell/module/microsoft.graph.identity.signins/new-mginvitation`.

Next, we'll shift gears to managing guest accounts once they are created in your tenant.

Managing External User Accounts in Microsoft Entra ID

Granting access to external users, customers, vendors, or partners can certainly be an important part of an organization's collaboration strategy. Managing external access after it has been granted is an equally important task, especially from the perspective of the Zero Trust model.

Administering and managing those external accounts can be done through several means—either by manipulating the objects manually in Entra ID, or through an entitlement management and access review process.

Managing Individual Objects

Guest objects are most easily managed through the Microsoft Entra admin center. In the **Identity | Users | All users** menu, you can apply a filter to identify guests in your tenant, as shown in *Figure 3.10*:

Figure 3.10: Identifying guest users in your tenant

Guest user objects can be manipulated like normal user objects, including updating properties, assigning to roles or groups, populating with custom attributes, and granting licenses.

Using Entitlement Management

From a scaling perspective, automation is key to ensuring that guest user identities don't outlive their needs. Just as it's important to de-provision identities for users who are no longer with your organization, it's just as important to manage the life cycle of guest accounts.

Microsoft Entra's entitlement management uses Microsoft Entra External ID B2B to manage the life cycle of external collaborators who need access to your organization's resources. External users authenticate through their home directory or identity provider but have a representation in your organization's directory, allowing them to be assigned access to resources. Entitlement management facilitates access requests for external individuals and automatically removes their digital identities when access is revoked.

> **Note**
>
> Entitlement management is a large topic and covers concepts around the entire identity governance and life cycle management area. This section will only specifically focus on access reviews, a component of entitlement management that can be used to automate the auditing process for external users.

Access reviews are a tool to help organizations track the resource access life cycle, including external guest access and membership. Some of the features of access reviews include the following:

- Performing ad hoc or scheduled reviews to evaluate who has access to resources (such as applications or groups)
- Delegating reviews to other administrators or individuals, including end users who can attest that they still need access to a resource
- Automating review outcomes, such as removing users from groups or teams
- Access reviews operate in a cycle, as shown in *Figure 3.11*:

Figure 3.11: Examining the access review cycle

The cycle begins with a review request sent to users and resource owners. The designated reviewers then review the current memberships for the users in scope for the review, assessing which users still require a certain role or group membership. The old or "stale" memberships can then be removed, and the status changes can be reported. This cycle repeats as needed.

Let's take a look at how we can plan an access review.

Planning Access Reviews

When planning out your access review strategy, you first need to decide what it is you're going to review—such as Microsoft 365 groups, Teams, or applications.

> **Exam tip**
>
> There are some caveats when selecting groups to review: groups must have at least one guest user assigned, and you can't choose dynamic groups or role-assignable groups as targets of an access review.

Depending on the type of review (groups or applications), you'll need to also understand who the reviewers will be.

For groups, the potential options are the following:

- **Group owner(s)**
- **Selected user(s) or group(s)**
- **Users review their own access**
- **Managers of users**

Applications have similar options for reviewing:

- **Selected user(s) or group(s)**
- **Users review their own access**
- **Managers of users**

You can also choose to configure single- or multi-stage reviews to help ensure only individuals with the appropriate level of authority are signing off on an access decision.

Another important factor when designing an access review strategy is specifying the recurrence. Your organization may have security, compliance, or other business requirements that dictate how often reviews should occur. You can specify a recurrence of **One time**, **Weekly**, **Monthly**, **Quarterly**, **Semi-annually**, or **Annually**, as well as start dates and ending parameters (**Never**, **End on a specific date**, or **End after a number of occurrences**).

Finally, you need to plan how to handle exceptions, default actions, and notifications.

With those options in mind, let's create an access review!

Implementing Access Reviews

Creating an access review is pretty straightforward.

Exam tip

In the least-privilege model, you should use either the User Administrator or Identity Governance Administrator role. Access reviews also support a new preview feature to allow the owner of a group to create an access review. Also, from a licensing perspective, access reviews are an Entra ID P2 feature.

To create an access review, follow these steps:

1. Navigate to the Microsoft Entra admin center (`https://entra.microsoft.com`). Expand **Identity** | **Identity Governance** | **Access reviews** and then choose **New access review**.

Figure 3.12: Creating a new access review page

2. On the **Review type** page, select the type of review to perform.

If you select **Teams + Groups**, you can choose either **All Microsoft 365 groups with guest users** or **Select Teams + groups**. If you specify **All Microsoft 365 groups with guest users**, you can choose which (if any) groups to exclude. The scope is automatically configured to **Guest users only**.

If you choose **Select Teams + groups**, then you can select a scope of either **Guest users only** or **All users**. In either group scenario, you can choose **Inactive users (on tenant level) only** along with a period of time for inactivity.

> **Important**
>
> If you want to create an ad hoc review, you must use the **Select Teams + groups** option and identify one or more groups that you want to include in the review. You cannot do an ad hoc review with the **All Microsoft 365 groups with guest users** option.

For **Applications,** you must select one or more enterprise applications that are currently configured in Entra. Select **Next.**

Figure 3.13: Selecting an access review type

3. On the **Reviews** tab, select whether you will be performing a single- or multi-stage review. For each review (or review stage), select who will be performing the review. Depending on the reviewer option selected, you may need to specify individual users or groups. You may also have the option to specify a fallback reviewer who will be contacted if for some reason the primary user no longer exists.

4. On the **Reviews** tab, configure the recurrence settings, as shown in *Figure 3.14*. The **Duration (in days)** value is how long you have to perform the review and **Review recurrence** is how often a review will be triggered. Click **Next.**

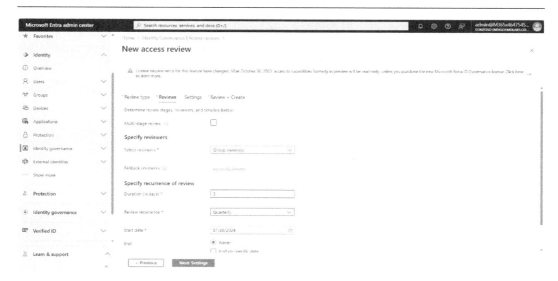

Figure 3.14: Configure reviewers and frequency

5. On the **Settings** tab, select the options. You can choose to auto-apply results to resources, as well as a default action if reviewers don't respond, as shown in *Figure 3.15*:

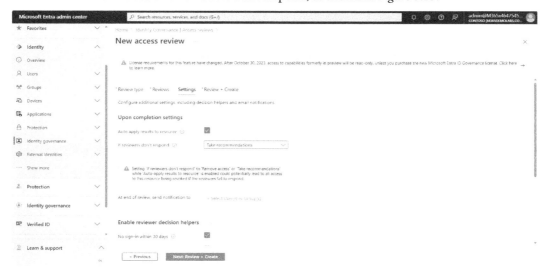

Figure 3.15: Access review – Upon completion settings

6. On the **Settings** tab, you can also configure an email notification that an access review is starting, reminder emails, and decision helpers to provide additional contextual information during the review.

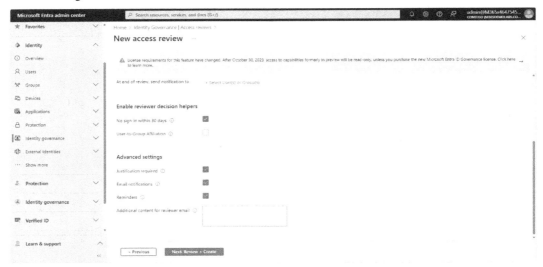

Figure 3.16: Configuring advanced settings for review

7. Click **Next**.

8. Enter a review name and description for the review and click **Create**.

Once an access review is created, it will adhere to your configured schedule. Users will be notified via email when they have pending actions to review.

When looking at the results of an access review, you'll be presented with options. Clicking on the value in the **Recommended action** column will present you with data on Entra's recommendations.

Figure 3.17: Viewing recommended actions for an access review

You can also view the history of performed access reviews by selecting **Review History** under **Access reviews**. You can create a new report of access reviews by selecting **New report** and choosing the date range and types of reviews and outcomes you want to inventory.

Next, we'll look at cross-tenant access features.

Implementing Cross-Tenant Access Settings

Microsoft Entra organizations can leverage cross-tenant access settings to effectively manage collaboration activity with other Microsoft Entra organizations and Microsoft Azure clouds via B2B collaboration and B2B direct connect. These cross-tenant access settings provide control over interactions with external Microsoft cloud-based entities. They are used to manage inbound access (dictating how others collaborate with you) and outbound access (determining how your users engage with external Microsoft Entra organizations).

Cross-tenant access settings can be used to manage features such as allowing you to trust **multi-factor authentication** (**MFA**) and device claims from other tenants. This essentially means that if a user meets the Conditional Access requirements in another tenant (such as MFA token or hybrid device join compliance), you can accept that as proof of MFA to access resources in your tenant.

Figure 3.18 depicts the control features of the cross-tenant access settings:

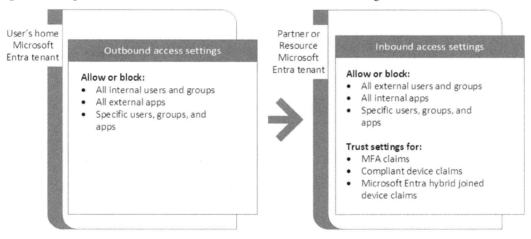

Figure 3.18: Understanding cross-tenant access flows

As you can see, the outbound access settings apply to a user's home tenant and are used to manage which users, groups, or apps are allowed to initiate external access requests. From the partner or resource Entra tenant's perspective, the inbound access settings control which external users, groups, and apps are allowed to participate. The inbound access settings also include the trust settings for MFA, device compliance, and Entra hybrid joined devices.

By default, B2B collaboration is enabled while B2B direct connect is blocked. You can modify the settings at the tenant level (which will affect your interactions with all tenants), or you can enable per-organization settings to provide more granular control.

> **Exam tip**
>
> Currently, Microsoft Teams shared channels are the only feature that supports B2B direct connect.

Outbound access settings manage how your users will interact with other Microsoft Entra tenants, while inbound access settings control how external Microsoft Entra tenants will be able to interact with you. In both cases, you can apply settings tenant-wide or to individuals, groups, and applications. On the inbound side, you can also configure whether you will accept another tenant's MFA attestations or proof so that external users don't need to complete another MFA challenge to access resources in your tenant.

Configuring Default Settings

When working with cross-tenant access settings, you can easily manage the defaults that will apply to all interactions with all external tenants. The defaults area is where you set the baselines for all of your external collaboration.

To locate the **Default settings** page, navigate to `https://entra.microsoft.com`, expand **Identity | External Identities**, and then select **Cross-tenant access settings**.

Figure 3.19: Viewing the default settings

From here, you can edit the defaults for **Inbound access settings**, **Outbound access settings**, and **Tenant restrictions**.

If you select **Edit inbound defaults**, you'll be presented with a series of tabs covering the settings for **B2B collaboration**, **B2B direct connect**, and **Trust settings**. See *Figure 3.20*.

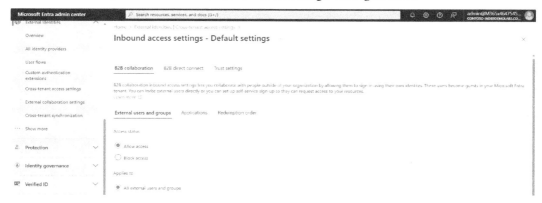

Figure 3.20: Configuring inbound access settings

By default, your tenant is configured to allow all access to everyone who is invited—that is, if someone in your organization invites an external user, there are no restrictions on who can be invited.

On the **B2B collaboration** page, if you flip the **Access status** from **Allow access** to **Block access** under **External users and groups**, all inbound access will be cut off and existing external users will receive the following message when they attempt to sign in:

Figure 3.21: Inbound access block

Exam tip

When you configure the user policy to block access, you'll also need to configure the application policy to block access.

Depending on your organization's security and collaboration needs, you may need to configure a "block all" policy and then add organizations separately to customize an allow policy for them. If you have a more specific policy (for example, a custom policy for a partner organization), that policy takes precedence over the tenant-wide configuration settings.

You can also manage what applications are available. By default, all tenant applications are available. You can use the **Applications** tab and then choose **Select applications** to identify which applications you want to allow external users to access. See *Figure 3.22*.

Figure 3.22: Configuring inbound application access

The **Redemption order** tab of the **B2B collaboration** page allows you to specify the order in which identity providers are tried for authenticating external users. When you send a B2B invitation, users present their own credentials to access your tenant. These credentials could be from another Entra tenant, an on-premises identity provider, or even a consumer system such as Facebook. You can enable, disable, and reorder the identity providers to meet your organization's needs. The default configuration and ordering are shown in *Figure 3.23*:

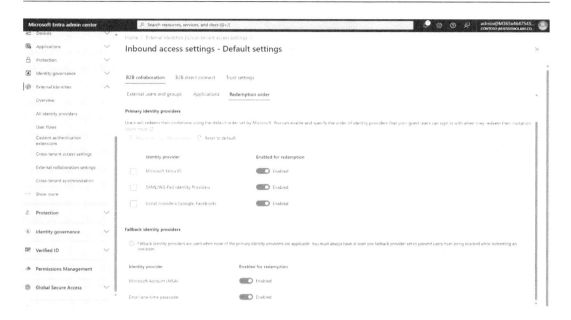

Figure 3.23: Viewing the redemption order

The **Trust settings** tab allows you to configure whether your tenant will accept security or MFA claims from users in remote tenants. See *Figure 3.24*:

Figure 3.24: Viewing trust settings

For the trust settings to take effect, you'll also need to configure one or more Conditional Access policies that include guest users in the scope. The controls configured on this page will require a corresponding Conditional Access policy control option (**Require MFA**, **Require compliant devices**, or **Require hybrid-joined devices**) to be selected in order to honor the external tenant's claims.

Just as you can edit **Inbound access settings** to manage what access others have to your tenant, you can also use the **Outbound access settings** options to control how your users can communicate with other tenants.

As with **Inbound access settings**, you can control the scope of users and applications that you want to allow to have access to cross-tenant features, as shown in *Figure 3.25*:

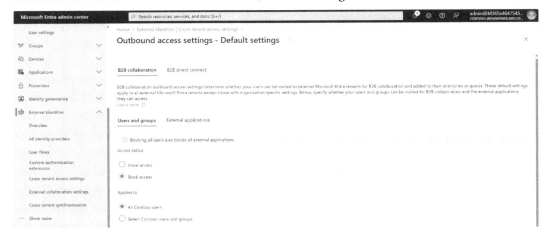

Figure 3.25: Managing outbound access

Once you've saved the tenant-wide settings to match your security use case, you should select which potential clouds your organization will collaborate with.

Configuring Cloud Settings

By default, organizations can only configure cross-tenant integration with other tenants in the same sovereign cloud set (such as Worldwide Commercial, Microsoft Azure Government, or Microsoft Azure China).

If you want to enable cross-cloud access, you'll need to enable the other tenant types. *Figure 3.26* shows the Microsoft cloud settings from a Worldwide Commercial tenant's point of view.

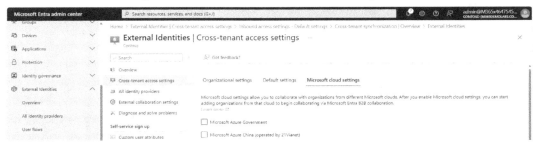

Figure 3.26: Viewing cross-cloud settings

Once the appropriate clouds have been enabled, you can begin configuring custom policies for other organizations.

Adding an Organization

The default cross-tenant access settings in the previous section cover every tenant for which a more specific policy is not configured. If you want to have more (or less) restrictive controls for partners, customers, vendors, or even competitors, you'll need to create a custom policy covering each of those tenant organizations.

To add a custom organization, simply select the **Organizational settings** tab from the **Cross-tenant access settings** page and then click **Add organization**.

Figure 3.27: Adding a custom organization

In the **Add organization** flyout, you can enter any of the domain names registered in the remote tenant (such as the initial domain or custom domain) or the tenant ID.

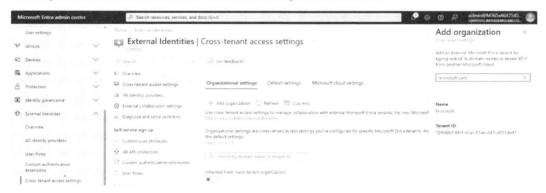

Figure 3.28: Locating a remote tenant

Once you have added a new tenant, it will appear in your organization list, as shown in *Figure 3.29*:

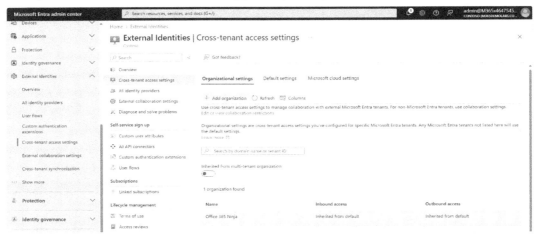

Figure 3.29: Viewing a newly added organization

From here, you'll notice that the current values for the **Inbound access**, **Outbound access**, and **Tenant restrictions** columns are **Inherited from default**. You can select either of those items to begin managing the inbound or outbound access settings specific to that organization.

The individual organization configuration pages are nearly identical to the tenant-wide settings, with the exception of radio buttons that allow you to choose either **Customize settings** or **Default settings**. Choosing **Customize settings** allows you to begin updating the access settings to meet your requirements. If at any time, you wish to revert to the original settings and inherit from your tenant's current default configuration, choose the **Default settings** button. You'll be presented with a prompt to discard your customizations. See *Figure 3.30*.

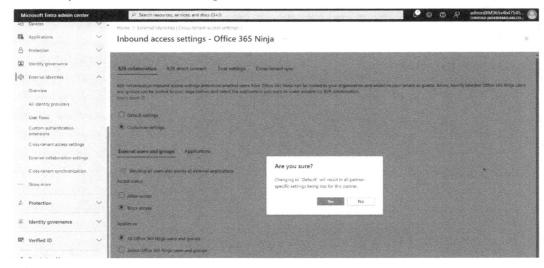

Figure 3.30: Reverting to default settings

Once reverted, users and domains that are part of this organization will go back to using the tenant default settings.

To delete an organization settings object, navigate to the **Organizational settings** tab of **Cross-tenant access settings** and select the **Delete** (trash can) icon.

With cross-tenant access settings under your belt, it's time to shift gears to a new cross-tenant feature: provisioning identities.

Implementing and Managing Cross-Tenant Synchronization

Cross-tenant synchronization is the result of many customer requests over the past several years, allowing organizations to automate provisioning, updating, and deleting Microsoft Entra B2B users across their tenants. The primary use case is to enable a more seamless integration for application and resource sharing for multi-tenant organizations.

> **Important**
>
> While there is nothing to stop you from doing it, Microsoft recommends that you only use the cross-tenant synchronization features with tenants that are part of your business organization. Microsoft states that the solution is currently not suitable for crossing organizational boundaries.

The core benefits of cross-tenant synchronization include the following:

- Automating B2B user account creation across tenant boundaries without custom scripting or full external identity management solutions

- Eliminating the need for users to accept invitation emails and processing consent prompts

- Keeping B2B user accounts up to date and de-provisioning them when the user leaves their home tenant

Cross-tenant synchronization enables a variety of business scenarios for organizations that maintain multiple tenants. Each tenant can be both a source of identities being synchronized to other tenants as well as a target for identities from other tenants.

> **Important**
>
> Cross-tenant synchronization can only synchronize internal member identities from the source tenant. External users or guests cannot be synchronized to partner tenants.

To configure cross-tenant synchronization, you'll need to make configuration changes in both the source and target tenants.

Configuring the Target Tenant

You'll perform the first part of the configuration in each of the target tenants (tenants that will receive synchronized B2B user objects). Much of this will seem familiar from the *Implementing Cross-Tenant Access Settings* section of this chapter:

1. Navigate to the Entra admin center (`https://entra.microsoft.com`) and log in with an account that has been assigned either the Global Administrator or Security Administrator role.

2. Expand **Identity | External Identities** and select **Cross-tenant access settings**.

3. Select the **Organizational settings** tab and then click **Add organization**.

4. Enter the tenant ID or one of the tenant's domains and click **Add**.

5. In the **Inbound access** column of the newly added organization, select **Inherited from default**.

6. Select the **Cross-tenant sync** tab.

7. On the **Cross-sent sync** tab, enable the **Allow users to sync into this tenant** checkbox and click **Save**.

Figure 3.31: Enabling inbound synchronization

8. If you are prompted with **Enable cross-tenant sync and auto-redemption?**, select **Yes**.

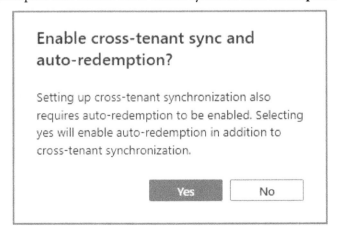

Figure 3.32: Enable cross-tenant sync and auto-redemption? dialog box

9. If you aren't prompted to automatically redeem invitations, select the **Trust settings** tab and enable the **Automatically redeem invitations with the tenant <tenant name>** checkbox. Click **Save**.

Next, we'll move on to setting up the source tenant configuration.

Configuring the Source Tenant

In this set of steps, we'll make the necessary configuration changes in the source tenant to support cross-tenant synchronization:

1. In the source tenant, repeat the steps to add an organization for cross-tenant access settings.

2. In the **Outbound access** column of the newly added organization, click **Inherited from default**.

3. On the **Trust settings** tab, select the **Automatically redeem invitations with the tenant <tenant name>** checkbox and click **Save**.

Figure 3.33: Enabling outbound access trust settings from cross-tenant synchronization

4. Select **Cross-tenant synchronization** under **Identity | External Identities**.

5. On the **Configurations** page, select **New configuration**.

Figure 3.34: Creating a new cross-tenant synchronization configuration

6. Enter a configuration name and click **Create**.

7. Select the newly created configuration.

8.   On the configuration's **Overview** page, click **Get started**.

Figure 3.35: Starting the cross-tenant synchronization configuration

9.   Under **Provisioning Mode**, select **Automatic**, as shown in *Figure 3.36*. This will expand the **Admin Credentials** section.

Figure 3.36: Choosing the provisioning mode

10. In the **Admin Credentials** section, enter the **Tenant Id** value from the target tenant and click **Test Connection**.

Figure 3.37: Entering the Tenant Id value

If you haven't enabled the cross-tenant synchronization inbound rules in the target tenant, you'll get the error shown in *Figure 3.38*, and you need to correct it.

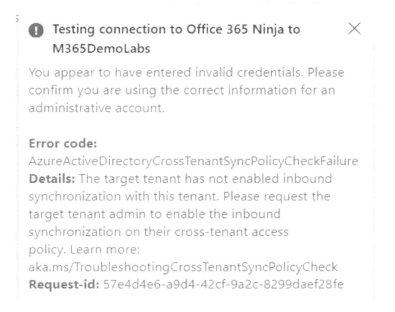

Figure 3.38: Viewing errors for cross-tenant synchronization

Once the cross-tenant access settings to allow synchronization and redemption have been updated, the **Test Connection** action should result in a success notification, similar to what is shown in *Figure 3.39*:

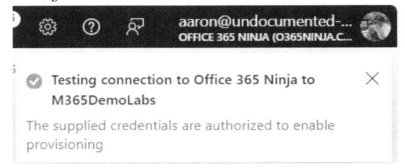

Figure 3.39: Viewing a successful test notification

11. Expand the **Settings** dropdown. If desired, enter a notification email for failures and set an **Accidental deletion threshold** value. When you're finished, click **Save**.

Figure 3.40: Configuring notifications and deletion threshold

At this point, you now have a basic configuration enabled that will allow you to synchronize users from the source tenant to the target tenant. Over the next few sections, we'll look at the settings you can use to customize the configuration.

Configuring Users and Groups

You can choose which users or groups to assign the configuration to by adding them to the **Users and groups** section of the configuration, as shown in *Figure 3.41*. This is a broad selection option for putting users in scope for configuration and can be further refined through the use of filters.

Figure 3.41: Viewing the users and groups assigned to the configuration

This setting works in conjunction with the **Settings | Scope** configuration on the **Provisioning** page and any scoping filters configured in the **Mapping** section. You can add users and groups to the configuration on the **Users and groups** page. Nested groups are not supported, though you can use dynamic groups.

Configuring Mappings

Entra has a default scoping and attribute mapping selection. This determines which objects will be synchronized (through scoping filters) as well as how attributes will be mapped between source and target users.

To edit the mappings and scoping filters, expand **Mappings** on the **Provisioning** page and select **Provision Microsoft Entra ID users**, as shown in *Figure 3.42*:

Figure 3.42: Viewing the Mappings setting

With the **Attribute Mapping** page open, you can select **Source Object Scope** to define filters for who will be included in the synchronization as well as what types of actions (**Create**, **Update**, and **Delete**) the synchronization service will take on objects synchronized to the target tenant.

Figure 3.43: Viewing the Attribute Mapping page

By selecting **All records** under **Source Object Scope**, you can create scoping filters that will be used to select which objects will be synchronized to the target tenant. You can create scoping filter groups to manage the rules. Inside the scoping filter groups, you can use any of the available attributes to create a filter. Multiple scoping filter groups are evaluated using *OR* logic, while multiple scoping clauses inside a filter group are evaluated using *AND* logic.

The following example scoping filter selects objects that are enabled (**accountEnabled IS TRUE**) and are cloud-only or not synchronized (**dirSyncEnabled IS FALSE**).

Figure 3.44: Configuring a scoping filter

As a reminder, these scoping filters work in conjunction with other scoping configuration options, so you'll want to test carefully to make sure all of the objects you intend to include in the synchronization are included.

> **Exam tip**
>
> By default, B2B identities provisioned through cross-tenant synchronization are provisioned as members in the target tenant. You can change to guest if desired by editing the user mapping. For more information, see `https://learn.microsoft.com/en-us/entra/identity/multi-tenant-organizations/cross-tenant-synchronization-configure#step-9-review-attribute-mappings`.

Configuring In-Scope Users

You can choose which users to synchronize from the source to the target tenant. By default, the scope is set to **Sync only assigned users and groups**. You can update this on the **Provisioning** page under **Settings**, as shown in *Figure 3.45*:

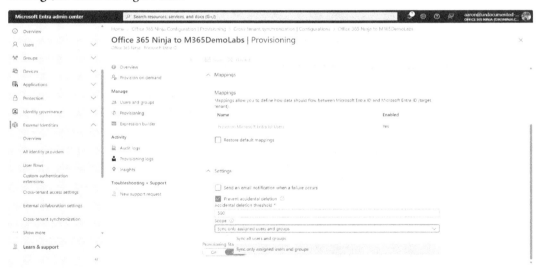

Figure 3.45: Viewing the default synchronization setting

The default setting, **Sync only assigned users and groups**, refers to users and groups assigned to the configuration on the **Users and groups** page.

> **Important**
>
> Microsoft recommends synchronizing the minimal number of users necessary and generally doesn't recommend selecting the **Sync all users and groups** option in the **Provisioning | Settings** area for large directories. Each of the scoping configuration options is reductive—that is to say, each further limits the objects that will be sent to the provisioning engine. In practice, you'll want to establish what users and groups are in scope and choose the best combination of scoping filters and inclusion groups to reduce any complexity in your provisioning rules.

Testing

Prior to enabling your configuration, it's good to test to make sure objects are successfully provisioned. You can use the **Provision on demand** feature to verify that provisioning is working cross-tenant.

Figure 3.46: Viewing the Provision on demand page

After selecting a user or group, you can initiate the engine by clicking **Provision**. If successful, you'll see a page similar to the one displayed in *Figure 3.47*:

Figure 3.47: Provision on demand success

You can click **View details** for each step to see the evaluation and execution process. See *Figure 3.48* for an example.

Figure 3.48: Viewing the results of an action

You can choose to provision another object for testing or then move on to enabling the synchronization process.

Enabling the Synchronization Engine

Once you've verified a successful provisioning action, you're ready to enable the synchronization engine.

In the cross-tenant synchronization configuration, under **Manage**, select **Provisioning** to open the **Provisioning** page. At the bottom of the page, slide the **Provisioning Status** toggle to **On** and click **Save**.

Figure 3.49: Enabling provisioning

The provisioning engine will begin evaluating in-scope objects and managing them in the target tenant.

You can view the results of any of the provisioning actions by looking at the provisioning logs in the **Activity** section. See *Figure 3.50*.

Figure 3.50: Viewing the provisioning logs

Next, we'll tackle the final topic in this chapter—configuring external identity providers.

Configuring External Identity Providers

Throughout this chapter, we've been exploring external identity concepts such as guest users, Entra B2B accounts, and cross-tenant synchronization.

From an identity architecture perspective, there are two distinct types of Entra tenants available: workforce tenants (which is the primary type of Entra tenant we've been working with up to this point—the kind that contains your internal users) and external tenants (tenants that contain non-organization users, such as customer-facing applications).

Using an external identity provider means that you as an application vendor or host of an environment are not responsible for maintaining an identity store for your users. You might think of this scenario as **bring your own identity**—users with well-known services (such as Google or Facebook) have already created an identity with those systems and you're just allowing them to reuse that identity with your tenant or system. You've seen this in action if you've ever gone to a website that had the option to sign in with your Google, Facebook, or Twitter/X account.

In this section, we'll touch on those types of scenarios—exploring how to create an external tenant and then configure an external identity provider such as Google to allow social identity sign-in.

Creating an External Tenant

Before we configure an external identity provider, we'll need an external tenant. The process is very straightforward:

1. Log in to the Microsoft Entra admin center (`https://entra.microsoft.com`) with an account that has either the Global Administrator or Tenant Creator role.

2. Expand **Identity** and select the **Overview** tab.

3. Select **Manage tenants**. This will launch a new page where you can view associated tenants or create new Microsoft 365 tenants.

Figure 3.51: Launching the Manage tenants page

4.   On the **Manage tenants** page, click **Create**.

5.   Select the **External** tile and click **Continue**.

Figure 3.52: Creating a new external tenant

6.   Depending on your licensing, you can choose to either start a trial or use an Azure subscription. In this instance, we'll select **Start a free trial**.

Figure 3.53: Choosing how to subscribe to the tenant

There is no difference in either the features or the functionality of the tenant, only the payment mechanism.

7. Expand the **Change settings** dropdown and update the settings. Here, you can select a new tenant name (the display name that appears in the Microsoft 365 admin center), domain name (initial tenant domain), and location.

Figure 3.54: Changing the external tenant settings

8. Click the checkbox to agree to the Microsoft Customer Agreement, and then click **Continue**.

After a few minutes, a new tenant will be provisioned. Now, it's time to start configuring external identity providers!

Configuring Google as an External Identity Provider

With your new tenant, we'll begin the process of configuring Google as an identity provider. This process will involve creating an application in the Google environment as well as configuration in the Entra external tenant.

> **Important**
>
> This example requires a Google account, which you can obtain from https://accounts.google.com/signup. While configuring an external app and social sign-up is outside the scope of the SC-300 exam, we thought it would be best to include a few examples to help expand the understanding of this topic.

Before you begin, you'll want to switch to your new external tenant if you're not already there by navigating to **Identity** | **Overview** | **Manage tenants** in the Microsoft Entra admin center, selecting your new external tenant, and then clicking **Switch**.

Figure 3.55: Switching tenants

Once you've switched, you'll want to capture both the **Tenant ID** and **Primary domain** values listed on the **Overview** page—you'll need them later.

Figure 3.56: Gathering tenant details

Creating a Google Application

Follow these steps to set up your external identity provider:

1. Sign in to `https://console.developers.google.com` with your Google account.

2. Expand **Projects** at the top, and then on the **Select a project** page, click **New project**. See *Figure 3.57*.

Figure 3.57: Creating a new Google project

3.   Populate **Project name** and then click **Create**.

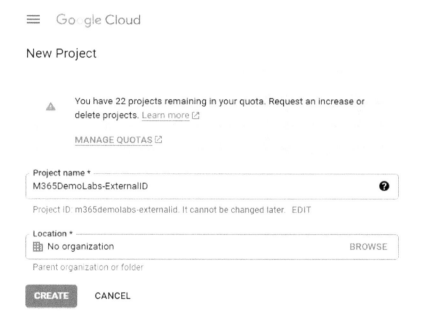

Figure 3.58: Editing the project details

4.   After your project has been created, click **Select project** in the notification window.

Figure 3.59: Switching to the new project

5. Expand **APIs & Services** and click **OAuth consent screen**, as shown in *Figure 3.60*.

Figure 3.60: Configuring OAuth settings

6. Select **External** and click **Create**.

7. Fill out the **App information** page details with your own information.

Figure 3.61: Configuring the app information

8. In the **Authorized domains** section, add any required domains (such as your personal or vanity domain) as well as `ciamlogin.com` and `microsoftonline.com`.

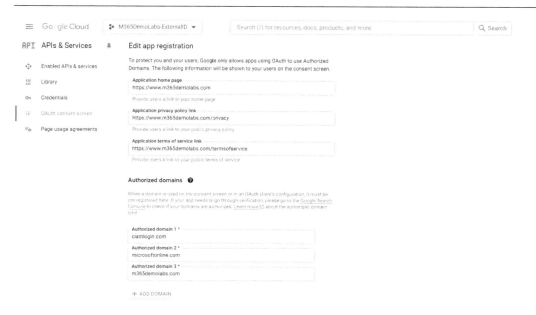

Figure 3.62: Configuring the authorized domains

9. When finished, click **Save and continue**.

10. Under **APIs & Services**, click **Credentials**.

11. Click **Create credentials** and then select **OAuth client ID**, as shown in *Figure 3.63*.

Figure 3.63: Creating credentials

12. Under **Application type**, select **Web application**. Specify a name.

13. Under **Authorized redirect URIs**, click **Add URI** and enter the following URIs—replacing `<tenant-id>` with the **Tenant ID** value you captured earlier and `<tenant-subdomain>` with the initial tenant domain value (before the `onmicrosoft.com` domain portion) of the **Primary domain** value you captured earlier:

- `https://login.microsoftonline.com`

- `https://login.microsoftonline.com/te/<tenant-ID>/oauth2/authresp`

- `https://login.microsoftonline.com/te/<tenant-subdomain>.onmicrosoft.com/oauth2/authresp`

- `https://<tenant-ID>.ciamlogin.com/<tenant-ID>/federation/oidc/accounts.google.com`

- `https://<tenant-ID>.ciamlogin.com/<tenant-subdomain>.onmicrosoft.com/federation/oidc/accounts.google.com`

- `https://<tenant-subdomain>.ciamlogin.com/<tenant-ID>/federation/oauth2`

- `https://<tenant-subdomain>.ciamlogin.com/<tenant-subdomain>.onmicrosoft.com/federation/oauth2`

See *Figure 3.64*.

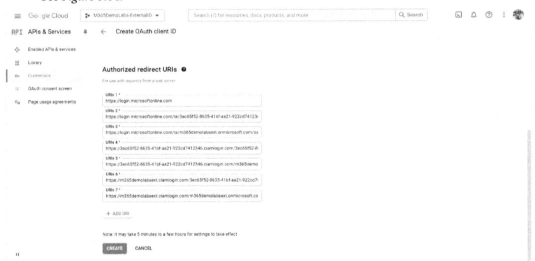

Figure 3.64: Adding the redirect URIs

14. Click **Create** when finished.

15. Store the **Client ID** and **Client secret** values displayed.

Next, we'll configure the federation settings in the external tenant.

Configuring Federation Settings

After switching to your external tenant, follow these steps to configure federation with the Google identity provider:

1. From the Microsoft Entra admin center, expand **Identity | External Identities** and select **All identity providers**.

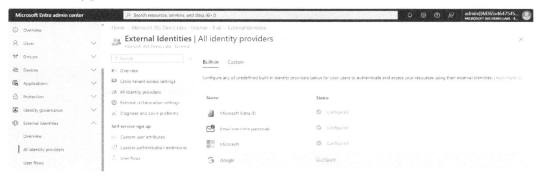

Figure 3.65: Navigating to All identity providers

2. Select **Configure** in the row corresponding to the Google provider.

3. Enter the **Client ID** and **Client secret** values that you copied when you created the Google application in the previous section.

Figure 3.66: Configuring the OAuth parameters

4. Click **Save**.

Next, we'll create some flows to pull it all together.

Creating Sign-Up and Sign-In User Flows

The sign-up and sign-in flows are necessary to provide the settings necessary to allow users to self-provision into your environment and access resources. Use the following process to create a simple set of flows:

1. Navigate to the Microsoft Entra admin center (`https://entra.microsoft.com`) for an external tenant.

2. Expand **Identity | External Identities** and select **User flows**.

3. On the **User flows** page, select **New user flow**.

Figure 3.67: Creating a new user flow

4. On the **Create** page, enter a name and select the identity providers you wish to use (such as the newly created Google IdP). Select any user attributes (such as name, address, or age) that you wish to collect during the sign-in process and click **Create**.

Figure 3.68: Configuring the flow parameters

Next, we'll create a single-page app to leverage the flows.

Registering an App

Creating or registering a new enterprise app will connect all of the parts of the solution together. In the real world, you'd have developers building an application that would leverage flows and connect to the application, but this will show how all of the pieces work:

1. Navigate to the Microsoft Entra admin center (`https://entra.microsoft.com`) for an external tenant.

2. Expand **Identity** | **Applications** and select **App registrations**.

3. Click **New registration**.

Figure 3.69: Creating a new app registration

4. On the **Register an application** page, enter a name and select the **Accounts in this organizational directory only** account type.

5. Under **Redirect URI**, select **Single-page application (SPA)** in the dropdown and enter http://localhost in the URL field.

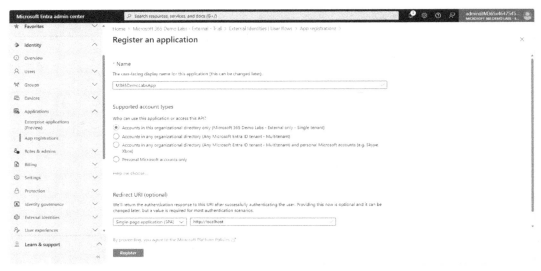

Figure 3.70: Configuring the app registration

6. Click **Register**.

7. If you're not redirected to the new app, select the app from the **App registrations** page and then click **Overview**.

8. Click **API permissions**.

9. Click **Grant admin consent**.

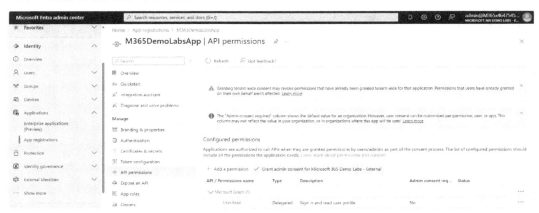

Figure 3.71: Managing API permissions

10. Click **Yes** in the **Grant admin consent confirmation.** dialog box.

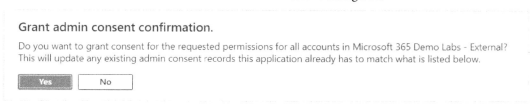

Figure 3.72: Confirming granting admin consent

At this point, you've set up the necessary application and configured permissions to allow users to interact with the app.

Next, you'll associate the app with the sign-up flow you created earlier.

Updating the Sign-Up Flow

In this section, you'll add the app to the sign-up flow, which will allow you to test the authentication process:

1. Navigate to the Microsoft Entra admin center (`https://entra.microsoft.com`) for an external tenant.

2. Expand **Identity | External Identities** and select **User flows**.

3. Select the user flow that you've created.

4. Under **Use**, select **Applications**, as shown in *Figure 3.73*.

Figure 3.73: Configuring the applications for a user flow

5. On the **Applications** page, click **Add application**.

6. Pick the newly registered application from the list and click **Select**.

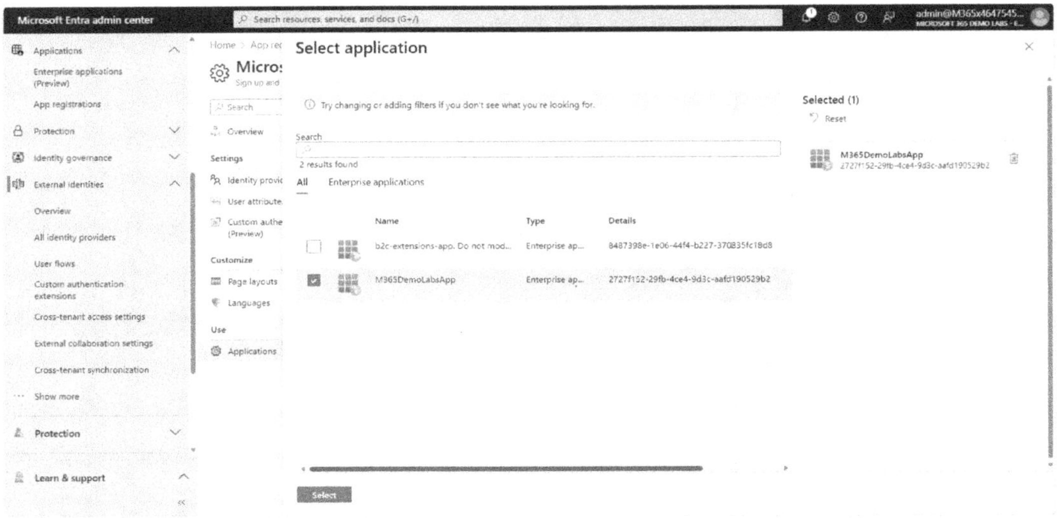

Figure 3.74: Choosing the applications to link to the user flow

Now all that's left is to test!

Testing the External ID App

While this example doesn't have a fully functioning app to test, you can make sure that the authentication flow is working correctly and learn how the integration with external identity works.

To test the external identity provider, follow these steps:

1. Navigate to the Microsoft Entra admin center (`https://entra.microsoft.com`) for an external tenant.

2. Expand **Identity | External Identities** and select **User flows**.

3. Select the user flow that you've created.

4. Under **Use**, select **Applications**.

5. Click **Run user flow**, as shown in *Figure 3.75*.

Figure 3.75: Running the user flow

6. On the **Run user flow** flyout, click **Run user flow**.

7. In the new browser tab that opens, click **Sign in with Google**.

Figure 3.76: Signing in with Google

You should be redirected to the Google identity picker at `https://accounts.google.com`.

8. Sign in to your Google identity.

9.   Click **Continue**.

Figure 3.77: Signing in to Microsoft Online with Google ID

10. If you chose to collect additional attributes in the user sign-up flow, you'll be prompted to enter them now. Enter the necessary fields and click **Next**.

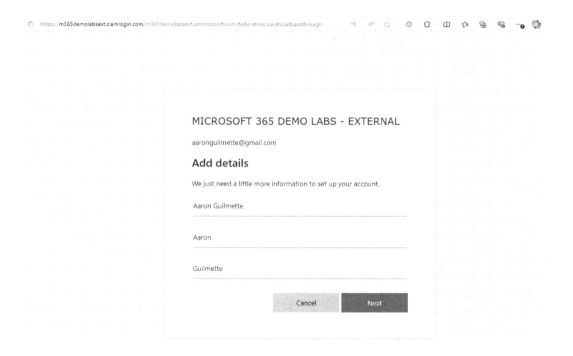

Figure 3.78: Entering additional fields

11. Choose whether to stay signed in by clicking **Yes** or **No**.

Since you likely don't have an application configured to listen at this stage, you won't get redirected further, but you will at least know that the Google authentication workflow is working correctly.

Configuring a Custom External Identity Provider

If you are working with an identity provider that wasn't listed, you can also configure a custom SAML/WS-Fed provider, as shown in *Figure 3.79*:

Figure 3.79: Configuring an external identity provider

When configuring an external provider, you'll need to provide the identity provider protocol (SAML or WS-Fed) and the domain name and specify how you'll populate the federation metadata information. The values that you need should be available from the custom directory's configuration page or documentation. See *Figure 3.80* for an example configuration.

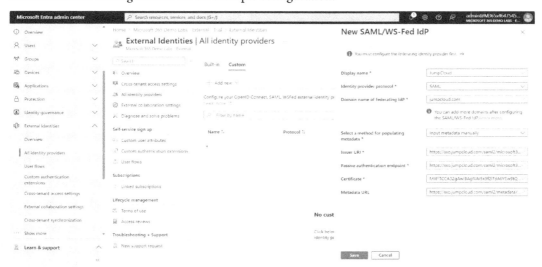

Figure 3.80: Completing a custom identity provider

Depending on the integration requirements, you may be able to use the existing user sign-up flows that you've already created, though you'll need to set up a new application.

Summary

In this chapter, you learned about a wide variety of external collaboration options and features, including guest users, B2B provisioning, and working with external Microsoft Entra tenants.

External identity capabilities enable organizations to collaborate with users outside the tenant boundary—whether those external users are part of an Entra organization or not. As an identity administrator, being able to configure both workforce and external tenants will be critical to your future success.

In the next chapter, we'll begin working with hybrid identity models.

Exam Readiness Drill – Chapter Review Questions

Apart from mastering key concepts, strong test-taking skills under time pressure are essential for acing your certification exam. That's why developing these abilities early in your learning journey is critical.

Exam readiness drills, using the free online practice resources provided with this book, help you progressively improve your time management and test-taking skills while reinforcing the key concepts you've learned.

HOW TO GET STARTED

- Open the link or scan the QR code at the bottom of this page

- If you have unlocked the practice resources already, log in to your registered account. If you haven't, follow the instructions in *Chapter 19* and come back to this page.

- Once you log in, click the START button to start a quiz

- We recommend attempting a quiz multiple times till you're able to answer most of the questions correctly and well within the time limit.

- You can use the following practice template to help you plan your attempts:

Working On Accuracy		
Attempt	**Target**	**Time Limit**
Attempt 1	40% or more	Till the timer runs out
Attempt 2	60% or more	Till the timer runs out
Attempt 3	75% or more	Till the timer runs out
Working On Timing		
Attempt 4	75% or more	1 minute before time limit
Attempt 5	75% or more	2 minutes before time limit
Attempt 6	75% or more	3 minutes before time limit

The above drill is just an example. Design your drills based on your own goals and make the most out of the online quizzes accompanying this book.

First time accessing the online resources? 🔒

You'll need to unlock them through a one-time process. **Head to *Chapter 19* for instructions**.

Open Quiz	
`https://packt.link/sc300ch3`	
OR scan this QR code →	

4

Implementing and Managing Hybrid Identity

As you've already seen throughout the first few chapters of this book, Microsoft 365 relies heavily on identity. When moving to Microsoft 365, many organizations may want to use their existing on-premises identity as a basis for populating their Entra ID directory.

Microsoft provides several identity and authentication mechanisms to help meet organizations' operational and security requirements.

In this chapter, we're going to step through basic configuration tasks for **Entra Connect Sync** and **Entra Cloud Sync**—Microsoft's two solutions for synchronizing and provisioning Entra identities.

This chapter covers the following exam objectives:

- Preparing for identity synchronization
- Implementing and managing Microsoft Entra Connect Sync
- Implementing and managing Microsoft Entra Connect cloud sync
- Implementing and managing password hash synchronization
- Implementing and managing pass-through authentication
- Implementing and managing seamless single sign-on
- Migrating from Active Directory Federation Services to other authentication and authorization mechanisms
- Implementing and managing Microsoft Entra Connect Health

By the end of this chapter, you should be able to describe the different hybrid identity technologies and features natively supported by the Entra platform.

Preparing for Identity Synchronization

Before we dive into the specific planning and deployment exercises for Entra Connect and Entra Cloud Sync, let's take a look at planning and preparation exercises that apply to both.

> **Exam tip**
>
> While identity preparation isn't explicitly called out in the SC-300 exam study guide, it's critical to understand some of the basic requirements and best practices to ensure that your single sign-on deployment works as intended.

Since the purpose of Entra Connect and Entra Connect cloud sync is to synchronize user, group, contact, and device objects to Entra, you'll need to make sure your on-premises objects are ready for their journey to the cloud.

Microsoft has guidance regarding the preparation of user objects for synchronization. Some attributes (specifically those that are used to identify the user throughout the system) must be unique throughout the organization. For example, you cannot have two users that have the same **userPrincipalName** value.

The following attributes should be prepared before synchronizing the directory to Azure **Active Directory (AD)**:

Attribute	Constraints	Must be unique	Required
displayName	≤ 256 characters	X	✓
givenName	≤ 64 characters	X	X
mail	≤ 113 characters ≤ 64 characters before the @ symbol Adheres to RFC 822/2822/5322 standard	✓	X
mailNickName	≤ 64 characters Cannot start with a "." Cannot contain certain characters such as &	✓	X
proxyAddresses	≤ 256 characters per value No spaces Diacritical marks are prohibited	✓	X

Attribute	Constraints	Must be unique	Required
sAMAccountName	≤ 20 characters	✓	✓
sn	≤ 64 characters	X	X
targetAddress	≤ 256 characters No spaces Includes prefix (such as SMTP:) Value after prefix adheres to RFC 822/2822/5322 standard after prefix	✓	X
userPrincipalName	≤ 113 characters Must use a routable domain name Unicode characters are converted to underscores	✓	✓

Table 4.1: Entra Connect attributes

As you can see, very few attributes are *required* for an object to synchronize. Several attributes (such as **mailNickname**, **userPrincipalName**, **mail**, **sAMAccountName**, and **proxyAddresses**) must contain unique values—that is, no other object in the directory of any type can share that value.

> **Further Reading**
>
> You can learn more about the required and supported values for attributes at `https://learn.microsoft.com/en-us/powershell/module/exchange/set-mailbox` and `https://learn.microsoft.com/en-us/microsoft-365/enterprise/prepare-for-directory-synchronization`.

IdFix is Microsoft's tool for detecting common issues with on-premises AD identity data. While it doesn't fix all possible errors (such as duplicate attribute values), it is able to identify and remediate data formatting errors.

To get started with the tool, follow these steps:

1. Navigate to `https://aka.ms/idfix`.
2. Scroll to the bottom of the page and click **Next**.

3. Review the prerequisites for the tool. Scroll to the bottom of the page and click **Next**.

4. Click **setup.exe** to download the file and start the installation.

5. After the installation wizard starts, click **Install**.

6. Acknowledge the IdFix privacy statement by clicking **OK**.

7. IdFix, by default, targets the whole directory:

   - You can select **Settings** to change the options for IdFix

   - You can edit the **Filter** options to scope to certain object types

   - You can select the **Search Base** option to specify a starting point for IdFix to begin its search.

8. After modifying the settings, click **OK**.

9. Click **Query** to connect to AD and begin the analysis.

After IdFix has analyzed the environment, results are returned to the application's data grid, as shown in *Figure 4.1*. The **DISTINGUISHEDNAME** column shows the full path to the object in question, while the **ATTRIBUTE** column shows the attribute or property impacted. The **ERROR** column shows what type of error was encountered (such as an invalid character or duplicate object value). The **VALUE** column shows the existing value and the **UPDATE** column shows any suggested value.

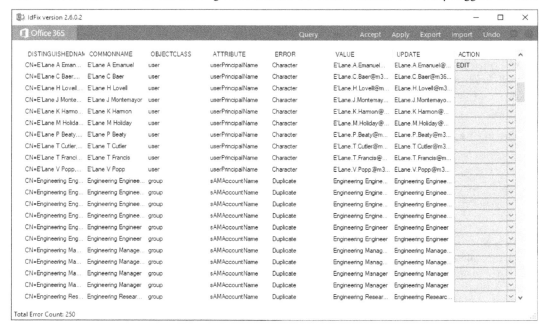

Figure 4.1: Reviewing the IdFix data

After you have investigated an object, you can choose to accept the suggested value (if one exists) in the **UPDATE** column. You can also choose to either enter or edit a new value in the **UPDATE** column.

Once you're done investigating or updating an object, you can use the dropdown in the **ACTION** column to mark an object:

- Selecting **EDIT** instructs IdFix to apply the value in the **UPDATE** column
- Selecting **COMPLETE** indicates you want to leave the object as is
- Selecting **REMOVE** instructs IdFix to clear the value

In addition, you can select **Accept** to accept any suggested values in the **UPDATE** column. Choosing this option will configure all objects with a value in the **UPDATE** column to **EDIT**, indicating that the changes are ready to be processed.

After you have configured an action for each object, select **Apply** to instruct IdFix to make the proposed changes.

Once you have ensured that your on-premises directory data is ready to synchronize to Entra, you can deploy and configure one of the Entra Connect synchronization products.

Implementing and Managing Microsoft Entra Connect Sync

The synchronization product we have today started off as DirSync, a software appliance based on **Forefront Identity Manager** (**FIM**), supporting the deployment of Microsoft **Business Productivity Online Suite** (**BPOS**) in 2007. The synchronization engine product's code base was upgraded from FIM to **Microsoft Identity Manager** (**MIM**) and renamed to **Azure AD Sync** in September 2014. It was then upgraded further and renamed to **Azure AD Connect** in June 2015. As of August 2023, the Azure AD Connect product has been renamed **Entra Connect**.

> Important
>
> While Microsoft has updated the product names in its marketing materials and documentation, many of the tools and user interfaces have yet to be fully refreshed. As such, you may see references to both the legacy names (Azure AD Connect or Azure AD Sync) and the current name (Entra Connect).

Entra Connect allows you to connect to multiple directory sources and provision those objects to Entra.

Planning and Sizing

Depending on your organization's requirements for onboarding to Microsoft 365 as well as additional features or services that are included with your subscription, you may want (or need) to enable or configure additional Entra Connect features.

Table 4.2 lists features that can be enabled through the Entra Connect setup:

Feature	Description
Device Writeback	This synchronizes Entra-joined devices back to on-premises AD
Directory Extensions	This enables the synchronization of additional on-premises attributes
Federation	This enables authentication federation with Microsoft AD FS or PingFederate
Hybrid Entra Join	This enables on-premises domain-joined devices to be synchronized and automatically joined to Entra
Password Hash SynchronizationPHS	This enables the hash of on-premises passwords to be synchronized to Entra; it can be used for authentication, as a backup option for authentication, or to support Leaked Credential Detection (an Entra Premium P2 feature)
PTAPass-through Authentication	An authentication method where passwords are validated on-premises through the Entra Connect service's connection to Entra Service Bus
Unified Group Writeback	This enables cloud-based Microsoft 365 groups to be written back to on-premises AD

Table 4.2: Entra Connect features

We'll explore two of these features (Password Hash SynchronizationPHS and PTAPass-through Authentication) in upcoming sections. There are additional features available post-installation for Entra Connect, such as managing **duplicate attribute resiliency** and **user principal name soft-matching**, both of which are used to control how Entra handles conflicts and links on-premises accounts to cloud objects.

> **Further reading**
>
> More detailed information about the Entra Connect optional features, such as duplicate attribute resiliency, is available here: `https://learn.microsoft.com/en-us/entra/identity/hybrid/connect/how-to-connect-syncservice-features`.

Installing the Synchronization Service

The first step to deploying Entra Connect is gathering the requirements of your environment. These requirements can impact the prerequisites for deployment (such as additional memory or a standalone SQL Server environment). As part of the planning process, you'll also want to identify which sign-in method will be employed (password hash synchronization, pass-through authentication, or federation—all of which are described later in this chapter in more depth).

While Entra Connect does come with its own SQL engine, it has scalability limits. If you are going to synchronize more than 100,000 objects, you will need to configure Entra to use a standalone SQL server.

> **Exam tip**
>
> To perform the express installation, you'll need an Enterprise Admin credential for the on-premises Active Directory forest so that the installer can create a service account and delegate the correct permissions.
>
> You'll also need an account that has either the Global Administrator or Hybrid Identity Administrator roles in Entra. The Entra Connect installer will create a cloud synchronization service account.

With that information in hand, it's time to start deploying Entra Connect:

1. On the server where Entra Connect will be deployed, download the latest version of the Entra Connect setup files (`https://aka.ms/aad-connect`) and launch the installer.

2.   Agree to the installation terms and select **Continue**.

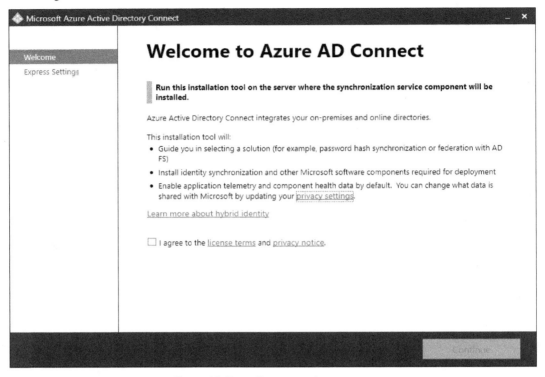

Figure 4.2: Entra Connect welcome page

3.   Review the **Express Settings** page shown in *Figure 4.3*. **Express Settings** automatically configures password hash synchronization. You can choose **Customize** if you want to configure Entra Connect to use pass-through or federated authentication methods, group-based filtering, or a custom SQL Server installation. While the sign-in methods and other features can be changed after installation, it is not possible to enable group-based filtering or change the SQL Server location after setup.

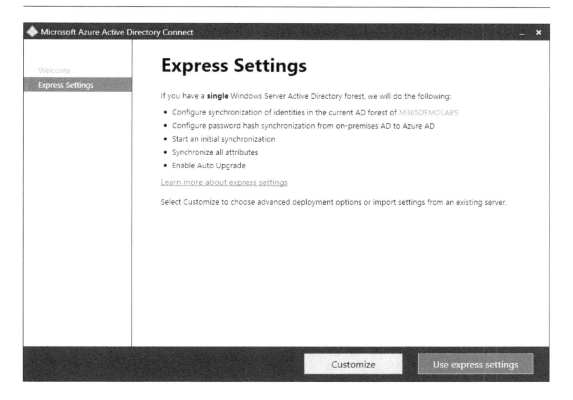

Figure 4.3: The Entra Connect Express Settings page

Installation notes

If you have other domains in your Active Directory forest, they must all be reachable from the Entra Connect server or express installation will fail. You can perform a custom installation to specify which domains to include in synchronization. You can check networking using a community tool located at https://aka.ms/AADNetwork.

4. On the **Connect to Entra** page, enter a credential that has either the Global Administrator or Hybrid Identity Administrator role in Entra. Click **Next**.

5. On the **Connect to AD DS** page, enter an Enterprise Administrator credential and click **Next**.

6. Verify the configuration options. By default, the Exchange hybrid scenario is not enabled. If you have an on-premises Exchange environment that you will be migrating to Microsoft 365, select the **Exchange hybrid deployment** option to include the Exchange-specific attributes. If you want to perform additional configuration tasks prior to synchronizing users, clear the **Start the synchronization process when configuration completes** checkbox, as shown in *Figure 4.4*.

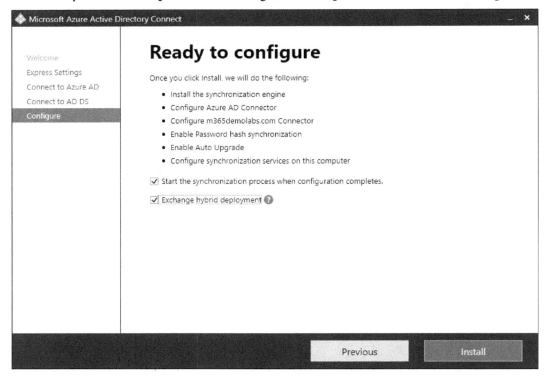

Figure 4.4: The Entra Connect Ready to configure page

> **Tip**
> If you don't have Exchange Server currently deployed or have upgraded your existing version of Exchange Server after your Entra Connect installation, you will need to re-run the setup to update the configuration. To avoid this, it's recommended to run the schema preparation for the latest available version of Exchange prior to running the Entra Connect setup.

7. Click **Install**.

8. Review the **Configuration complete** page and click **Exit**. If presented with recommendations (such as enabling the Active Directory recycle bin), you can click the link to learn more about them.

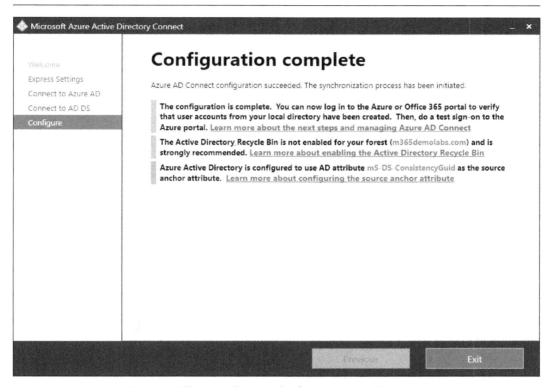

Figure 4.5: The Entra Connect Configuration complete page

If you select the **Start the synchronization process when configuration completes** checkbox, you can review the Entra portal to verify that users have been synchronized.

> **Exam tip**
>
> Filtering isn't explicitly called out or included in the SC-300 exam study guide, but it's good to know that it's available to reduce the scope of objects being synchronized. Many organizations use filtering to help align with Zero Trust principles.

Configuring Entra Connect Filters

If you need to exclude objects from Entra Connect's synchronization scope, you can do that through three methods:

- Domain- and **organizational unit** (**OU**)-based filtering
- Group-based filtering (only available during initial Entra Connect setup)
- Attribute-based filtering

Let's quickly examine these.

Domain- and Organizational Unit-Based Filtering

With this method, you can select large portions of your directory by modifying the list of domains or organizational units included in synchronization. While there are several ways to do this, the easiest way is through the Entra Connect setup and configuration tool:

1. To launch the Entra Connect setup tool, double-click the Entra Connect icon on the desktop of the server where Entra Connect is installed. After it launches, click **Configure**.

2. On the **Additional tasks** page, select **Customize synchronization options** and then click **Next**.

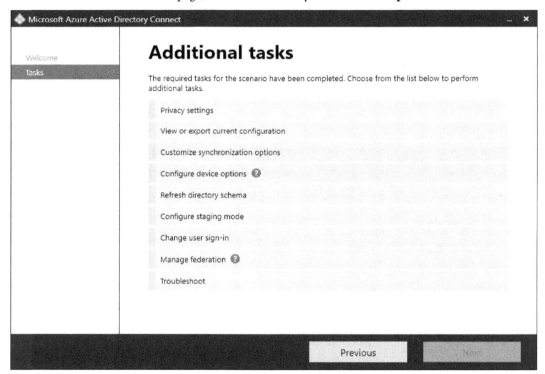

Figure 4.6: The Additional tasks page

3. On the **Connect to Entra** page, enter a credential with either the Global Administrator or Hybrid Identity Administrator role and click **Next**.

4. On the **Connect your directories** page, click **Next**.

5.   On the **Domain and OU filtering** page, select the **Sync selected domains and OUs** radio button and then either select (or clear) domains and organizational units to include or exclude from synchronization. From a Zero Trust perspective, you should consider excluding organizational units that contain service accounts or other privileged objects.

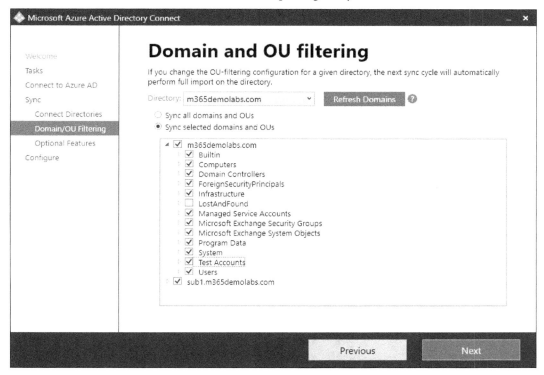

Figure 4.7: The Entra Connect Domain and OU filtering page

6.   Click **Next**.

7.   On the **Optional features** page, click **Next**.

8.   On the **Ready to configure** page, click **Configure**.

After synchronization completes, verify that only objects from in-scope OUs or domains that were selected are present in Azure AD.

Group-Based Filtering

Entra Connect only supports the configuration of group-based filtering if you choose to customize the Entra Connect setup. It is not available if you perform an express installation.

If you've chosen a custom installation, you can choose to set the synchronization scope to a single group. On the **Filter users and devices** page of the wizard, select the **Synchronize selected** radio button and then enter either the name or **distinguishedName (DN)** of a group that contains the users and devices to be synchronized to Entra, as shown in *Figure 4.8*.

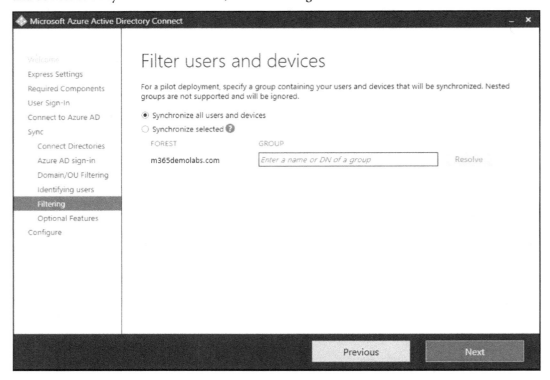

Figure 4.8: The Filter users and devices page

With group-based filtering, only direct members of the group are synchronized. Users, groups, contacts, or devices nested inside other groups are not resolved or synchronized.

> **Important**
>
> Microsoft recommends group-based filtering for piloting purposes only. For long-term or production filtering needs, use either domain- and organizational unit-based filtering or attribute-based filtering.

Attribute-Based Filtering

You can also control the synchronization scope by using an attribute filter. This method is more complex than other methods. It requires creating one or more custom synchronization rules in the Entra Connect Synchronization Rules Editor.

To create an attribute-based filtering rule, select an attribute that isn't currently being used by your organization for another purpose. This attribute will be a scoping filter to exclude objects.

The following procedure can be used to create a simple filtering rule:

1. On the server running Entra Connect, launch the Synchronization Rules Editor.
2. Under **Direction**, select **Inbound** and then click **Add new rule**. See *Figure 4.9.*

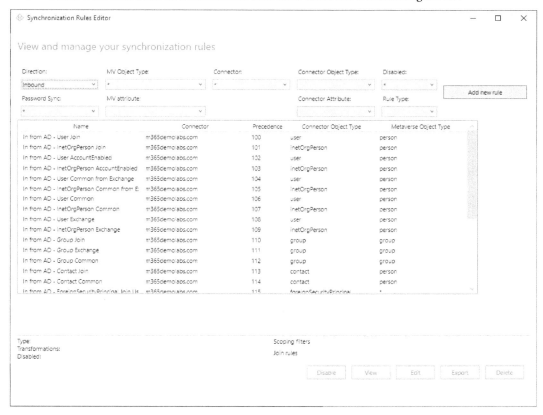

Figure 4.9: Synchronization Rules Editor

3. Provide a name and a description for the rule.

4. Under **Connected System**, select the object that represents your on-premises AD forest.

5. Under **Connect System Object Type**, select **user**.

6. Under **Metaverse Object Type**, select **person**.

7. Under **Link Type**, select **Join**.

8. In the **Precedence** text field, enter an unused number that is lower than any of the existing rules (such as **50**). See *Figure 4.10*. Click **Next**.

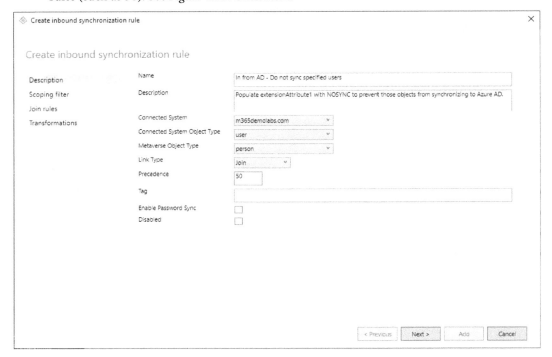

Figure 4.10: Creating a new inbound synchronization rule

9. On the **Scoping filter** page, click **Add group** and then click **Add clause**.

10. Under **Attribute**, select **extensionAttribute1** (or whichever unused attribute you have selected).

11. Under **Operator**, select **EQUAL**.

12. In the **Value** text field, enter NOSYNC and then click **Next**.

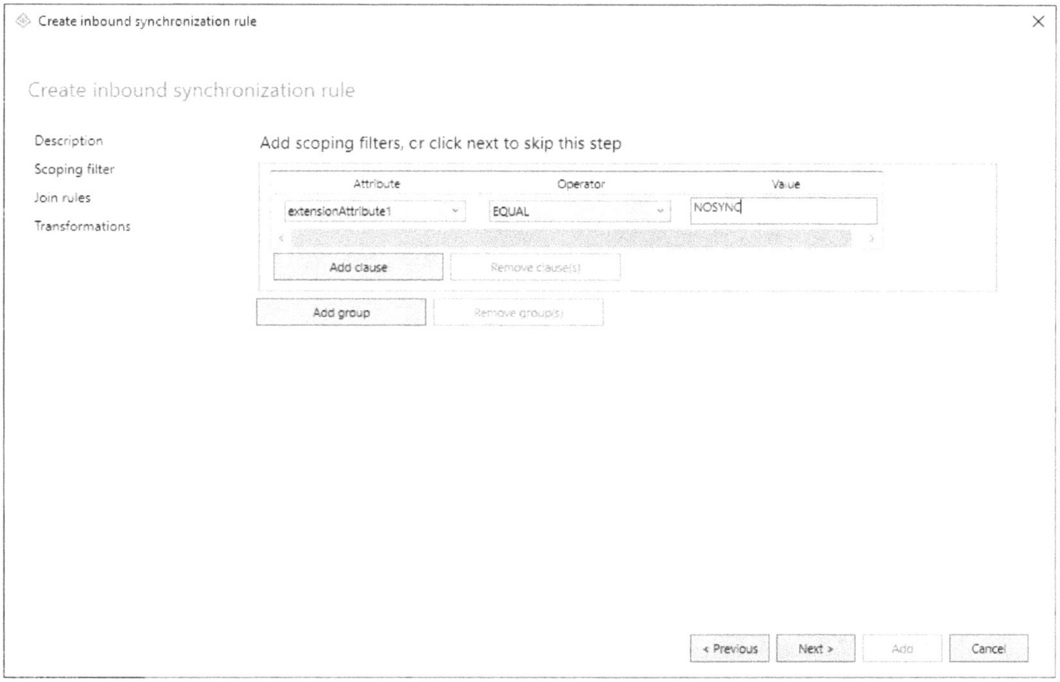

Figure 4.11: Configuring a scoping filter for extensionAttribute1

Tip

It's important to note that the value *itself* doesn't really matter. What matters is that the value you configure here *must* match the value you'll populate in the user's attribute, including case sensitivity.

---

13. On the **Join rules** page, do not add any parameters or values. Click **Next**.

14. On the **Transformations** page, click **Add transformation**.

15. Under **FlowType**, select **Constant**.

16. Under **Target Attribute**, select **cloudFiltered**.

17. In the **Source** text field, enter the value **True**.

18. Click **Add**.

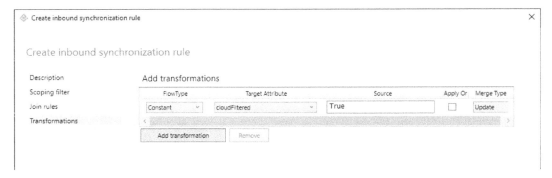

Figure 4.12: Adding a transformation for the cloudFiltered attribute

19. Acknowledge the warning that a full import and synchronization cycle will be required by clicking **OK**.

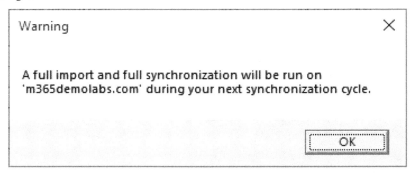

Figure 4.13: Warning for full import and synchronization

After creating, deleting, or modifying a synchronization rule, Entra Connect will require a full import and full synchronization. You don't have to perform any special steps, however; Entra Connect is aware of the change and will automatically perform the necessary sequences.

> **Further reading**
>
> There's a lot that can be done with the declarative rules engine in Entra Connect. In addition to filtering, you can use rules to transform or dynamically create attribute values in both Entra and Active Directory. For more information, see `https://learn.microsoft.com/en-us/entra/identity/hybrid/connect/concept-azure-ad-connect-sync-declarative-provisioning-expressions`.

Next, we'll shift gears to the lightweight Microsoft Entra Connect cloud sync provisioning agent.

Implementing and Managing Microsoft Entra Connect Cloud Sync

While the Entra Connect synchronization engine runs and processes identity on-premises, **Entra Connect cloud sync** is a synchronization platform that allows you to manage directory synchronization directly from the Entra portal. Depending on your organization's goals and environments, Entra Connect cloud sync may be a viable, flexible option that allows you to begin directory synchronization quickly.

> **Exam tip**
>
> To perform the installation, you'll need either a domain admin or Enterprise Admin credential for the on-premises AD forest so that the installer can create the **group Managed Service Account (gMSA)**. You'll also need an account that has either the Global Administrator or Hybrid Identity Administrator role in Entra.
>
> Microsoft recommends configuring a unique identity in Entra with the Hybrid Identity Administrator role for Entra Connect cloud sync.

Entra Connect cloud sync relies on a lightweight on-premises agent to detect changes in Active Directory. The agent is not responsible for any of the provisioning logic—that's handled by the Entra Connect cloud sync service.

Installing the Provisioning Agent

To begin installing Entra Connect cloud sync, follow these steps:

1. Log on to a server where you wish to install the Entra Connect cloud sync provisioning agent.

2. Navigate to the Microsoft Entra admin center (`https://entra.microsoft.com`), expand **Identity**, and select **Hybrid management| Microsoft Entra Connect**.

Figure 4.14: Entra Connect in the Microsoft Entra admin center

3. From the menu, select **Cloud sync**.

4. Under **Monitor**, select **Agents**.

5. Select **Download on-premises agent**, as shown in *Figure 4.15*.

Figure 4.15: Download on-premises agent for Entra Connect cloud sync

6. On the **Entra Provisioning Agent** flyout, select **Accept terms & download** to begin the download.

7. Open the **AADConnectProvisioningAgentSetup.exe** file to begin the installation.

8. Agree to the licensing terms and click **Install** to deploy the Microsoft Entra Connect Provisioning Agent package.

9. After the software installation is complete, the configuration wizard will launch automatically. Click **Next** to begin the configuration.

10. On the **Select Extension** page, choose the **HR-driven provisioning (Workday and SuccessFactors) / Microsoft Entra Cloud Sync** radio button and click **Next**.

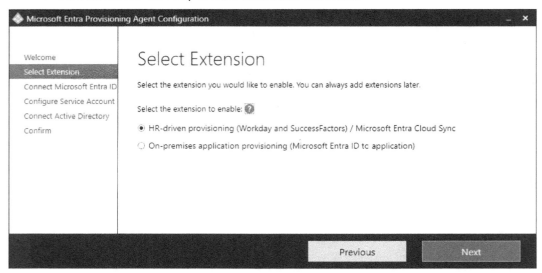

Figure 4.16: The Entra Connect cloud sync Select Extension page

11. On the **Connect Entra** page, click **Authenticate** to sign into Entra.

12. **On the Configure Service Account** page, select the **Create gMSA** radio button to provision a new group managed service account, as shown in *Figure 4.17*. Enter either a domain admin or Enterprise Admin credential and click **Next**.

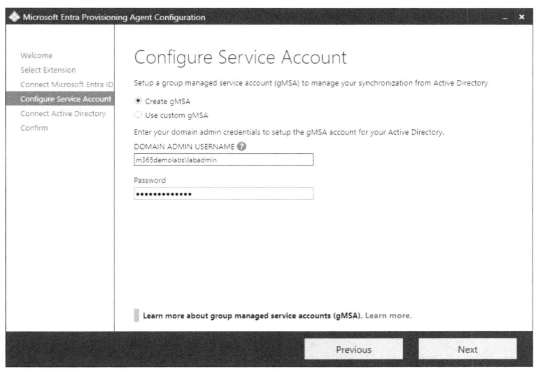

Figure 4.17: Configure Entra Connect cloud sync service account

13. On the **Connect Active Directory** page, click **Add directory** and provide the credentials to add the directory to the configuration. When finished, click **Next**.

Figure 4.18: Adding a directory to Entra Connect cloud sync

14. Review the details on the **Agent configuration** page and click **Confirm** to deploy the provisioning agent. When finished, click **Exit**.

After the agent has been deployed, you will need to continue the configuration in the Entra portal.

Configuring the Provisioning Service

In order to complete the Entra Connect cloud sync deployment, you'll need to set up a new configuration in the Microsoft Entra admin center:

1. Navigate to the Microsoft Entra admin center (`https://entra.microsoft.com`), expand **Identity**, and select **Hybrid management| Microsoft Entra Connect**.

2. Select **Cloud sync** from the menu and then, on the **Configurations** tab, click **New configuration** and then select **AD to Microsoft Entra ID sync**.

3. On the **New cloud sync configuration** page, select which domains you would like to synchronize to Entra. If desired, select the **Enable password hash sync** checkbox. If you don't select this option, users will need to maintain two separate passwords (one on-premises and one in Entra). See *Figure 4.19*.

> **Exam tip**
>
> Entra Connect cloud sync does not support using password hash sync for **InetOrgPerson** objects.

Figure 4.19: Creating a new Entra Connect cloud sync configuration

4. Scroll to the bottom of the page and click **Create** to complete the basic configuration.

The Entra Connect cloud sync configuration has been completed but it is not yet enabled and ready to start provisioning users. In the next series of steps, you can customize the service before fully enabling it.

Customizing the Provisioning Service

Like the on-premises Entra Connect service, Entra Connect cloud sync features the ability to perform scoping (including or excluding objects from synchronization) as well as attribute mapping (though, as mentioned before, no precedence rules or combining attributes from multiple domains or forests are supported).

After creating a new configuration, you should be redirected to the properties of the configuration, as shown in *Figure 4.20*:

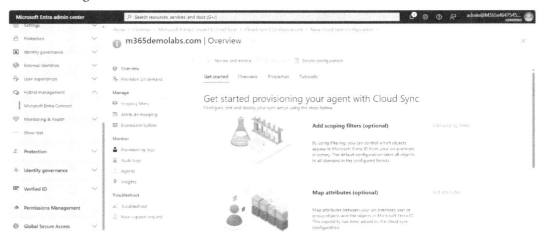

Figure 4.20: Provisioning agent overview page

From this page, you can set up the scoping filters and attribute mappings for customizing your environment. By default, Entra Connect cloud sync will include all objects in the connected forest and domains for synchronization.

Scoping Filters

By selecting **Scoping filters** under **Manage**, you can configure what objects should be included for synchronization to Entra. You can specify a list of security groups or select organizational units, but not both. See *Figure 4.21*.

Figure 4.21: Entra Connect cloud sync scoping filters

There are a few important caveats to keep in mind when using scoping filters with Entra Connect cloud sync:

- When using group-based scoping, nested objects beyond the first level will not be included in the scope
- You can only include 59 individual OUs or security groups as scoping filters

It's also important to note that using security groups to perform scoping is only recommended for piloting scenarios.

Attribute Mapping

Another customization option available involves mapping attribute values between on-premises and cloud objects. As with Entra Connect, you can configure how cloud attributes are populated—whether it's from a source attribute, a constant value, or some sort of expression.

Like Entra Connect, Entra Connect sync has a default attribute mapping flow, as shown in *Figure 4.22*:

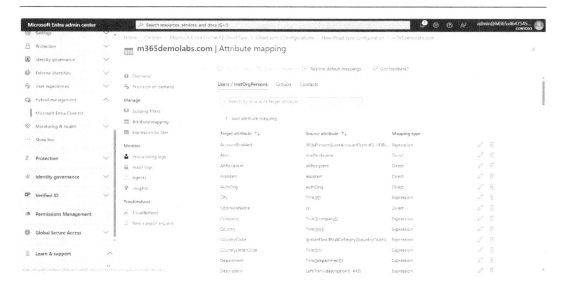

Figure 4.22: Entra Connect cloud sync attribute mappings

You can select an existing attribute to modify or create a new attribute flow.

Entra Connect cloud sync also features an expression builder, allowing you to create your own custom attribute flows.

> **Exam tip**
>
> Unlike Entra Connect, attribute mappings and expressions *cannot* be used to merge attributes from different domains or forests. Entra Connect cloud sync also does not support synchronization rule or attribute flow precedence. If you need to have a very specific ordering of rules and attribute flows, you should choose Entra Connect instead of Entra Connect cloud sync.

Once you have finished customizing scoping filters and attribute flows, you can return to the **Overview** page and enable synchronization by selecting **Review and enable**.

Next, we'll look at the various Entra authentication methods.

Implementing and Managing Password Hash Synchronization

Password hash synchronization (PHS) is one of the three sign-in methods that support hybrid identity. Microsoft Entra Connect synchronizes a hashed version of a user's password from an on-premises Active Directory to the Entra instance associated with an organization's tenant.

This feature builds on the directory synchronization capability provided by Microsoft Entra Connect Sync. It allows you to use the same password for signing into Microsoft Entra services, such as Microsoft 365, as you use for your on-premises Active Directory.

Password hash synchronization is Microsoft's recommended authentication method when deploying Entra Connect.

If you deploy Entra Connect using the **Express** option, PHS is automatically configured. If you choose a custom installation and choose another authentication option, you can easily switch back to PHS using the Entra Connect setup wizard.

From a security perspective, it's important to understand how Microsoft protects user password data as part of the password hash synchronization process. The following high-level overview captures the steps that Entra Connect performs to ensure your organization's password data is safe:

1. Every two minutes, the password hash synchronization agent on the Entra Connect server requests stored password hashes (stored in the `unicodePwd` attribute) from a **domain controller (DC)**. This request is made via the standard **Directory Replication Service Remote Protocol (MS-DRSR)**, which is typically used for synchronizing data between DCs.

2. Before transmission, the DC encrypts the MD4 password hash using a key derived from an MD5 hash of the RPC session key and a salt. This encrypted hash is then sent to the PHS agent over **Remote Procedure Call (RPC)**. The DC also transmits the salt to the synchronization agent using the DC replication protocol, enabling the agent to decrypt the envelope.

3. Once the PHS agent receives the encrypted envelope, it utilizes `MD5CryptoServiceProvider` and the provided salt to generate a decryption key, restoring the data to its original MD4 format. It's important to note that the agent never has access to the clear text password. The use of MD5 by the PHS agent is solely for compatibility with the DC's replication protocol and is limited to on-premises interactions between the DC and the agent.

4. The PHS agent converts the 16-byte binary password hash into a 32-byte hexadecimal string, and then back into a 64-byte binary format using UTF-16 encoding.

5. To further secure the original hash, the PHS agent appends a per-user salt, which is 10 bytes in length, to the 64-byte binary hash.

6. The agent then combines the MD4 hash with the per-user salt and processes it through the `PBKDF2` function, using 1,000 iterations of the `HMAC-SHA256` keyed hashing algorithm.

7. The resulting 32-byte hash is concatenated with both the per-user salt and the number of `SHA256` iterations (for use by Microsoft Entra ID), and this concatenated string is transmitted from Microsoft Entra Connect to Microsoft Entra ID over **Transport Layer Security (TLS)**.

8. When a user attempts to sign in to Microsoft Entra ID and inputs their password, the password undergoes the same *MD4+salt+PBKDF2+HMAC-SHA256* process. If the resulting hash matches the one stored in Microsoft Entra ID, the user is authenticated, confirming that the correct password was entered.

Next, we'll look at the steps to deploy password hash synchronization.

Deploying Password Hash Synchronization

Password hash synchronization requires two advanced Active Directory rights (*Replicating Directory Changes* and *Replicating Directory Changes All*) in order to be able to access the password data. During initial setup, these rights are granted automatically. If you configure Entra Connect to use a dedicated service account or change accounts on the deployed connectors later, you will need to delegate those rights manually. You can use the Active Directory **Delegation of Control** wizard to accomplish that task, as shown in *Figure 4.23*:

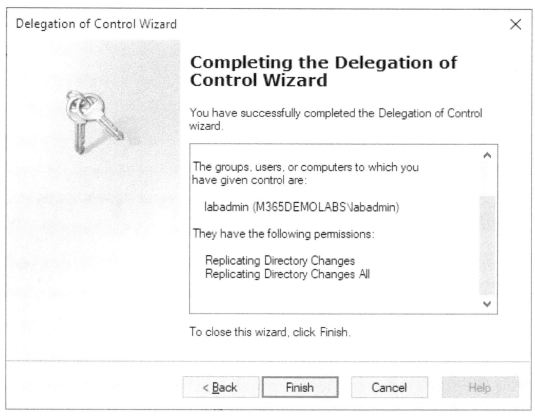

Figure 4.23: Using the Delegation of Control wizard

To update your deployment to use PHS, follow these steps:

1. On the server running Microsoft Entra Connect, double-click the icon labeled **Azure AD Connect** on the desktop to launch the setup wizard, as shown in *Figure 4.24*.

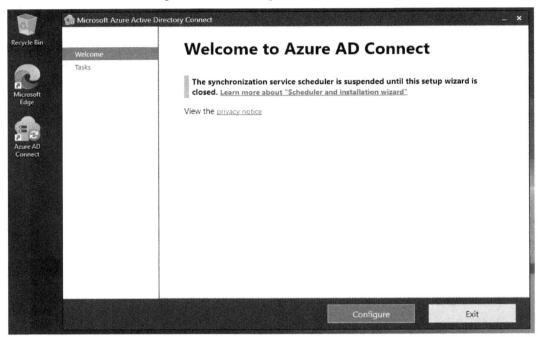

Figure 4.24: Launching the Entra Connect setup wizard

2. Click **Configure**.
3. Select **Change user sign-in** from the list of options and click **Next**. See *Figure 4.25*.

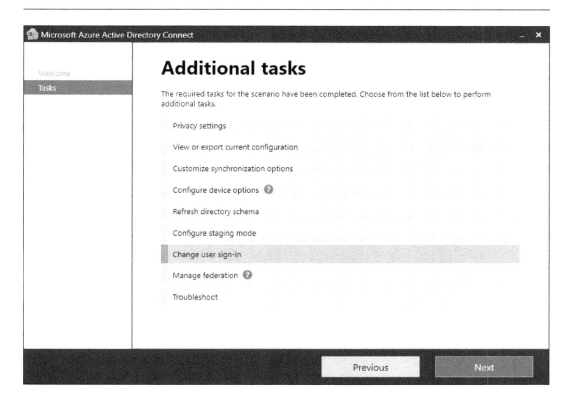

Figure 4.25: Selecting Change user sign-in

4. On the **Connect to Azure AD** page, enter a credential that has either the Global Administrator or Hybrid Identity Administrator role in your Entra tenant and click **Next**.

5. On the **User sign-in** page, select **Password Hash Synchronization** and click **Next**.

Figure 4.26: Selecting Password Hash Synchronization

6. On the **Ready to configure** page, click **Configure**.

After the setup has been completed, password hash data will be synchronized to Entra and the sign-in method for users will be set to password hash synchronization (which is essentially a cloud-based authentication).

Managing Password Hash Synchronization

There are a few password hash synchronization features that can be managed, either directly through Entra Connect configurations or through the on-premises directory. In this section, we'll look at the following:

- Password complexity
- Password expiration
- Temporary passwords
- Selective password hash synchronization

Let's look at each of these areas.

Password Complexity

When password hash synchronization is enabled, the password complexity policies from your on-premises AD take precedence over those in the cloud for synchronized users. This allows you to use all valid passwords from your on-premises AD to access Microsoft Entra services.

Password Expiration

If a user is in the scope for password hash synchronization, their synchronized cloud account password is set to **Never Expire**.

Users can continue to sign on to Microsoft 365 or Entra services using a password that may be expired in your on-premises directory. This may be a compliance risk for some organizations—especially if remote users never log on to the domain. If you need your cloud users to comply with your on-premises password expiration policy, you can run the following Microsoft Graph script:

```
Connect-MgGraph -Scopes "OnPremDirectorySynchronization.ReadWrite.All"
$OnPremSync = Get-MgDirectoryOnPremiseSynchronization
$OnPremSync.Features.CloudPasswordPolicyForPasswordSyncedUsersEnabled
= $true
Update-MgDirectoryOnPremiseSynchronization
-OnPremisesDirectorySynchronizationId $OnPremSync.Id -Features
$OnPremSync.Features
```

If any users have the **DisablePasswordExpiration** value already set on their user account, this won't take effect until they change their on-premises password unless you also run `Update-MgUser -UserID <User Object ID> -PasswordPolicies "DisablePasswordExpiration"` for each affected user.

> **Best practices**
>
> If you're going to enable `CloudPasswordPolicyForPasswordSyncedUsersEnabled`, you should do it prior to initializing password hash synchronization. That way, users get configured correctly as soon as PHS has been enabled. It's also best practice to set your Active Directory and Entra password expiration policies the same. You can update the Entra password policy using the `Update-MgDomain` cmdlet.

Temporary Passwords

If your organization uses the **User must change password at next logon** option when resetting on-premises passwords for users, you may want to enable support for temporary passwords for Entra synchronized users.

You can do this using the following command:

```
Connect-MgGraph -Scopes "OnPremDirectorySynchronization.ReadWrite.All"
$OnPremSync = Get-MgDirectoryOnPremiseSynchronization
$OnPremSync.Features.UserForcePasswordChangeOnLogonEnabled = $true
Update-MgDirectoryOnPremiseSynchronization
-OnPremisesDirectorySynchronizationId $OnPremSync.Id -Features
$OnPremSync.Features
```

Once updated, when a user is forced to change their password on-premises, the flag to change passwords will be synchronized to Entra as well.

> **Best practice**
>
> It's recommended to also deploy **self-service password reset** (**SSPR**) along with this feature to ensure the updated cloud password is written back on-premises if the user changes their Entra password first.

Selective Password Hash Synchronization

While most organizations will deploy password hash synchronization across all users, there may be instances where your organization does not want to synchronize password hashes for every user that is in synchronization scope.

You can manage that through a feature called **selective password hash synchronization**. Configuring selective PHS involves three core steps:

1. Duplicating the default **In from AD – User AccountEnabled** rule, disabling PHS on the rule, and configuring a scoping filter to *include* objects with a certain attribute value set.

2. Duplicating the default **In from AD – UserAccountEnabled** rule, enabling PHS on the rule, and configuring a scoping filter to *exclude* objects with a certain attribute value set.

3. Updating objects in AD to include or exclude them from PHS.

It may sound a bit tricky at first, but it's easily configurable using the following process:

1. On the server running Entra Connect, launch an elevated PowerShell console.

2. In the PowerShell console window, run **Set-ADSyncScheduler -SyncCyleEnabled $False** to disable scheduled synchronizations while the rules are being modified.

3. Launch the **Synchronization Rules Editor**.

4. In the **Password Sync** dropdown, select **On**. Under **Rule Type**, select **Standard**. This will filter for the default **In from AD – User AccountEnabled** rule.

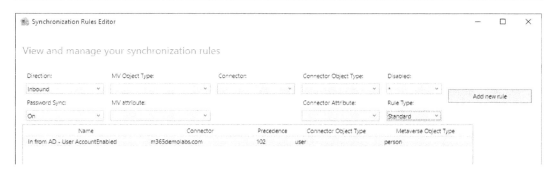

Figure 4.27: Filtering for the In from AD – UserAccountEnabled rule

5.   Select the rule and then click **Edit**.

6.   Click **Yes** to create a copy of this rule and disable the original rule.

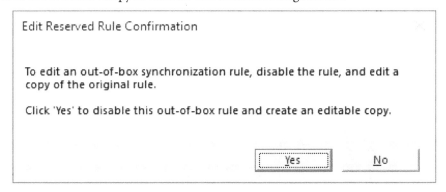

Figure 4.28: Disabling the original rule

Deep dive

The out-of-the-box rule is automatically cloned and disabled in modern versions of Entra Connect to ensure survivability during the upgrade process. During an Entra Connect upgrade, if there are rule changes to the default rules, they are automatically applied. This could have negative effects if you had edited a default rule and kept it active. With the clone-and-disable process, your changes to default rules are preserved in a new rule while the disabled out-of-the-box rule is upgraded.

7.   Edit the name and description of the rule to indicate that this rule will be scoped to users who have PHS *disabled*. Additionally, set the **Precedence** value to a number lower than the default rules, such as 95.

8. Ensure both the **Enable Password Sync** and **Disabled** checkboxes are clear. See the completed configuration page in *Figure 4.29*.

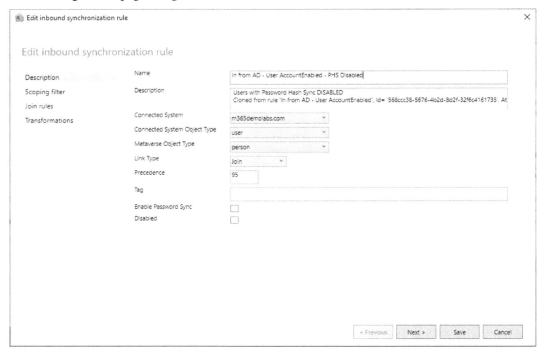

Figure 4.29: Creating the password hash synchronization disabled rule

9. Click **Next**.

10. On the **Scoping filter** page, click **Add clause**.

11. In the **Attribute** column dropdown, select an unpopulated attribute. In this example, we'll use the **adminDescription** attribute, but you can use any attribute that you want to designate for this filter.

12. In the **Operator** column dropdown, select **EQUAL**.

13. In the **Value** field, enter **PHSFiltered** (or whatever value you like to indicate that these users will be excluded from PHS). See *Figure 4.30*.

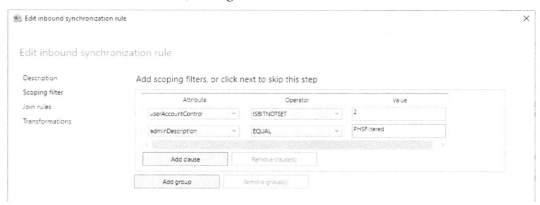

Figure 4.30: Configuring the scoping filter for the PHS-disabled users

14. Click **Save** to complete the rule.

15. Click **OK** to acknowledge the warning that a full synchronization task will be run on the next Entra Connect synchronization cycle.

16. On the Synchronization Rules Editor home page, the default **In from AD – User AccountEnabled** rule should still be highlighted. Select it and click **Edit**.

17. Click **Yes** to create another clone of this rule.

18. Edit the name and description of the rule to indicate that this rule will be scoped to users who have PHS *enabled*. Additionally, set the **Precedence** value to a number lower than the cloned rule you just created, such as 93.

19. Ensure **Enable Password Sync** is selected and that the **Disabled** checkbox is clear. See the completed configuration page in *Figure 4.31*.

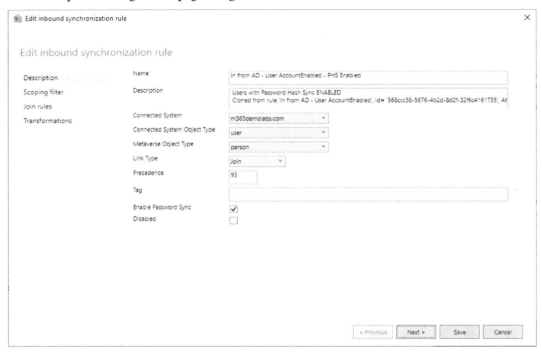

Figure 4.31: Creating the PHS-enabled rule

20. Click **Next**.

21. On the **Scoping filter** page, click **Add clause**.

22. In the **Attribute** column dropdown, select the attribute you chose in step 15. In this example, we'll use the **adminDescription** attribute, since that is what was used previously.

23. In the **Operator** column dropdown, select **NOTEQUAL**.

24. In the **Value** field, enter `PHSFiltered` (or whatever value you specified in step 17). See *Figure 4.32*.

Figure 4.32: Configuring the scoping filter for the PHS-enabled users

25. Click **Save** to complete the rule.

26. Click **OK** to acknowledge the warning that a full synchronization task will be run on the next Entra Connect synchronization cycle.

27. Close the Synchronization Rules Editor.

28. In the PowerShell console window, run `Set-ADSyncScheduler -SyncCyleEnabled $True` to re-enable scheduled synchronizations.

After the configurations are complete, you can then begin editing the properties for users that you want to be excluded from password hash synchronization. You can use PowerShell or **Active Directory Users and Computers (ADUC)**.

See *Figure 4.33* for an example of disabling password hash synchronization for a user. In this case, the scoping filter attribute is **adminDescription** and the scoping filter value is **PHSFiltered**.

Figure 4.33: Disabling password hash synchronization for a user

Important

Users excluded from having their passwords synchronized will be out of scope for self-service password reset operations that write back to on-premises Active Directory.

As a best practice, Microsoft recommends using password hash synchronization to both simplify hybrid identity deployments as well as to enable advanced E5 features such as leaked credential detection. Next, we'll look at a sign-in method that validates passwords using the on-premises domain controller infrastructure.

Implementing and Managing Pass-Through Authentication

Pass-through authentication (**PTA**) is a type of sign-in method where cloud authentication requests are processed by on-premises domain controllers. For organizations that aren't able to or don't want to deploy PHS, PTA is an alternative that minimizes the amount of on-premises infrastructure needed for authenticating requests and helps honor things such as password policies.

> Exam tip
>
> One of the benefits of PTA over a solution such as federation is that PTA only requires outbound connectivity from the servers running the pass-through authentication agent. Like federation, however, PTA depends on the availability of on-premises infrastructure to process user logons. If the on-premises infrastructure isn't available, then users will be unable to log onto the service.

If pass-through authentication is enabled on the tenant, the authentication process follows these steps:

1. The user tries to access a resource or application authenticated by Entra.

2. If the user is not already signed in or using single sign-on, the user is redirected to the Microsoft Entra ID user sign-in page where they enter their username and password and then attempt to sign in.

3. Microsoft Entra ID places the username and password (encrypted using the public key of each of the configured Entra PTA agents) in a queue on Entra Service Bus.

4. An on-premises PTA agent retrieves the username and encrypted password from the queue and the agent decrypts the password by using its private key.

5. The agent validates the username and password against Active Directory.

6. The on-premises domain controller evaluates the request and returns the appropriate response (success, failure, password expired, or user locked out) to the agent.

7. The authentication agent delivers this response back to Microsoft Entra.

8. Microsoft Entra evaluates the reply and delivers the response to the user (such as granting access, requesting a password change, or prompting for multifactor authentication).

9. If the user sign-in is successful, the user can access the resource or application.

Next, let's see how it can be implemented!

Deploying Pass-Through Authentication

Like password hash synchronization, PTA can be enabled from the Entra Connect setup wizard. To configure your Entra tenant to use PTA, follow these steps:

1. On the server running Microsoft Entra Connect, double-click the icon labeled **Azure AD Connect** on the desktop to launch the setup wizard.

2. Click **Configure**.

3. Select **Change user sign-in** from the list of options and click **Next**.

4. On the **Connect to Azure AD** page, enter a credential that has either the Global Administrator or Hybrid Identity Administrator role assigned and then click **Next**.

5. On the **User sign-in** page, select **Pass-through Authentication** and click **Next**.

6. On the **Ready to configure** page, click **Configure**.

After the setup has been completed, the tenant will be configured to use pass-through authentication.

Identifying Limitations

It's important to note that there are some limitations and supported scenarios with regard to pass-through authentication:

*   Leaked credentials detection does not work, since it requires PHS.

*   Microsoft Entra Domain Services requires PHS to be enabled on the tenant. Using PTA only is not supported.

*   Signing in to Microsoft Entra joined devices with a temporary or expired password is not supported for PTA users. These users must sign in to a browser first to update their temporary password.

*   PTA is not currently integrated with Microsoft Entra Connect Health.

It is supported, however, to have pass-through authentication configured as the sign-in method and then have password hash synchronization enabled as an additional option in the Entra Connect setup. This will enable a workaround for the first three scenarios, though Entra Connect Health will still not work for the PTA sign-in method.

Implementing and Managing Seamless single sign-on

Seamless single sign-on (sometimes written as **Seamless SSO** or **SSSO**) automatically signs the user into Entra ID when they are using corporate devices connected to their organization's corporate network. After it's been enabled, users no longer need to enter their passwords when browsing Entra-authenticated services such as Microsoft 365.

Seamless SSO is available to be combined with either PHS or PTA sign-in methods. Seamless SSO works by leveraging Kerberos delegation to an AD computer account (named AZUREADSSOACC in the local directory). Through the sign-in process, a native application or browser session requests a Kerberos ticket for the AZUREADSSOACC computer account and presents it to Entra ID along with the user's identity as proof that the user is logged into an AD domain.

Seamless SSO can be enabled through the Entra Connect setup wizard using the following process:

1. On the server running Microsoft Entra Connect, double-click the icon labeled **Azure AD Connect** on the desktop to launch the setup wizard.

2. Click **Configure**.

3. Select **Change user sign-in** from the list of options and click **Next**.

4. On the **Connect to Azure AD** page, enter a credential that has either the Global Administrator or Hybrid Identity Administrator role assigned and then click **Next**.

5. On the **User sign-in** page, select the **Enable single sign-on** checkbox and click **Next**.

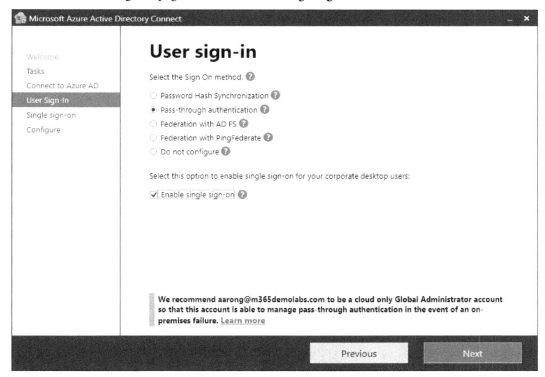

Figure 4.34: Enabling seamless single sign-on

6. On the **Enable single sign-on** page, click **Enter credentials** and provide a domain administrator or Enterprise Administrator credential. This credential is not stored and is only used to configure the **AZUREADSSOACC** computer account.

Figure 4.35: Providing a credential

7. Click **Next**.

8. On the **Ready to configure** page, click **Configure**.

That's all the configuration necessary to enable seamless SSO!

> **Further information**
>
> Configuring seamless SSO through the Entra Connect setup wizard provides support for Windows 7 and later devices. However, if your environment is using Windows 10 or later and Entra Hybrid Join, seamless SSO is achieved through **Primary Refresh Tokens (PRTs)**.

Next, we'll tackle migrating from Active Directory Federation Services to either password hash synchronization or pass-through authentication.

Migrating from AD FS to Other Authentication and Authorization Mechanisms

Over the past several years, Microsoft has made significant investments in both PHS and PTA, as well as security features that can easily be layered on top of them such as risk-based Conditional Access policies and multifactor authentication.

While AD FS can still be configured (either standalone or through the Entra Connect setup wizard), Microsoft recommends migrating to either PHS or PTA.

When migrating to PHS or PTA from AD FS, there are two main approaches: **cutover** and **staged rollout**. We'll briefly look at the steps for each of them.

Performing a Cutover Migration

From a process perspective, a **cutover migration** has the fewest steps and is the easiest to execute. In a cutover migration, everyone is migrated at one time from federated authentication to one of the cloud authentication methods.

Follow these steps to perform a cutover migration:

1. On the server running Microsoft Entra Connect, double-click the icon labeled **Azure AD Connect** on the desktop to launch the setup wizard.

2. Click **Configure**. Select **Change user sign-in** from the list of options and click **Next**.

3. On the **Connect to Azure AD** page, enter a credential that has either the Global Administrator or Hybrid Identity Administrator role assigned and then click **Next**.

4. On the **User sign-in** page, select either the **Password Hash Synchronization** or **Pass-through authentication** radio buttons. Click **Next**.

5. On the **Ready to configure** page, click **Configure**.

That's it! It couldn't be easier to perform a cutover migration.

However, if your organization isn't ready for a full cutover and you want to do it in phases, you can go through the staged rollout process.

Performing a Staged Rollout

Both password hash synchronization and pass-through authentication support staged rollouts when migrating from a federated authentication method. You can scale up as your organization becomes more comfortable with PHS and PTA.

Preparing for Staged Rollout

In order to support migrating users between federated and cloud authentication methods, you'll need to prepare one or more groups. Users that will be migrated to cloud authentication are placed in the staged rollout groups. Groups can be either cloud-based or synchronized from on-premises.

In addition, if your organization is going to move to PHS, you'll need to enable the **Password hash synchronization** optional feature (separate from the sign-in method configuration) in the Entra Connect setup using the following process:

1. On the server running Microsoft Entra Connect, double-click the icon labeled **Azure AD Connect** on the desktop to launch the setup wizard.

2. Click **Configure**.

3. Select **Customize synchronization options** from the list of options and click **Next**.

4. On the **Connect to Azure AD** page, enter a credential that has either the Global Administrator or Hybrid Identity Administrator role assigned and then click **Next**.

5. On the **Connect your directories** page, click **Next**.

6. On the **Domain and OU** filtering page, click **Next**.

7. On the **Optional features** page, select the checkbox for **Password hash synchronization**, as shown in *Figure 4.36*, and click **Next**.

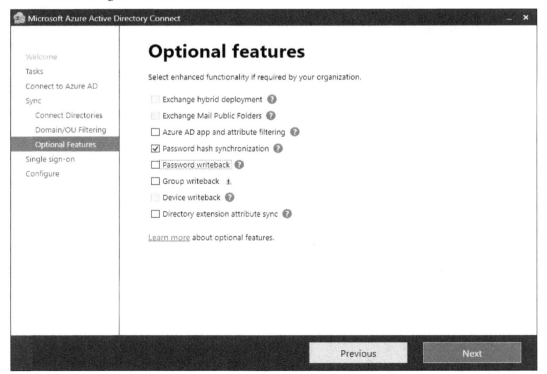

Figure 4.36: Enabling the Password hash synchronization optional feature

8. On the **Ready to configure** page, click **Configure**.

> **Important**
>
> It's important to note that this configuration selection *does not* update the sign-in method for your Entra tenant—this is just essentially configuring the **Enable password sync** checkbox on each of your organization's Active Directory connectors. The service account used by the connectors will need to have the *Replicating Directory Changes* and *Replicating Directory Changes All* rights granted if they haven't been granted.

Deploying a Staged Rollout Configuration

After you've gotten your staged rollout groups created (and enabled the password hash synchronization optional feature, if you're migrating to PHS), it's time to perform the cloud configuration. Follow these steps to complete the configuration:

1. Navigate to the Microsoft Entra admin center (`https://entra.microsoft.com`), expand **Identity**, and select **Hybrid management| Microsoft Entra Connect**.

2. Under **Staged rollout of cloud authentication**, select **Enable staged rollout for managed user sign-in**. See *Figure 4.37*.

Figure 4.37: Enabling staged rollout

3.   Select which sign-in method you wish to enable by moving the corresponding slider to **On**.

Figure 4.38: Choosing a managed authentication method

4.   When prompted, click **OK** to confirm.

5.   After a staged rollout method has been enabled, click **Manage groups**.

Figure 4.39: Managing groups for staged rollout

6.   Click **Add Groups** and then select up to 10 groups that will be used to manage users enabled for staged rollout.

Figure 4.40: Adding groups to staged rollout

After the groups have been added, you can begin placing users in the staged rollout groups to migrate them from federated authentication to either PHS or PTA.

Finalizing the Migration

Once your organization is comfortable with its adoption of PHS or PTA authentication, you can follow the steps outlined previously to enable either PHS or PTA as the sign-in method through the Entra Connect setup wizard.

Implementing and Managing Microsoft Entra Connect Health

Entra Connect Health is a premium feature of the Entra license. Entra Connect Health is used to provide monitoring information for components of your hybrid identity environment, including domain controllers, Entra Connect servers, and federation server infrastructure (if federation is used).

Entra Connect Health has separate agent features for Entra Connect, Entra Health for **Directory Services (DS)**, and Entra Health for AD FS.

Entra Connect Health

You can access the Entra Connect Health page by navigating to the Microsoft Entra admin center (`https://entra.microsoft.com`), expanding **Identity** | **Hybrid Management** | **Microsoft Entra Connect** | **Connect Sync**, and selecting **Microsoft Entra Connect Health** under **Health and analytics**, as shown in *Figure 4.41*:

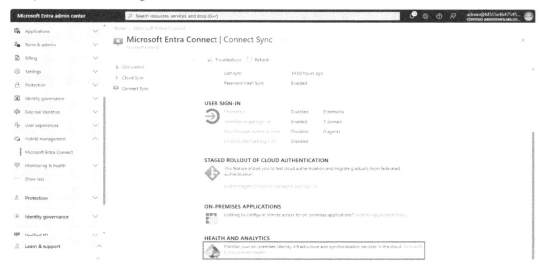

Figure 4.41: Navigating to Entra Connect Health

> **Other ways to get there**
>
> You can also browse to the Entra Connect Health page using this short link: `https://aka.ms/aadconnecthealth`. This link currently goes to the legacy Azure portal. The menu nodes and pages are exactly the same, though the external navigation menu from the Entra admin center is not displayed.

From there, you will be able to view basic details about your environment as well as obtaining agent installation packages.

While the Entra Connect Health for Sync agent is included in the Entra Connect installation, the health agents for DS and AD FS are separate installations and must be downloaded separately:

- **Entra Connect Health agent for DS**: `https://go.microsoft.com/fwlink/?LinkID=820540`

- **Entra Connect Health agent for AD FS**: `https://go.microsoft.com/fwlink/?LinkID=518973`

If you do not have AD FS deployed in your environment, you do not need to deploy the AD FS agents.

Entra Connect Health for Sync

The core health product, Entra Connect Health for Sync, shows the current health of your synchronization environment, including object synchronization problems and data-related errors.

You can view the health status and identified errors by selecting **Sync errors** under **Microsoft Entra Connect (Sync)**.

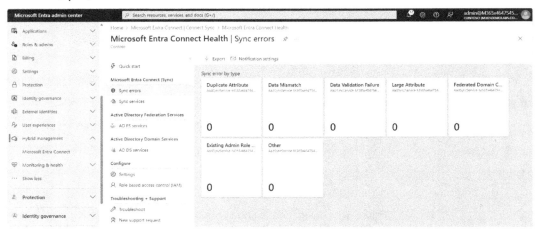

Figure 4.42: Entra Connect Health | Sync errors

If any errors exist, you can view them by selecting one of the tiles. This will allow you to drill down into individual errors. In the example in *Figure 4.43*, Entra Connect Health has detected two objects with the same address:

Figure 4.43: Entra Connect Health error details

You can use this information to identify and troubleshoot on-premises objects.

Entra Connect Health for Directory Services

Microsoft recommends deploying Entra Connect Health for Directory Services agents on all domain controllers you want to monitor or at least one for each domain.

The Entra Connect Health agent deployment is relatively straightforward. After downloading and installing the agent, you may be prompted to sign into your Entra tenant so that the agent can register with the health service.

Once the installation has been completed, you can review details about your domain controller health in the Entra Connect Health portal.

From the **Entra Connect Health** page, under **Active Directory Domain Services**, select **AD DS services**, as shown in *Figure 4.44*, and then select a domain to view details.

Figure 4.44: Entra Connect Health AD DS services

The health service agent displays a variety of details about the environment, including replication errors, LDAP bind operations, NTLM authentication operations, and Kerberos authentication operations, as shown in *Figure 4.45*:

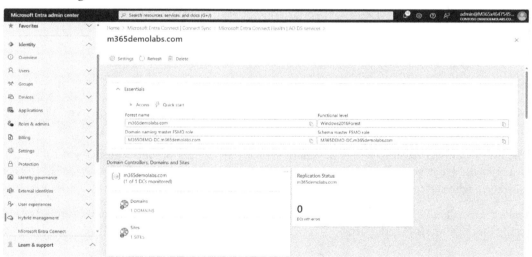

Figure 4.45: Entra Connect Health for Directory Services detail page

Errors that are detected here should be resolved in your on-premises AD environment.

Entra Connect Health for Active Directory Federation Services

In addition to gathering and reporting information for your on-premises Active Directory and synchronization services, Entra Connect Health also supports Active Directory Federation Services.

To get the most out of Entra Connect Health for AD FS, you'll need to enable auditing. Auditing is required to generate additional diagnostic log data, which the Entra Connect Health agents process and use to determine the health of the system. Enabling auditing involves three steps:

1. Ensure the AD FS farm service account has been granted the **Generate security audits** right in the security policy (**Local Policies | User Rights Assignment | Generate security audits**).

2. From an elevated command prompt, run the following command:

    ```
    auditpol.exe /set /subcategory:{0CCE9222-69AE-11D9-
    BED3-505054503030} /failure:enable /success:enable
    ```

3. On the AD FS primary farm server, open an elevated PowerShell prompt and run the following command:

    ```
    Set-AdfsProperties -AuditLevel Verbose
    ```

Then, you can deploy the agents to your servers.

After deploying the agents to your federation and proxy servers, you will see information reported in the Entra Connect Health portal under **Active Directory Federation Services**.

In addition to diagnostic information, the health services for AD FS can also provide usage analytics and performance monitoring, as well as failed logins and information regarding risky sign-ins.

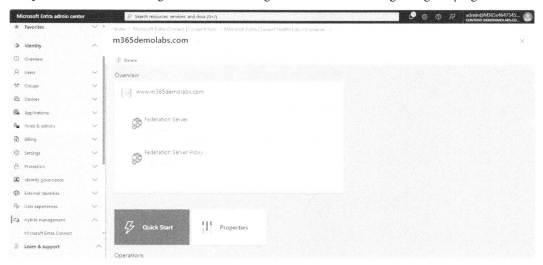

Figure 4.46: Entra Connect Health for AD FS

Entra Connect Health is a valuable premium service that can help keep you on top of the health and performance aspects of your hybrid identity deployment.

Summary

In this chapter, we explored many hybrid identity concepts, including directory synchronization through Entra Connect and Entra Connect cloud sync and authentication methods such as password hash synchronization (PHS) and pass-through authentication (PTA). Additionally, we also explored migrating from federated identity to one of the modern managed authentication methods and monitoring for the Entra Connect platform.

Selecting a hybrid identity model is an important part of an organization's security and is directly related to the Zero Trust Identity pillar.

In the next chapter, you'll learn how to manage advanced Entra authentication concepts such as Windows Hello for Business and multifactor authentication.

Exam Readiness Drill – Chapter Review Questions

Apart from mastering key concepts, strong test-taking skills under time pressure are essential for acing your certification exam. That's why developing these abilities early in your learning journey is critical.

Exam readiness drills, using the free online practice resources provided with this book, help you progressively improve your time management and test-taking skills while reinforcing the key concepts you've learned.

HOW TO GET STARTED

- Open the link or scan the QR code at the bottom of this page

- If you have unlocked the practice resources already, log in to your registered account. If you haven't, follow the instructions in *Chapter 19* and come back to this page.

- Once you log in, click the START button to start a quiz

- We recommend attempting a quiz multiple times till you're able to answer most of the questions correctly and well within the time limit.

- You can use the following practice template to help you plan your attempts:

Working On Accuracy		
Attempt	**Target**	**Time Limit**
Attempt 1	40% or more	Till the timer runs out
Attempt 2	60% or more	Till the timer runs out
Attempt 3	75% or more	Till the timer runs out
Working On Timing		
Attempt 4	75% or more	1 minute before time limit
Attempt 5	75% or more	2 minutes before time limit
Attempt 6	75% or more	3 minutes before time limit

The above drill is just an example. Design your drills based on your own goals and make the most out of the online quizzes accompanying this book.

First time accessing the online resources? 🔒
You'll need to unlock them through a one-time process. **Head to** *Chapter 19* **for instructions**.

Open Quiz	
https://packt.link/sc300ch4 OR scan this QR code →	

Planning, Implementing, and Managing Microsoft Entra User Authentication

The ability to effectively plan, implement, and manage user authentication is necessary to do your job well. This chapter delves into the comprehensive strategies and best practices for managing Microsoft Entra user authentication.

This chapter will guide you through the various facets of Microsoft Entra ID user authentication, from the foundational concepts to the details of configuration and management.

The objectives and skills we'll cover in this chapter include the following:

- Planning for authentication
- Implementing and managing authentication methods
- Implementing and managing tenant-wide **multi-factor authentication (MFA)** settings
- Managing per-user MFA settings
- Configuring and deploying self-service password reset
- Implementing and managing Windows Hello for Business
- Disabling accounts and revoking user sessions
- Deploying and managing password protection and smart lockout
- Enabling Microsoft Entra Kerberos authentication for hybrid identities
- Implementing certificate-based authentication in Microsoft Entra

By the end of this chapter, you'll be well equipped to plan, implement, and manage Microsoft Entra ID user authentication, ensuring a secure and efficient environment for your organization. This information will be valuable whether you're preparing for the SC-300 exam or seeking to improve your identity management practices.

This is one of the largest exam objectives, so let's get going!

Planning for Authentication

Planning for authentication within Microsoft Entra ID and Azure is critical for any **identity and access management** (**IAM**) administrator.

To begin with, you must evaluate the organization's security requirements and risk profile. This involves identifying sensitive resources, understanding potential threats, and determining the level of security needed for different user groups. For instance, high-risk users such as administrators or those with access to sensitive data may require more robust authentication methods such as MFA or biometric verification through **Windows Hello for Business** (**WHFB**). Conversely, regular users might be adequately protected with less stringent methods, provided they do not access critical systems.

A key aspect of planning for authentication is selecting and implementing appropriate authentication methods. Microsoft Entra ID supports a variety of methods, including password-based authentication, MFA, and passwordless options such as WHFB and FIDO2 security keys. Each method has its advantages and trade-offs. For example, while passwords are easy to implement, they are also susceptible to phishing and brute-force attacks. MFA significantly enhances security by requiring additional verification steps, such as a code sent to a mobile device or biometric verification. On the other hand, passwordless methods offer a seamless user experience and reduce the risk associated with password management.

Real-world scenarios highlight the importance of tailored authentication strategies. Consider a global enterprise with a diverse workforce, including remote employees and contractors. Implementing MFA for all users might be challenging in such a scenario due to varying levels of access to technology and internet connectivity. You could adopt a hybrid approach, deploying MFA for high-risk users and passwordless authentication for others, ensuring security and usability. You must also plan for contingencies, such as account recovery options and support for users who may have problems with the chosen authentication methods.

Another critical component of authentication planning is configuring tenant-wide settings and policies. This includes defining MFA policies, configuring Conditional Access rules, and setting up **self-service password reset** (**SSPR**) options. For example, an organization might enforce MFA for all users accessing cloud applications from outside the corporate network while allowing single-factor authentication within the network. Conditional Access policies can be fine-tuned to balance security and user experience, such as requiring a stronger MFA method only for high-risk sign-ins or when accessing sensitive applications.

Furthermore, you must stay informed about the latest updates and best practices in authentication. Microsoft regularly updates its authentication capabilities, and staying current with these changes is crucial for maintaining a secure environment. For instance, updates might include new MFA methods, enhanced reporting and monitoring tools, or improved integration with third-party identity providers.

Let's move on to the next critical aspect of user authentication: implementing and managing authentication methods.

Implementing and Managing Authentication Methods

Implementing and managing various authentication methods involves deploying and maintaining a range of authentication mechanisms to ensure secure access to resources while providing a seamless user experience. As you prepare for the SC-300 exam, you must understand the technical details and best practices for managing these methods, including **certificate-based authentication** (**CBA**), **temporary access passes** (**TAPs**), OAuth tokens, Microsoft Authenticator, and FIDO2 security keys.

CBA leverages digital certificates to verify user identities. This method is highly secure as it uses **public key infrastructure** (**PKI**) to authenticate users without relying on passwords. Implementing CBA involves issuing certificates to users and configuring Entra ID to trust these certificates. For example, an organization might use CBA to access sensitive applications or systems, ensuring that only devices with valid certificates can connect. This method is beneficial in environments where high security is essential, such as financial institutions or government agencies.

Implementing Temporary Access Passes

A TAP is a time-limited, one-time passcode that can be used for authentication. TAPs are beneficial for scenarios where users need temporary access to resources or when setting up new devices. For instance, an IAM administrator might issue a TAP to a new employee to allow them to sign in and set up their MFA methods. This approach simplifies the onboarding process and enhances security by limiting the validity of the passcode.

Here is a step-by-step guide on how to implement a TAP in Microsoft Entra ID:

1. Navigate to the Entra admin center (`https://entra.microsoft.com`) and sign in with an account that has either the Global Administrator or Authentication Policy Administrator role.

2. Expand **Identity** | **Protection** and select **Authentication methods**.

3. Under **Manage**, select **Policies**.

4. From the list of available authentication methods, click **Temporary Access Pass**. See *Figure 5.1*.

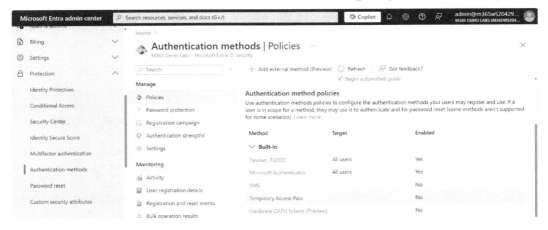

Figure 5.1: Selecting an authentication method policy

5. Under **Enable and Target**, toggle the slider for **Enable**. On the **Include** and **Exclude** tabs, choose which users or groups to include or exclude from the policy.

Figure 5.2: Configuring policy targeting

6. Optionally, select the **Configure** tab to modify default settings such as **Minimum lifetime**, **Maximum lifetime, Default lifetime**, whether an access pass is single-use or can be used multiple times, and **Length**. Click **Save** to apply the policy.

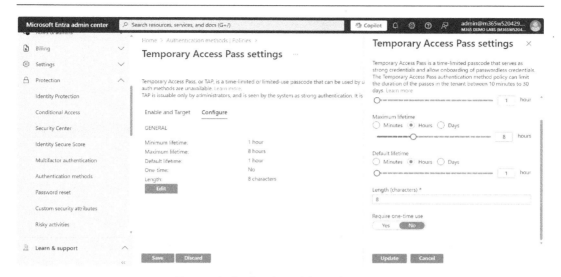

Figure 5.3: Configuring additional settings

7. With the policy established, now you can apply it to an individual user. Under **Identity**, expand **Users** and select **All users**.

8. Choose the user you want to create a TAP for. Click on **Authentication methods**.

9. On the **Authentication methods** page, select **Add authentication method**.

10. On the **Add authentication method** flyout, under **Choose method**, select **Temporary Access Pass**.

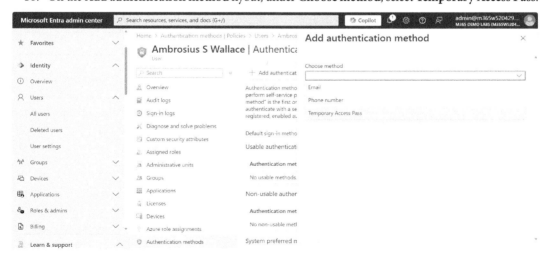

Figure 5.4: Adding a TAP authentication method

11. Configure any additional TAP properties (for example, **Delayed start time**, **Activation duration**, and whether it is single-use or multiple-use).

12. Click **Add** to generate the TAP.

13. Share the details of the TAP with the user securely via a secure messaging app, encrypted email, or a phone call. When you have finished, click **OK**.

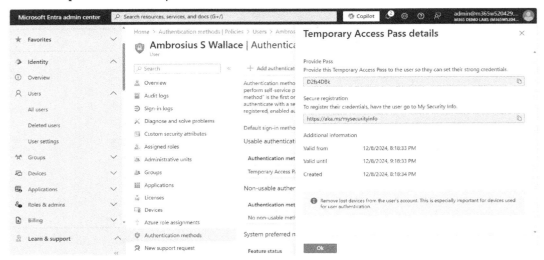

Figure 5.5: Confirming TAP details

The user can now use the TAP to sign in and register passwordless authentication methods such as Microsoft Authenticator, FIDO2 security keys, or WHFB. You can use Entra ID logs to monitor the issuance and usage of TAPs.

It's important to regularly review and update your TAP policies to ensure they meet your organization's security requirements. Use tools such as audit logs, SIEM systems, and monitoring dashboards to keep track of the effectiveness and security of your TAP setup. By continuously monitoring and refining your policies, you can maintain a robust security posture and respond swiftly to any potential issues.

Implementing OAuth

OAuth tokens are widely used to authorize resource access without exposing user credentials. OAuth is an open standard for access delegation, commonly used for granting websites or applications limited access to user information. In Entra ID, OAuth tokens authenticate and authorize access to APIs and services. For example, a mobile app might use OAuth tokens to access user data stored in Entra ID without requiring the user to enter their credentials repeatedly. Managing OAuth tokens involves configuring application registrations, defining permissions, and ensuring secure token handling practices.

Here's a step-by-step guide on how to implement OAuth tokens in Azure and Microsoft Entra ID.

Creating an App Registration

First, you'll need to create an app registration:

1. Navigate to the Entra admin center (`https://entra.microsoft.com`).
2. Expand **Identity** and select **App registrations**.
3. Click **New registration**.

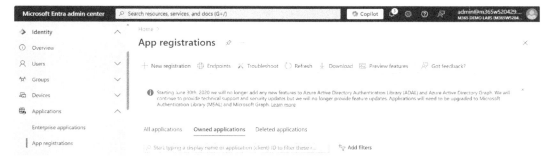

Figure 5.6: Adding a new app registration

4. Enter a name for your application.
5. Select the supported account types for this app registration (for example, **Single tenant**, **Multitenant**, **Multitenant and personal Microsoft accounts**, or **Personal Microsoft accounts only**).
6. Enter the redirect URI (for example, `https://localhost:5001/signin-oidc` for a web app).
7. Click **Register**.
8. In the app registration, go to **API permissions**.
9. Click **Add permission**.
10. Select **Microsoft Graph** (or another API your fictional app needs to access).
11. On the **Request API permissions** page, choose which type of permissions you will assign (such as **Delegated permissions** for user context or **Application permissions** for app-only context).

12. Locate the permissions you want to assign and select them. When you have finished, click **Add permissions**.

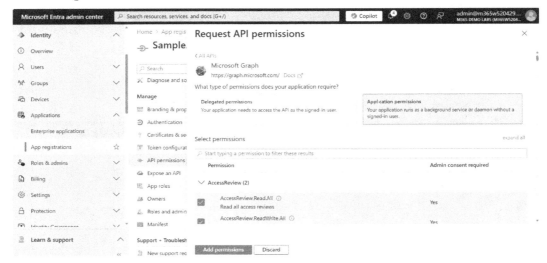

Figure 5.7: Configuring permissions

If you assigned app permissions, you also have the opportunity to grant admin consent to ensure that users don't get prompted.

13. Under **Manage**, select **Authentication**.

14. Click **Add a platform**. Select the platform type (for example, **Web**). Enter the redirect URIs and configure implicit grant settings if needed. Click **Configure**.

15. In the app registration, go to **Certificates & secrets**.

16. Click **New client secret**.

17. Enter a description and select an **Expiration** period. Click **Add**.

18. Copy the client secret value and store it securely.

Next, you need to implement an OAuth 2.0 authorization code flow.

Implementing an Authorization Code Flow

An authorization code flow is used to request access to data on a protected resource. It starts with a client application (such as the app registration you just created).

> **Don't worry!**
>
> The SC-300 exam doesn't test you on application development technologies or techniques. This process is really just to help you understand how the process works, but you shouldn't see any detailed questions on it. You'll want to be familiar with the keywords presented, but you won't have to set up an OAuth authorization code flow.

The basic process for creating an authorization code flow is as follows:

1. Depending on your application type, install the appropriate **Microsoft Authentication Library (MSAL)** (such as `MSAL.js` for JavaScript or `MSAL.NET` for .NET).

2. Configure the library using the client ID, tenant ID, and client secret from your app registration.

3. Redirect users to the Microsoft identity platform authorization endpoint. To do this, your application must send an authorization request to the Microsoft Entra authorization endpoint. This request typically includes several key parameters:

   - **Client ID**: Issued to your app during its registration

   - **Redirect URI**: The URL to which the authorization server will send the user after granting authorization

   - **Scope**: Specifies the level of access your app is requesting

   - **Response Type**: Indicates that your app expects an authorization code

   - **State**: A unique string that ensures the response matches the user's request and helps prevent cross-site request forgery attacks

   Handle the authorization response and exchange the authorization code for an access token. To do this, extract this code from the URL and prepare a POST request to the token endpoint, including the authorization code, client ID, client secret, redirect URI, and grant type (`authorization_code`). Ensure this request is sent using HTTPS. The server responds with a JSON object containing the access token and possibly a refresh token. Store the access token securely and use it in the authorization header of your HTTP requests, formatted as `Bearer {access_token}`, to access protected resources on behalf of the user.

4. Use the access token to call protected APIs.

Next, we'll shift focus to implementing Microsoft Authenticator.

Implementing Microsoft Authenticator

Microsoft Authenticator is a mobile app that provides an additional layer of security through MFA. Users can receive push notifications, generate **time-based one-time passcodes (TOTPs)**, or use biometric verification to authenticate. Implementing Microsoft Authenticator involves configuring Entra ID to support the app and guiding users through the setup process. For instance, an organization might require all employees to use Microsoft Authenticator for MFA, enhancing security by leveraging the app's robust authentication capabilities.

Here is a step-by-step guide on how to implement Microsoft Authenticator in Microsoft Entra ID:

1. Navigate to the Entra admin center (`https://entra.microsoft.com`). Ensure you have the necessary permissions; you need to have either the Global Administrator or Authentication Policy Administrator role to configure Microsoft Authenticator.

2. Expand **Protection** and select **Authentication methods**.

3. Select **Policies**.

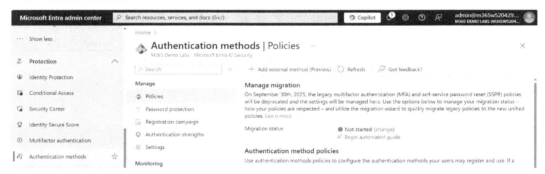

Figure 5.8: Authentication methods blade

4. Select **Microsoft Authenticator** from the list of available authentication methods.

5. On the **Enable and Target** tab, toggle the slider for **Enable** and select the users or groups to include in the policy.

6. Select the **Configure** tab.

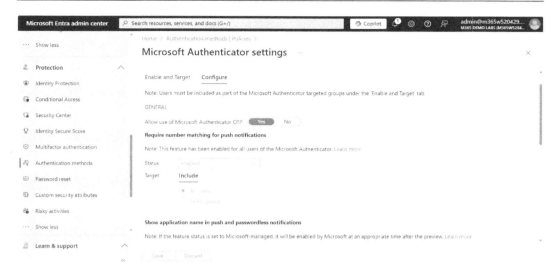

Figure 5.9: Configuring the Microsoft Authenticator settings

On this tab, there are several configuration settings that you can enable or disable, including the following:

- **Require number matching for push notifications**: This enhances security by requiring users to enter a number displayed on the sign-in screen into their Authenticator app when approving an MFA request. This setting has been enabled for all Microsoft Authenticator users and cannot be changed.

- **Show the application name in push and passwordless notifications**: This enhances security by displaying the name of the application requesting approval when a user receives a passwordless phone sign-in or MFA push notification.

- **Show geographic locations in push and passwordless notifications**: This adds an extra layer of security by displaying the geographic location of the sign-in attempt based on the IP address. When a user receives a passwordless phone sign-in or MFA push notification, they will see the location from which the sign-in request originated.

- **Enable Microsoft Authentication on companion applications**: This allows users to complete MFA using companion applications, such as Outlook mobile, without needing the full Microsoft Authenticator app.

7. Click **Save** to apply the policy.

You can instruct users to download the Microsoft Authenticator app from the Apple App Store or Google Play Store to their mobile devices. Users can sign in to the My Sign-Ins page (https://mysignins.microsoft.com) and then add their mobile device as a soft authentication token to their Microsoft 365 account. See *Figure 5.10*.

Figure 5.10: The My Sign-Ins page

Once that is all done, you will need to test the configuration. Use a non-administrator user account to test the Microsoft Authenticator setup. Ensure the user is prompted to use the Authenticator app during sign-in. Confirm that push notifications, verification codes, and passwordless sign-in are working correctly.

Implementing FIDO2

FIDO2 security keys offer a passwordless authentication method that uses public key cryptography to authenticate users. FIDO2 keys are hardware devices users can plug into their computers or connect via Bluetooth or **near-field communication** (**NFC**). Implementing FIDO2 involves configuring Entra ID to support these keys and distributing them to users. For example, an organization might deploy FIDO2 keys to employees who frequently travel or work remotely, providing a secure and convenient authentication method that reduces the risk of phishing attacks.

Here is a step-by-step guide on how to implement FIDO2 security keys in Azure and Microsoft Entra ID:

1. Navigate to the Entra admin center (https://entra.microsoft.com). Ensure you have the necessary permissions; you need to have either the Global Administrator or Authentication Policy Administrator role to configure Microsoft Authenticator.

2. Expand **Protection** and select **Authentication methods**.

3. Select **Policies**.

4. From the list of available authentication methods, select **Passkey (FIDO2) Security Key**.

5. On the **Enable and Target** tab, toggle the slider for **Enable** and select the users or groups to include in the policy.

6. Select the **Configure** tab.

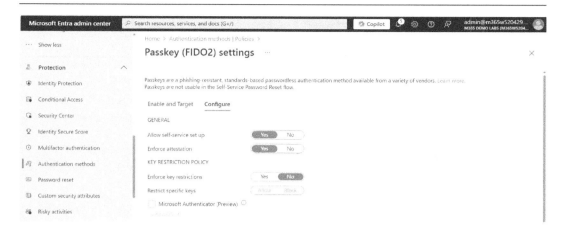

Figure 5.11: Configuring FIDO2 settings

On this tab, there are several configuration settings that you can enable or disable, including the following:

- **Allow self-service set up**: This enables users to set up FIDO2 keys.

- **Enforce attestation**: Require the attestation GUID to be provided during registration. Microsoft provides a list of compatible keys here: `https://learn.microsoft.com/en-us/entra/identity/authentication/concept-fido2-hardware-vendor#fido2-security-keys-eligible-for-attestation-with-microsoft-entra-id`.

- **Enforce key restrictions**: This parameter controls which specific FIDO2 security keys can be used for authentication. You can choose to block or allow keys based on the Authenticator attestation GUIDs.

7.   Click **Save** to apply the settings.

You can instruct users to download the Microsoft Authenticator app from the Apple App Store or Google Play Store to their mobile devices. Users can sign in to the My Sign-Ins page (`https://mysignins.microsoft.com`) and then add their mobile device as a soft authentication token to their Microsoft 365 account.

Consider a multinational corporation with a mix of on-site and remote employees. The organization might implement CBA for on-site access to critical systems, use a TAP for onboarding new employees, leverage OAuth tokens for mobile app access, require Microsoft Authenticator for MFA, and deploy FIDO2 keys for remote workers. This comprehensive approach ensures robust security while accommodating the diverse needs of the workforce.

IAM administrators can create a robust and flexible authentication framework by carefully configuring CBA, TAPs, OAuth tokens, Microsoft Authenticator, and FIDO2 security keys. With a solid understanding of these methods, you're ready to start learning about the next aspect of user authentication: implementing and managing tenant-wide MFA settings.

Implementing and Managing Tenant-Wide MFA Settings

MFA significantly enhances security by requiring users to provide multiple verification forms before accessing resources, reducing the risk of unauthorized access due to compromised credentials. This is an important part of the SC-300 exam.

Enabling tenant-wide MFA involves configuring policies that apply to all users within the organization. This can be achieved through the Entra admin center through the configuration of Conditional Access policies. These policies allow administrators to define specific conditions under which MFA is required, such as when users access sensitive applications, sign in from untrusted locations, or exhibit risky sign-in behaviors.

You can follow these steps to implement tenant-wide MFA settings in Microsoft Entra ID using a basic Conditional Access policy:

1. Navigate to the Entra admin center (`https://entra.microsoft.com`).

2. Expand **Protection** and select **Conditional Access**.

Figure 5.12: Configuring a Conditional Access policy

3. On the **Select a template** page, locate a template policy, such as **Require multifactor authentication for all users**.

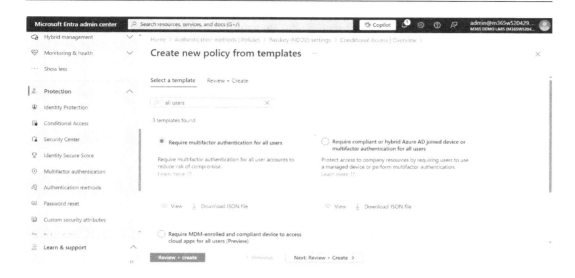

Figure 5.13: Selecting a policy template

4.   Click **Review + create**.

5.   On the **Review + Create** page, set **Policy state** to **On**.

A critical aspect of managing tenant-wide MFA settings is configuring Conditional Access policies. These policies enable administrators to enforce MFA based on various conditions, ensuring that security measures are applied dynamically based on the context of the sign-in attempt. For example, an organization might require MFA for all users accessing cloud applications from outside the corporate network while allowing single-factor authentication within the network. This approach balances security and user convenience, ensuring that high-risk activities are adequately protected without imposing unnecessary burdens on users.

We'll look at password reset options next.

Configuring and Deploying Self-Service Password Reset

Configuring and deploying SSPR in Microsoft Entra ID is an important tool in reducing the administrative burden associated with password management.

Enabling SSPR involves configuring the necessary settings in the Entra admin center. This includes defining which users or groups can use SSPR, typically by selecting specific Entra ID groups or enabling it for all users. Additionally, you must configure the authentication methods required for password reset.

In order for users to be able to have access to SSPR, they must register their authentication methods before they can use it. You can facilitate this process by prompting users to register during their next sign-in or sending targeted communications with instructions. For example, an organization might send an email campaign to all employees, guiding them through the registration process and highlighting the benefits of SSPR.

To enable SSPR, follow these steps:

1. Navigate to the Entra admin center (https://entra.microsoft.com).

2. Expand **Protection** and select **Password reset**.

3. Under **Manage**, choose the scope for users to enable for SSPR. You can choose **None**, **Selected** (and specify a group), or **All**.

4. Click **Save** when finished.

While that enables SSPR to be used, you must also configure which authentication methods will be available to users when they register. There are currently two ways to set up the authentication methods: the legacy **Authentication methods** page and the newer **Combined registration** process. We'll look at each.

Legacy Authentication Methods

The legacy authentication methods configuration is configured on the **Password reset** page. To walk through this process, follow these steps:

1. Navigate to the Entra admin center (https://entra.microsoft.com).

2. Expand **Protection** and select **Password reset**.

3. Under **Manage**, choose **Authentication methods**.

4. Choose an option for **Number of methods required to reset passwords** (either **1** or **2**; **2** is the recommended best practice), as well as which methods are available to users, such as **Mobile app notification**, **Mobile app code**, **Email**, **Mobile phone**, **Office phone**, or **Security questions**. If you select **Security questions**, you must also specify the number of questions users are required to answer for both registration and reset attempts. See *Figure 5.14*.

Figure 5.14: Configuring SSPR authentication methods

5.   Click **Save**.

6.   Under **Manage**, select **Registration**.

Figure 5.15: Configuring registration

7.   Enable the **Require users to register when signing in?** option and set the number of days before users are asked to reconfirm their authentication information. Click **Save**.

By following these steps, you can effectively configure SSPR in Microsoft Entra ID, reducing helpdesk calls and improving user productivity.

Combined Registration

Microsoft will be deprecating the legacy authentication methods configuration for SSPR and MFA on September 30, 2025, in favor of the new combined registration. Combined registration can be forced through the configuration of a Conditional Access policy that targets the registration action.

You can live in the future and configure the settings now (instead of waiting until the deprecation date) by following this process:

1. Navigate to the Entra admin center (`https://entra.microsoft.com`).

2. Expand **Protection** and select **Conditional Access**.

3. Select **Create new policy from templates**.

Figure 5.16: Creating a new Conditional Access policy

4. On the **Select a template** tab, choose the **Secure foundation** subtab. You can select **Securing security info registration** or search for *registration*.

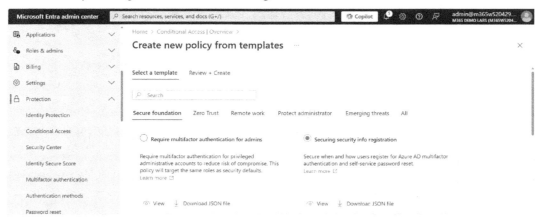

Figure 5.17: Locating the template

5. Click **View** to see the pre-built configuration displayed in the **Policy summary** flyout panel. Under **Cloud apps or actions**, note that the **User actions** item shows **Register security information**. Close the flyout.

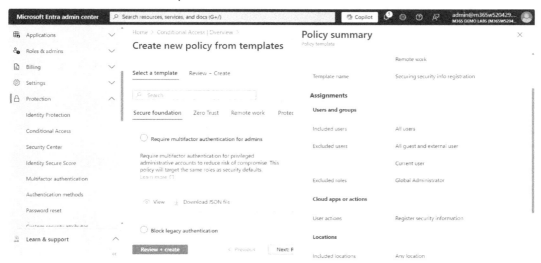

Figure 5.18: Viewing the registration policy

6. Click **Review + create**.

7. On the **Review + Create** tab, set **Policy state** to **On**. Note that the current user is excluded by default from template policies. For this policy template, external users and the Global Administrator role are also excluded.

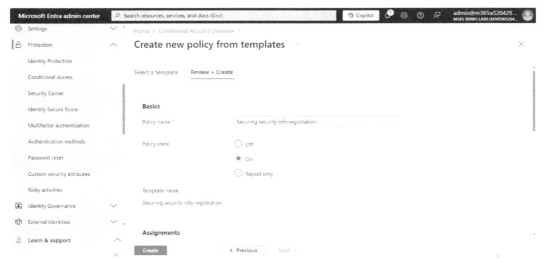

Figure 5.19: Updating the policy state

8. Click **Create**.

9. In the Entra admin center, expand **Protection** and select **Authentication methods**.

10. Under **Manage**, select **Policies**.

11. Under **Manage migration**, select **Begin automated guide** to start converting legacy policies and methods to the modern combined registration settings.

Figure 5.20: Updating to combined registration

12. On the **Authentication method settings migration** page, review your tenant's current configuration and click **Next**.

13. Review the settings. You can click the pencil icon next to an authentication method to update both the **Status** and **Targeted users & groups** options. When finished, click **Migrate**.

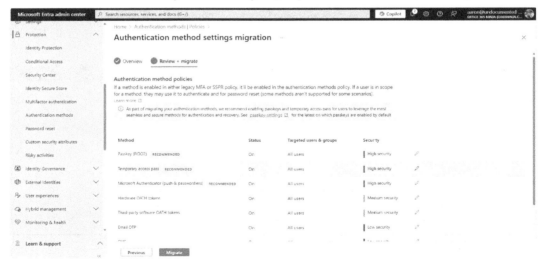

Figure 5.21: Completing the authentication methods migration

14. Click **Continue** to confirm the migration.

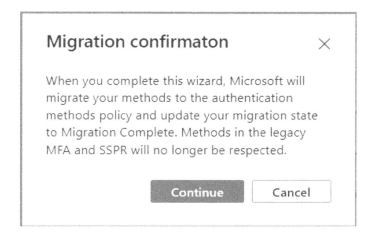

Figure 5.22: Confirming migration

15. On the **Policies** tab, confirm that **Migration status** is **Complete**.

Consider a large enterprise with thousands of employees where password reset requests are expected. Without SSPR, the IT helpdesk may be overwhelmed with password reset requests, leading to increased operational costs and reduced productivity. By implementing SSPR, the organization can significantly reduce the number of helpdesk tickets related to password resets, allowing IT staff to focus on more strategic tasks.

In the **Monitoring** section, you can review which users have registered with different types of authentication methods.

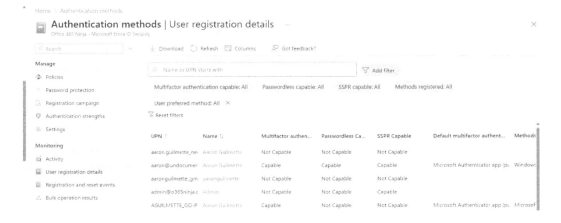

Figure 5.23: Reviewing user registration information

The next section will guide you through the essential steps and best practices to effectively implement and manage WHFB in your organization.

Implementing and Managing Windows Hello for Business

Implementing and managing WHFB provides users with a robust, passwordless authentication method that enhances security and the user experience. Microsoft recommends WHFB as the ideal solution for passwordless authentication, particularly for users with their own dedicated PCs. To log in, users simply present a biometric credential or enter a PIN to unlock the device.

> **Note**
> Since WHFB requires either a PIN or a biometric unlock that is device specific, it is treated as an MFA method.

WHFB supports various biometric methods, including facial recognition and fingerprint scanning. Devices configured with WHFB can be easily identified by the signature WHFB smiley face greeting displayed at the top, as shown in *Figure 5.24*.

Figure 5.24: WHFB-enabled sign-on

WHFB is unique from regular passwords in that the biometric and PIN used to unlock the device are unique to that device. They're protected by the device's **Trusted Platform Module (TPM)**.

> **Under the hood**
>
> If you were to examine how WHFB works, it's based on public key cryptography (the combination of a private and public key). WHFB is essentially a biometric-enhanced version of a PKI solution.

WHFB uses the following process to unlock or sign in to a device:

1. The user signs in with either a biometric identifier or PIN (if the configured biometric input method can't be accessed), which unlocks the WHFB private key. The key is then sent to the **cloud authentication security support provider** (sometimes referred to as the **Cloud AP**) part of the on-device security package.

2. The Cloud AP requests a **nonce** (a single-use randomly generated number) from Entra.

3. Entra sends the nonce to the Cloud AP on the device.

4. The Cloud AP signs the nonce with the user's private key and returns the signed nonce to Entra.

5. Entra decrypts and validates the signed nonce with the user's public key. After it's validated, Entra issues a **primary refresh token (PRT)** with a session key, encrypts it using the device's public transport key, and sends that back to the Cloud AP on the device.

6. The Cloud AP decrypts the PRT and session key data using the device's transport private key and then stores it in the device's TPM.

7. The Cloud AP returns a success response to Windows, allowing the user to complete the sign-in process.

WHFB is available to be deployed as either a cloud-only or hybrid identity solution and can be used for both Windows logon and logon to Microsoft 365 services. As mentioned earlier, WHFB authentication is unique to each device, meaning you have to set it up individually for each device that you will be using.

The easiest way to deploy WHFB is the cloud-only model, since your Microsoft 365 organization is set up for it automatically. WHFB can also be enabled via Group Policy for hybrid deployments. Microsoft is focusing on the cloud-only model, which is what we'll demonstrate in this section.

During the **out-of-the-box experience (OOBE)**, users are prompted for credentials. After providing Entra credentials, if the Intune enrollment policy has not been configured to block WHFB, the user will be prompted to enroll with their biometric data (such as a fingerprint or facial scan with a compatible camera) and set a PIN (which will be used if the biometric input is not available).

Devices will be joined to Entra during the initial sign-in process and WHFB will be enabled.

Microsoft recommends creating a WHFB policy to configure settings for your organization. To set up a WHFB policy in Intune, follow these steps:

1. Navigate to the Intune admin center (`https://intune.microsoft.com` or `https://endpoint.microsoft.com`).

2. Expand **Devices** and, under **Device onboarding**, select **Enrollment**.

3. Select **Windows Hello for Business**.

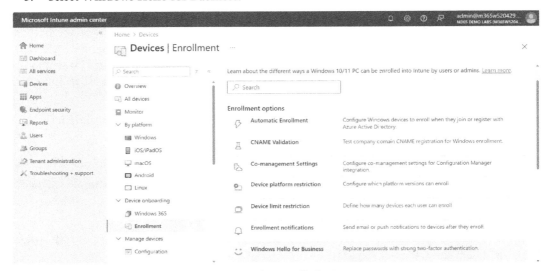

Figure 5.25: Windows Hello for Business

4. On the **Windows Hello for Business** flyout, under **Assigned to**, select a group (if you are scoping the enrollment policy to a subset of users) or select **All users**.

5. Under **Configure Windows Hello for Business**, select **Enabled** to display the configuration options. Configure the options for WHFB (bold options are the default settings for the enrollment policy):

    *   **Use a Trusted Platform Module (TPM)**: Choose **Required** or **Preferred**

    *   **Minimum PIN length**: Configure a numeric value between 4 and 127

    *   **Maximum PIN length**: Configure a numeric value between 4 and 127

- **Lowercase letters in PIN**: Choose **Not allowed**, **Allowed**, or **Required**

- **Uppercase letters in PIN**: Choose **Not allowed**, **Allowed**, or **Required**

- **Special characters in PIN**: Choose **Not allowed**, **Allowed**, or **Required**

- **PIN expiration (days)**: Choose **Never** or a numeric value between 1 and 730

- **Remember PIN history**: Choose **Never** or a numeric value between 1 and 50

- **Allow biometric authentication**: Choose **Yes** or **No**

- **Use enhanced anti-spoofing, when available**: Choose **Not configured**, **Yes**, or **No**

- **Allow phone sign-in**: Choose **Yes** or **No**

- **Use security keys for sign-in**: Choose **Not configured**, **Enabled**, or **Disabled**

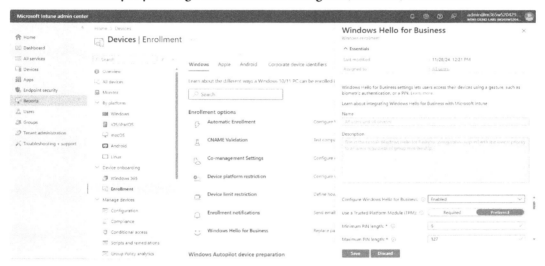

Figure 5.26: Configuring the Windows Hello for Business enrollment policy

6.   Click **Save** to complete the policy configuration.

With the policy configured, new device enrollments will receive the WHFB setup prompt to begin WHFB configuration, as shown in *Figure 5.27*:

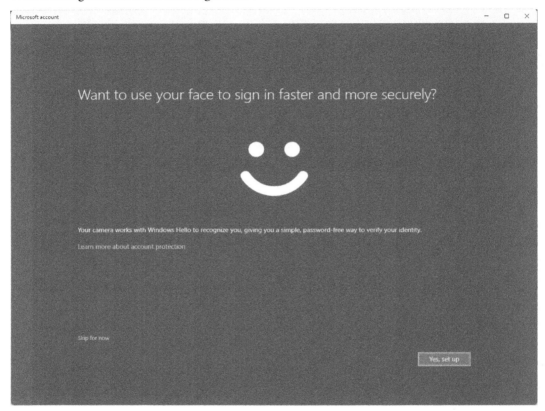

Figure 5.27: Windows Hello for Business enrollment

After completing enrollment, users will be able to unlock and log in to devices using supported biometrics or their PIN. The biometric or PIN can also be used to unlock other protected storage areas such as browser passwords or other saved credentials.

Users that are already connected to Entra can also trigger the Windows Hello setup wizard by either navigating to the **Account protection** blade in the Windows settings app or by pressing *Win + R* and entering ms-cxh://nthaad in the **Run** dialog box.

Disabling Accounts and Revoking User Sessions

Disabling accounts and revoking user sessions are essential for maintaining security and compliance, especially in scenarios involving compromised accounts, employee terminations, or insider threats.

Revoking user sessions involves invalidating the user's access tokens and refresh tokens, which are used to maintain authenticated sessions. It's important to note that access tokens typically have a lifespan of one hour, so revoking tokens ensures that the user is signed out of all sessions within this timeframe. For immediate effect, administrators can also reset the user's password, which forces a sign-out from all active sessions.

Here is a step-by-step guide on how to disable accounts and revoke user sessions in Microsoft Entra ID:

1. Navigate to the Entra admin center (`https://entra.microsoft.com`) and sign in.
2. Expand **Identity** | **Users** and select **All users**.
3. Locate and select the user account you want to disable.
4. Under **Account status**, click **Edit**.
5. Clear the checkbox for **Account enabled** and click **Save**.
6. Click the **Revoke sessions** button for the user.

Figure 5.28: Revoking user sessions

7. Click **Yes** to confirm.

If an employee is terminated, the IAM administrator must quickly deactivate the account and revoke all sessions to prevent unauthorized access to sensitive information. This process might involve additional steps in a hybrid environment, such as deactivating the account on-premises **Active Directory** (**AD**) and ensuring all associated sessions are terminated across cloud and on-premises resources.

Another scenario could involve a compromised account where an attacker has gained access. In this case, the administrator must act swiftly to deactivate the account, revoke all sessions, and investigate the breach to understand the extent of the compromise. Continuous monitoring and regular audits of user activities can help identify such incidents early and mitigate potential damage.

In the next section, you will go through the steps and best practices for implementing and managing password protection in your organization.

Deploying and Managing Entra Password Protection and Smart Lockout

Entra ID Password Protection helps prevent users from creating weak or easily guessable passwords by enforcing a global banned password list and allowing administrators to define custom banned password lists. This feature is crucial in mitigating password spraying attacks, where attackers attempt to gain access by trying common passwords across many accounts.

We'll start with the smart lockout settings.

Custom Smart Lockout

The smart lockout settings determine how Entra ID handles failed login attempts. **Lockout threshold** is the number of times in a row a user can enter a bad password before getting locked out. By default, **Lockout threshold** is set to **10** in the Microsoft 365 Worldwide (sometimes referred to as Commercial or Public) cloud and Microsoft 365 China 21Vianet tenants, while it is set to **3** for Azure US Government customers.

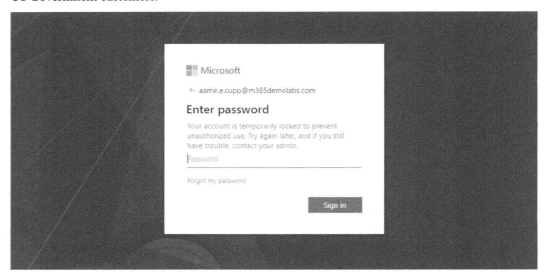

Figure 5.29: Account lockout

Lockout duration only specifies the initial lockout duration after the lockout threshold has been reached. Each subsequent lockout increases the lockout duration. Microsoft does not publish the rate at which the duration increases.

The smart lockout settings are now located in the **Password protection** section of **Authentication methods** (in the Entra ID admin center, click on **Identity | Protection | Authentication methods**), as shown in *Figure 5.30*:

Figure 5.30: Configuring smart lockout settings

Next, we'll look at configuring custom banned passwords.

Custom Banned Passwords

While Microsoft recommends moving toward passwordless authentication as a primary mechanism, passwords are still required to be configured in many scenarios. To help minimize using well-known, weak, or easy-to-guess passwords, you can specify a custom list of words that you want to exclude from being used as passwords. For example, you may wish to include your organization's name, products or services offered by your organization, cities or landmarks in your area, or local sports teams.

To enable the custom banned password feature, set **Enforce custom list** to **Yes**, and then add up to 1,000 banned words in the **Custom banned password list** text area. The list is not case-sensitive. Entra ID automatically performs common substitutions (such as *0* and *o* or *3* and *e*), so you do not need to generate multiple permutations for a word you want to ban.

> **Tip**
> Each entry in the custom banned password list must be at least four characters.

Like the smart lockout settings, the custom banned password settings are available on the **Authentication methods** page, as shown in *Figure 5.31*:

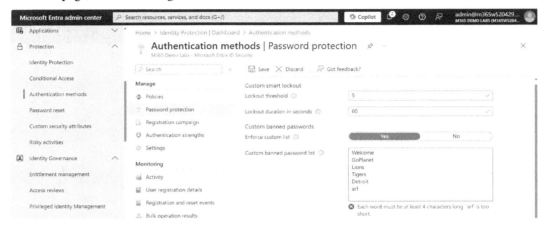

Figure 5.31: Configuring a custom banned password list

Next, we'll look at configuring password policy integration with AD.

Password Protection for Windows Server Active Directory

This settings area allows you to extend the custom banned password list to your on-premises infrastructure. There are two components:

- **Microsoft Entra Password Protection DC agent**, which must be installed on domain controllers.
- **Microsoft Entra Password Protection proxy**, which must be installed on at least one domain-joined server in the forest. As a security best practice, Microsoft recommends deploying on a member server since it requires internet connectivity.

In this configuration, the Entra Password Protection proxy servers periodically retrieve the custom banned password list from Entra ID. The DC agents cache the password policy locally and validate password change requests accordingly.

This integration is also configured on the **Password protection** page of **Authentication methods** by toggling **Enable password protection on Windows Server Active Directory** to **Yes**.

After enabling this toggle, you can then choose what mode to process password change requests. They can be processed in **Audit** mode (where changes are logged but the policy is not implemented) or **Enforced** mode, where password resets are actively evaluated against the banned password list and rejected if they do not meet the requirements.

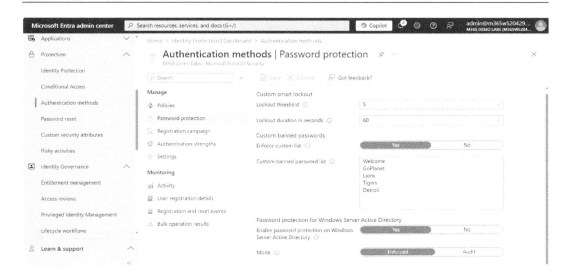

Figure 5.32: Enabling on-premises password protection

After enabling these settings in the Entra admin center, you can install the DC and proxy agents in your on-premises environment.

Further reading

To view detailed steps for deploying password protection on-premises, see `https://learn.microsoft.com/en-us/entra/identity/authentication/howto-password-ban-bad-on-premises-deploy`. Both the DC agent and the proxy are available on the Microsoft download center: `https://www.microsoft.com/en-us/download/details.aspx?id=57071`.

Next, we will look at enabling Microsoft Entra Kerberos authentication for hybrid identities.

Enabling Microsoft Entra Kerberos Authentication for Hybrid Identities

Enabling Microsoft Entra Kerberos authentication for hybrid identities allows hybrid users with on-premises AD identities synchronized to Microsoft Entra ID to access Azure resources such as Azure file shares using Kerberos authentication. This setup is particularly beneficial for organizations that must maintain seamless access to both on-premises and cloud resources without compromising security.

Kerberos integration is only available for hybrid identity environments synchronized with Microsoft Entra Connect.

Enabling Microsoft Entra Kerberos authentication requires several different steps:

1. Configuring an Azure resource that supports Kerberos authentication, such as a storage account

2. Granting access to the Azure resource's corresponding enterprise application

3. Updating Conditional Access policies to exclude the resource from MFA (if necessary)

4. Deploying a policy to configure endpoints to use the cloud Kerberos ticket

Let's review each of them.

Configuring a Storage Account

First, we'll enable an Azure resource for Kerberos authentication. The easiest one is a storage account that can be used to host Azure file shares. To configure Azure file shares to support Kerberos authentication, follow these steps:

1. Navigate to the Azure portal (`https://portal.azure.com`) and sign in with an account that has the Global Administrator role.

2. In the Azure portal, go to **Storage Accounts** and select the storage account you want to configure. If you don't already have a storage account, you will need to create one.

Figure 5.33: Storage accounts blade

3. Under **Data storage**, select **File shares**.

4. Configure AD by clicking **Not configured** next to **Identity-based access**. See *Figure 5.34*:

Figure 5.34: File share settings

5.   Under **Microsoft Entra Kerberos**, select **Set up**.

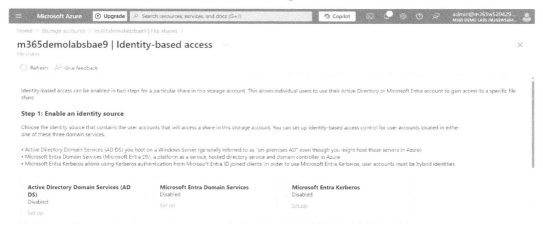

Figure 5.35: Microsoft Entra Kerberos option

6. In the **Microsoft Entra Kerberos** flyout, populate your on-premises domain name and the domain's `ObjectGuid` value (which you can derive from running `(Get-ADDomain).ObjectGuid`) in a PowerShell console session).

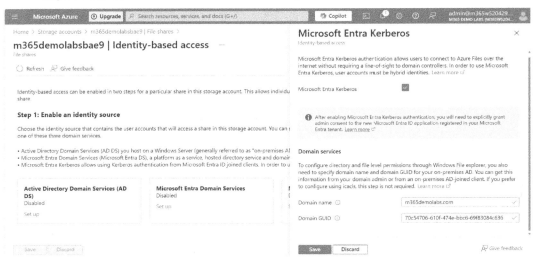

Figure 5.36: Adding the domain name and GUID to the Microsoft Entra Kerberos configuration

7. Click **Save**.

8. You can optionally configure default share-level permissions for authenticated users and groups by selecting **Enable permissions for all authenticated users and groups** and then choosing a role (such as **Storage File Data SMB Share Contributor**). If you configure this option, then you'll need to click **Save** again.

Next, we'll configure the permissions for the storage account's enterprise application.

Configuring Enterprise Application Permissions

Follow these steps:

1. Navigate to the Entra admin center (`https://entra.microsoft.com`).

2. Under **Identity**, expand **Applications** and select **Enterprise applications**.

3. Under **Manage**, select **All applications**. Clear the filter and then select the application with the name matching **[Storage Account] <your-storage-account-name>.file.core.windows. net**, as shown in *Figure 5.37*:

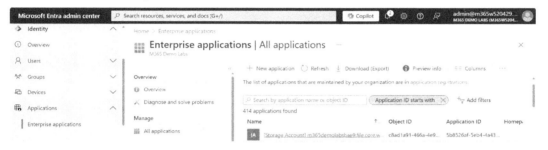

Figure 5.37: Selecting the [Storage Account] enterprise application

4. Under **Security**, select **Permissions**.

5. Click **Grant admin consent**.

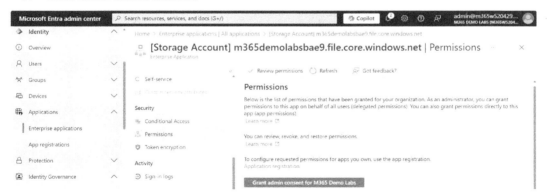

Figure 5.38: Permissions blade for the storage account

6.   On the popup, click **Accept**.

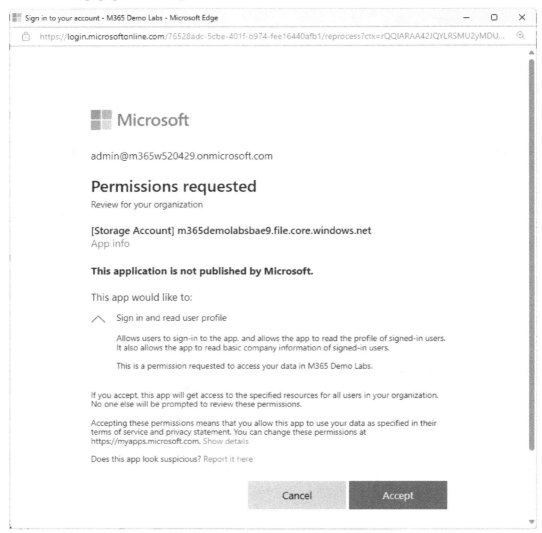

Figure 5.39: Granting admin consent

7.   Refresh the **Permissions** blade and verify that the Microsoft Graph API for the **openid**, **profile**, and **User.Read** claim values have been added.

Next, you'll want to update any Conditional Access policies that require MFA.

Updating Conditional Access Policies

In our example configuration, Azure file shares don't support integration with MFA. You'll need to review each Conditional Access policy and exclude the Kerberos-enabled Azure applications.

You'll also need to update Conditional Access policies to exclude target resources such as the storage account that don't support MFA. You can do this by going to the Entra admin center, to **Protection** | **Conditional Access**, and then selecting the enterprise application object corresponding to the Azure resource under **Target resources**. See *Figure 5.40* for an example of excluding the [**Storage Account**] resource from an MFA policy.

Figure 5.40: Configuring an MFA exclusion

Repeat this process for each Conditional Access policy that requires MFA.

Finally, you'll need to enable endpoints to request cloud Kerberos tickets.

Updating Endpoint Settings

Update client settings to ensure client devices are configured to retrieve Kerberos tickets from Entra ID. You can do this through either Group Policy or Intune.

Since Microsoft's position is cloud-first deployments, we'll start with configuration through Intune.

Configuring an Intune Configuration Service Provider

You can configure an Intune **configuration service provider** (**CSP**) to update your Intune-enrolled devices to retrieve a cloud Kerberos ticket using the following steps:

1. Navigate to the Intune admin center (`https://intune.microsoft.com`).
2. Select **Devices** | **Windows**.
3. Under **Manage devices**, select **Configuration**.
4. On the **Policies** tab, click **Create** | **New Policy**.

5. On the **Create a profile** flyout, under **Platform**, select **Windows 10 and later**.

6. On the **Create a profile** flyout, under **Profile type**, select **Templates** and then choose the **Custom** template.

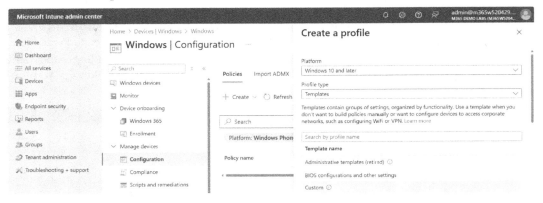

Figure 5.41: Creating a custom configuration profile

7. On the **Basics** tab, enter a name such as `Kerberos configuration` and click **Next**.

8. On the **Configuration settings** tab, click **Add**.

Figure 5.42: Adding an OMA-URI configuration

9. On the **Add Row** flyout, add a **Name** value of `CloudKerberosTicketRetrievalEnabled`.

10. Optionally, add a description.

11. In the **OMA-URI** field, enter the following value (be sure to include the leading period):

```
./Device/Vendor/MSFT/Policy/Config/Kerberos/
CloudKerberosTicketRetrievalEnabled
```

12. Under **Data type**, select **Integer**.

13. In the **Value** field, enter 1. See *Figure 5.43*.

Figure 5.43: Configuring the OMA-URI setting

14. Click **Save**.

15. Click **Next**.

16. Under **Included groups**, either add a device group or select **Add all devices**. Click **Next**.

17. On the **Applicability Rules** tab, click **Next**.

18. Click **Create**.

After the Intune refresh cycle, enrolled devices should request a cloud Kerberos ticket.

Configuring Group Policy

You can also configure devices to retrieve a cloud Kerberos ticket through Group Policy:

1. Log in to a server with the Group Policy management console.

2. Create or edit a **Group Policy object** (**GPO**) that applies to the client devices you want to configure.

3. Within the GPO, navigate to **Computer Configuration | Policies | Administrative Templates | System | Kerberos**.

4. Enable Kerberos authentication by double-clicking on **Define host name-to-Kerberos realm mapping** and setting it to **Enabled**.

5. In the **Value name** field, enter the Kerberos realm for Entra ID:

    KERBEROS.MICROSOFTONLINE.COM

6. Add an entry.

7.   In the **Value** field, enter your internal domain name. If you have multiple domains or subdomains, separate them with a semicolon. See *Figure 5.44*:

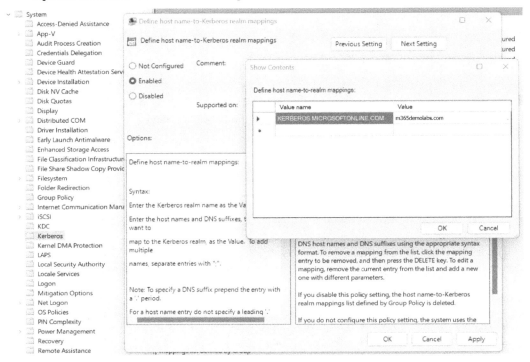

Figure 5.44: Configuring the hostname-to-Kerberos realm mapping

8.   Click **OK** to close the dialog box.

9.   Click **Apply** and then **OK** to save the settings. Ensure the GPO is linked to the appropriate **organizational unit (OU)** containing the client devices.

Verify the configuration using tools such as `klist` to confirm that Kerberos tickets are being issued correctly. Using the `klist` command-line tool, you can display the list of cached Kerberos tickets. Look for the Kerberos tickets associated with the Entra ID realm to confirm they are being issued correctly.

Use Entra ID logs to monitor authentication events and troubleshoot issues. Periodically review and update your Kerberos authentication settings to maintain security.

A real-world scenario where this configuration is beneficial involves a company with a hybrid environment where employees must access Azure file shares from both on-premises and remote locations. By enabling Microsoft Entra Kerberos authentication, the company can ensure that users authenticate seamlessly using their existing AD credentials, enhancing security and user experience. This setup also simplifies the management of access controls and permissions across different environments.

As we transition from deploying internet access to more advanced security measures, the next critical topic is implementing CBA in Microsoft Entra.

Implementing Certificate-Based Authentication in Microsoft Entra

Implementing CBA enhances security by allowing users to authenticate using X.509 certificates issued by a trusted PKI. This method is particularly effective in providing phishing-resistant, passwordless authentication. Administrators must ensure that their environment meets specific prerequisites to implement CBA, such as having a configured PKI with at least one **certification authority (CA)** and any intermediate CAs uploaded to Microsoft Entra ID.

Enabling CBA in your Microsoft 365 tenant requires several steps:

- Configuring certificates
- Updating authentication methods
- Configuring Entra Connect

Over these next sections, we'll look at enabling CBA with hybrid identity and an on-premises enterprise PKI solution. We'll be using Active Directory Certificate Services.

Configuring Certificates

The first step is to upload your PKI system's CA root certificate (and any subordinate or intermediate certificates) to Entra. You can export them from your PKI solution. The certificates should be in `.cer` format. Follow these steps:

1. Navigate to the Entra admin center (`https://entra.microsoft.com`).

2. In the **Search** bar, enter `certificate` and select **Certificate Authorities** or navigate to **Protection | Security Center** and select it from the list.

Figure 5.45: Searching for Certificate Authorities

3. Click **Upload**.

4. On the **Upload certificate file** flyout, under **Certificate**, browse to your certificate file.

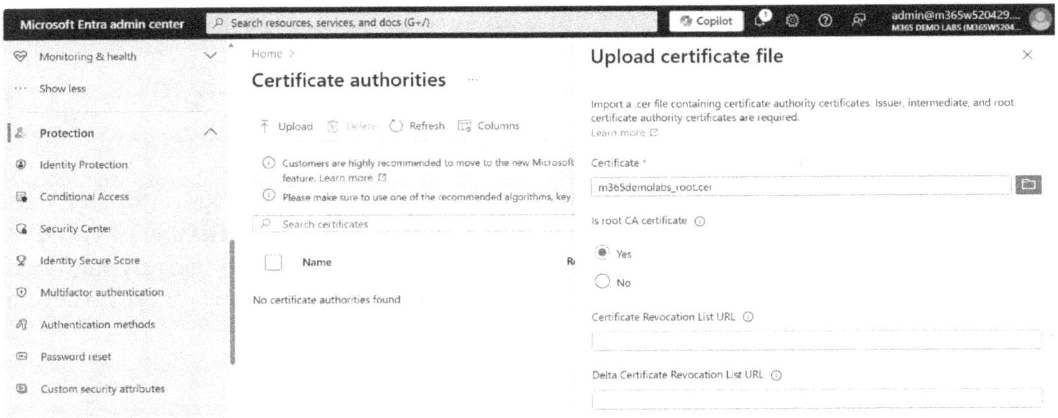

Figure 5.46: Uploading a certificate

5. Select **Yes** under **Is root CA certificate** if it is a root or **No** if it is an intermediate CA certificate.

6. If available, enter a publicly available **certificate revocation list** (**CRL**) URL. If this URL is not populated with a CRL, Entra will not be able to verify whether certificates have been revoked (which presents a security risk).

7. Click **Add**.

8. Repeat for any additional certificates.

> **Tip**
>
> Create an Entra group that will contain users with certificates. If you enable CBA and scope it to include users who don't have certificates, they won't be able to log in.

Next, you'll need to update the authentication methods to support CBA.

Updating Authentication Methods

In order to use CBA to access Microsoft 365 services, you'll need to enable it as an authentication method by following these steps:

1. In the Entra admin center, expand **Identity | Protection | Authentication methods** and select **Policies**.

2. Click **Certificate-based authentication**.

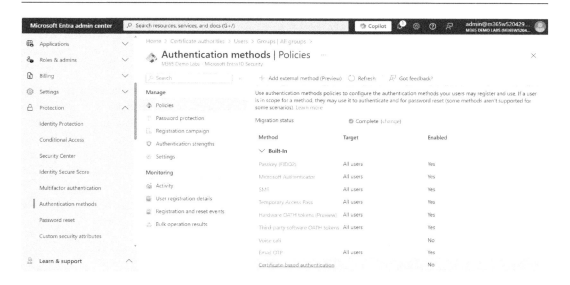

Figure 5.47: Configuring the CBA method

3. On the **Enable and Target** tab, move the **Enable** slider to **On**.

4. On the **Include** tab, click **Select groups** and then add your security group that contains users with valid certificates.

5. Click **I Acknowledge** to acknowledge that users without certificates will be unable to log in.

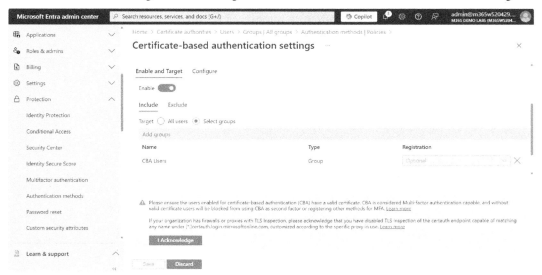

Figure 5.48: Configuring group targeting

6. Select the **Configure** tab.

7. If you published an external CRL when uploading your certificates, you can select **Require CRL validation (recommended)**.

8. Under **Authentication binding**, you can select **Multi-factor authentication** as the protection level if you want to treat CBA as equal to MFA. If you select the **Multi-factor authentication** option, this will cause all users authenticating to the tenant with CBA to have their logon sessions stamped as *X.509 Multifactor logins* for the duration of their session. The default value is **Single-factor authentication**. For this demonstration, you can select either.

9. Click **Save**.

Next, you'll need to update the Entra Connect configuration to synchronize the CertificateIds property.

Updating Entra Connect

As previously mentioned, you'll need to make sure users have certificates added to their accounts. If you have configured an enterprise PKI platform such as Active Directory Certificate Services with user auto-enrollment and are using Entra Connect, on-premises certificates will be added to the users in the directory (which is half of the solution).

In this section, you'll configure Entra Connect to map the certificate IDs:

1. Launch the Entra Connect Synchronization Rules Editor.

2. In the **Direction** dropdown, select **Outbound**.

3. Locate the **Out to AAD - User Identity** rule and click **Edit**.

4. Click **Edit**.

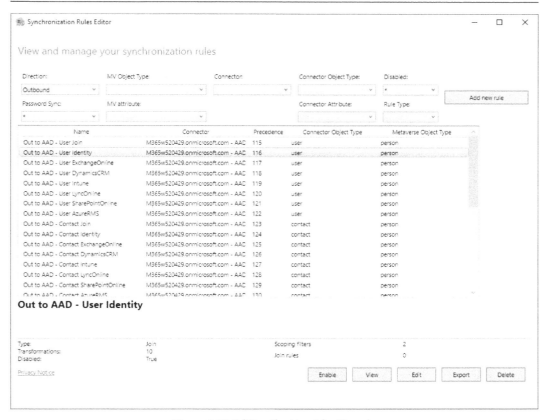

Figure 5.49: Editing the User Identity rule

5. When prompted, click **Yes** to disable the default rules and create a duplicate.

6. In the **Precedence** field, add a high value (such as `155`) and click **Next**.

7. On the **Scoping filter** page, click **Next**.

8. On the **Join rules** page, click **Next**.

9. On the **Transformations** page, click **Add transformation**. Scroll to the bottom of the transformations window to select the blank rule.

10. Under **FlowType**, select **Expression**.

11. Under **Target Attribute**, select **certificateUserIds**.

12. In the **Source** field, enter the following value:

```
"X509:<PN>"&[userPrincipalName]
```

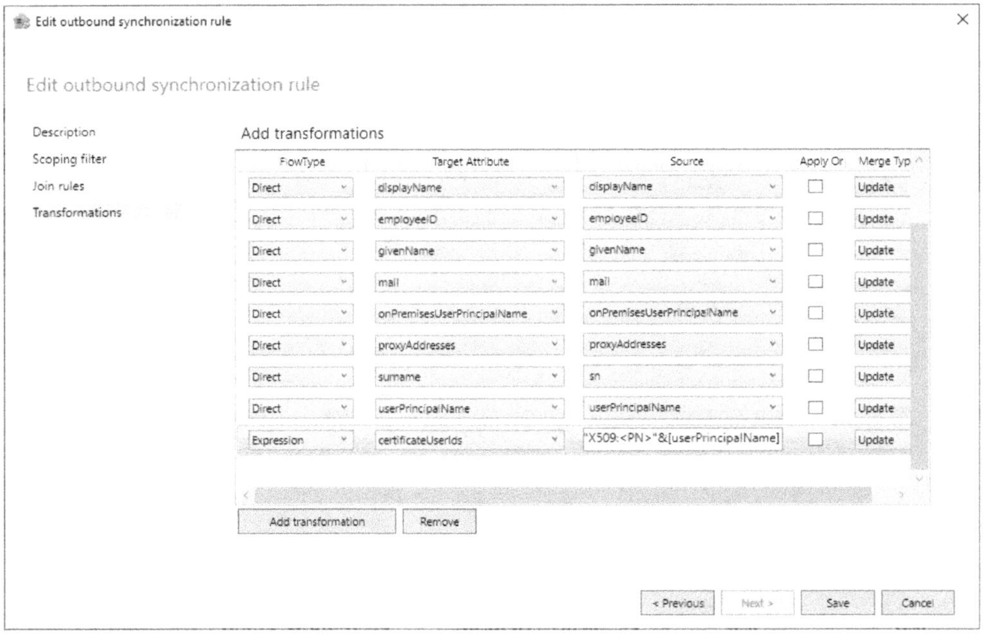

Figure 5.50: Adding a transformation

13. Click **Save**.

14. Click **OK** to acknowledge the full synchronization.

To see whether the **certificateUserIds** transform has populated the property in Entra, check a user by looking in the Entra admin center. Select a user and then click **View** next to **Authorization info**:

Figure 5.51: Viewing the certificateUserIds value

> **Further reading**
>
> Entra Connect configuration can be complex. There are a number of different **certificateUserIds** mapping options available, depending on how your enterprise PKI is configured. For an in-depth review of more configuration choices, see `https://learn.microsoft.com/en-us/entra/identity/authentication/concept-certificate-based-authentication-certificateuserids`.

Next, let's see what CBA looks like from the end user perspective.

Logging In with CBA

To validate that CBA is working, you can log in with a user identity that has a certificate assigned. To use CBA, the user will need to have the certificate saved in their local certificate store. You can view the certificate using the Certificates MMC snap-in:

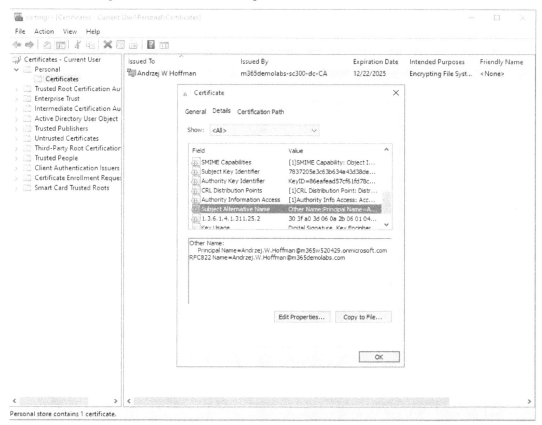

Figure 5.52: Viewing a local user certificate

The **Subject Alternative Name** certificate property should have the **Principal Name** value listed. This value should match the value stored in the **certficateUserIds** value in Entra.

To test CBA, follow these steps:

1. Navigate to `https://www.microsoft365.com` and click **Sign in**.
2. Enter or select your identity.
3. Enter your password and click **Sign in**. If the user is in the CBA group, they will be directed to select a certificate for login.

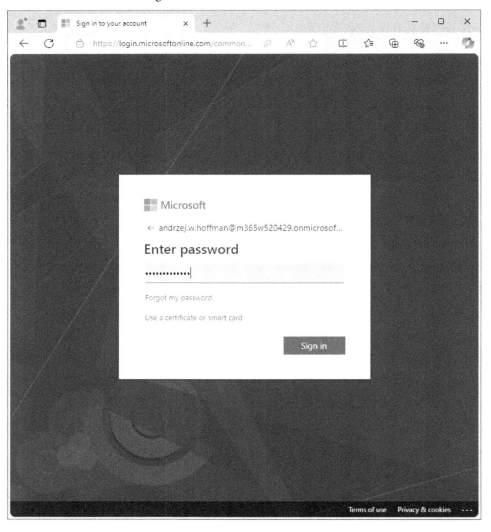

Figure 5.53: Logging in to Microsoft 365

4. When prompted, choose your user certificate and click **OK**.

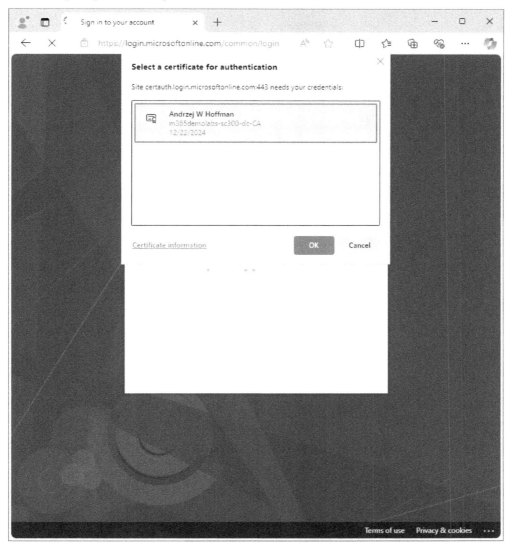

Figure 5.54: Selecting the certificate

5. If this is the first time logging in and your tenant has combined registration enabled, you'll need to register MFA by clicking **Next**.

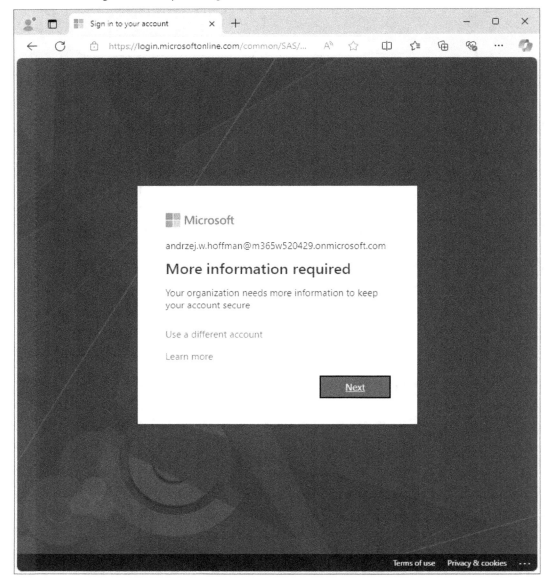

Figure 5.55: Starting the combined registration wizard

6. Proceed to register MFA by following the on-screen prompts.

Congratulations! You've configured CBA!

Monitoring Logins

After a successful login, you can view the sign-in logs for a user to see whether they were successfully validated using CBA:

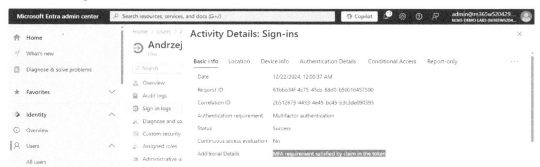

Figure 5.56: Verifying CBA sign-in details

Monitor authentication logs using Entra ID to track authentication events and troubleshoot issues. You should plan on setting reminders to renew your enterprise root and subordinate CA certificates, both in the on-premises environment and by uploading the new certificates to Entra.

When thinking about scenarios where CBA might be useful, consider a financial institution that wants to require high-security authentication for brokers using an Entra ID-integrated stock trading application to ensure users are only accessing the application from an organization-owned asset enrolled in the enterprise PKI solution. In addition to providing a highly secure environment, this configuration simplifies the user experience by eliminating the need for passwords. Additionally, in a hybrid environment, CBA can provide seamless authentication for users accessing both on-premises and cloud resources.

Summary

This chapter guided you through the various facets of Microsoft Entra ID user authentication, including planning for authentication, implementing and managing authentication methods, implementing and managing tenant-wide MFA settings, managing per-user MFA settings, configuring and deploying SSPR, implementing and managing WHFB, disabling accounts and revoking user sessions, deploying and managing password protection and smart lockout, enabling Microsoft Entra Kerberos authentication for hybrid identities, and implementing CBA in Microsoft Entra.

You should be well equipped to plan, implement, and manage Microsoft Entra ID user authentication, ensuring a secure and efficient environment for your organization. Whether you're preparing for the SC-300 exam or seeking to improve your identity management practices in your Azure and Entra ID tenants, this information will be valuable.

In the next chapter, we will explore Microsoft Entra Conditional Access.

Exam Readiness Drill – Chapter Review Questions

Apart from mastering key concepts, strong test-taking skills under time pressure are essential for acing your certification exam. That's why developing these abilities early in your learning journey is critical.

Exam readiness drills, using the free online practice resources provided with this book, help you progressively improve your time management and test-taking skills while reinforcing the key concepts you've learned.

HOW TO GET STARTED

- Open the link or scan the QR code at the bottom of this page

- If you have unlocked the practice resources already, log in to your registered account. If you haven't, follow the instructions in *Chapter 19* and come back to this page.

- Once you log in, click the START button to start a quiz

- We recommend attempting a quiz multiple times till you're able to answer most of the questions correctly and well within the time limit.

- You can use the following practice template to help you plan your attempts:

Working On Accuracy		
Attempt	**Target**	**Time Limit**
Attempt 1	40% or more	Till the timer runs out
Attempt 2	60% or more	Till the timer runs out
Attempt 3	75% or more	Till the timer runs out
Working On Timing		
Attempt 4	75% or more	1 minute before time limit
Attempt 5	75% or more	2 minutes before time limit
Attempt 6	75% or more	3 minutes before time limit

The above drill is just an example. Design your drills based on your own goals and make the most out of the online quizzes accompanying this book.

First time accessing the online resources? 🔒

You'll need to unlock them through a one-time process. **Head to** *Chapter 19* **for instructions.**

Open Quiz	
https://packt.link/sc300ch5 OR scan this QR code →	

Planning, Implementing, and Managing Microsoft Entra Conditional Access

Microsoft Entra Conditional Access can support and manage the balance between security and a seamless user experience. It provides a robust set of tools designed to tailor access requirements to each organization's unique needs. This chapter digs into the strategic planning, implementation, and ongoing management of Conditional Access policies within Microsoft Entra, providing administrators with the knowledge and skills necessary to enforce their security posture.

This chapter will take you from the foundational concepts of planning Conditional Access policies, where you'll learn how to craft policies that align with organizational goals and compliance standards, to the practical aspects of implementing Conditional Access policy assignments and controls that dictate how and when access is granted or denied.

You'll explore the intricacies of implementing session management, ensuring that user sessions are secure yet flexible enough to support productivity. Then, you will explore implementing device-enforced restrictions, a critical aspect of modern security that ensures only compliant devices can access corporate data.

As threats become more sophisticated, you need to be able to respond in real time. You'll discover how to implement continuous access evaluation to adjust user access rights dynamically in response to changing conditions. Lastly, this chapter will equip you with the skills to efficiently deploy policies using best practices by creating a Conditional Access policy from a template.

The objectives and skills we'll cover in this chapter are as follows:

- Planning Conditional Access policies
- Implementing Conditional Access policy assignments
- Implementing Conditional Access policy controls

- Implementing session management
- Implementing device-enforced restrictions
- Implementing Continuous Access evaluation
- Creating a Conditional Access policy from a template

By the end of this chapter, you will understand how to effectively plan, implement, and manage Conditional Access within Microsoft Entra, ensuring that the right people have suitable access at the right time, all while maintaining the highest security standards.

Planning Conditional Access Policies

When planning **Conditional Access** policies, it's essential to have a comprehensive understanding of the components and strategies that ensure robust security within Azure environments. Conditional Access policies are `if-then` statements that evaluate certain conditions before granting access to resources. These policies are composed of three main components:

- **Assignments**: Identify the users (or groups), networks and locations, and the cloud apps affected or in scope for the policy
- **Conditions**: The signals, such as risk (user, sign-in, and insider), networks and locations, and device platforms that the policy is targeting
- **Access controls**: The actions taken when the combination of assignments and conditions are met, such as granting access or requiring **multi-factor authentication (MFA)**

Conditional Access policies are crucial to implementing a zero-trust model in your Azure environment. The principle of zero trust is "never trust, always verify." Conditional Access policies enforce this principle by controlling access to resources based on the various conditions that are configured.

A successful strategy begins with identifying which applications contain sensitive information and thus require stricter access controls. Users should be classified based on their access needs and risk profiles, with high-risk users subjected to more stringent policies. When setting conditions, contextual factors such as location, device compliance, and access time should be considered. Policies should be granular, targeting specific scenarios to ensure security without hindering productivity. It is also crucial to plan for emergency access accounts that bypass policies to ensure access during critical situations.

High-risk users are categorized in identity protection, which evaluates various risk factors to determine the likelihood that an account has been compromised. Some factors include risky sign-in, where an account signs in from multiple unfamiliar locations, or atypical travel patterns, including signing in from multiple countries within a short time frame without the necessary travel time between them.

For environments that leverage either cloud identity or hybrid identity with password hash synchronization, Microsoft detects that the user credentials were part of a data breach. The presence of a leaked credential adds to the risk calculation.

> **Exam Tip**
> Checking for leaked credentials is an Entra ID P2 subscription feature.

Password spray attacks are another factor that could cause a user account to be at high risk; this is when access to an account is continuously attempted using common passwords. The last factor is if there are unfamiliar sign-in properties when the account sign-ins significantly differ from the user's typical behavior. Identity protection uses these factors to assign a risk level (low, medium, or high) to an account based on the likelihood that the account is compromised.

Default or templated policies can be used for common scenarios, while policies for specific needs can be based on a template and then customized. Understanding how multiple policies combine and their order of precedence is essential. Monitoring and reporting are also vital for tracking policy usage and effectiveness, with regular reviews of sign-in logs and audit trails recommended.

Testing and troubleshooting are essential steps in the implementation process. Rest assured; you're not alone on this journey. Pilot testing with a small group of users can validate the policy's impact before full deployment. Troubleshooting tools such as Entra ID's sign-in logs, Conditional Access report-only mode, and the *Conditional Access Insights* workbook are valuable resources at your disposal.

The following section will explore implementing Conditional Access policy assignments.

Implementing Conditional Access Policy Assignments

Implementing Conditional Access policy assignments is critical for managing identity and access within Azure and Microsoft Entra. As an administrator, it's essential to understand how to effectively assign these policies to protect resources without negatively impacting user experience.

Conditional Access policy assignments define who is subject to the policy. These assignments can be directed at individual users, groups, or organizational roles. When planning your assignments, consider the principle of least privilege, create Conditional Access policies for each resource that needs to be secured, and only assign the approved policy to those who need it to perform their job functions. This minimizes potential attack surfaces and reduces the risk of unauthorized access.

In the context of the SC-300 exam, you'll need to demonstrate proficiency in creating and managing these assignments. This includes understanding the distinctions of including and excluding specific users or groups in your policy assignments. For instance, you might include all users but exclude certain administrative accounts requiring unrestricted troubleshooting access. Follow these steps to create and manage Conditional Access assignments:

1. Navigate to the Entra admin center (`https://entra.microsoft.com`). Log in with an account that has the Global Administrator, Security Administrator, or Conditional Access Administrator role assigned.

2. Expand **Protection** and select **Conditional Access**.

3.   Select **Policies**.

4.   Click **New policy** to create a new Conditional Access policy. This will bring up the screen shown in *Figure 6.1*.

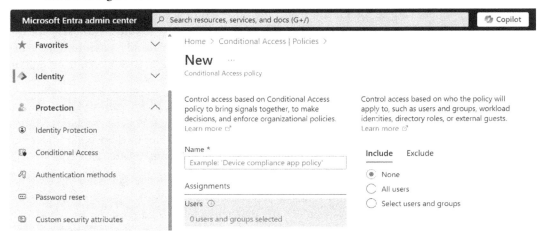

Figure 6.1: Conditional Access user assignments

5.   Name your policy by entering a descriptive name in the **Name** field.

6.   Use the **Assignments** area to define the in- and out-of-scope users, applications, and networks. Click the link for **Users and groups** and select the **Include** subtab.

7.   On the **Include** subtab, select the **All users** radio button to bring all users in the tenant into the scope of the policy. You can also choose the **Select users and groups** radio button to expose options for choosing **Users and groups**, **Guest or external users**, or **Directory roles**, as shown in *Figure 6.2*.

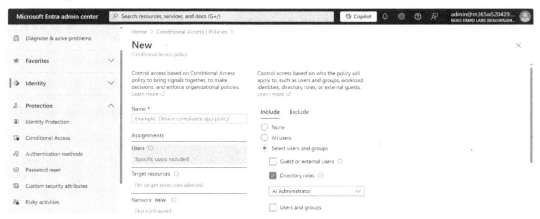

Figure 6.2: Selecting the users' assignment scope

8. If you've chosen the **Select users and groups** option, choose the users or groups you want to include in this policy by searching for their names or email addresses. Click **Done** to confirm your selections.

    You can also select the **Exclude** subtab to configure exclusions to the assignment. Exclusions are configured using the same options for **Guest or external users**, **Directory roles**, and **Users and groups**.

9. Select the **Cloud apps or actions** tab to specify which applications or actions the policy applies to.

10. Under **Target resources**, click **No target resources selected** to begin configuring the assignment if scoping is needed. See *Figure 6.3*.

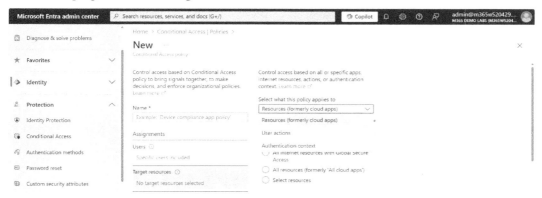

Figure 6.3: Configuring target resources

You can choose from the following options:

- **Resources (formerly cloud apps)**: Choose **All resources (formerly 'All cloud apps')** to select all Entra ID-federated applications in the tenant. **All internet resources with Global Secure Access** will control access for users who are in scope for GSA and have GSA configured. The **Select resources** option will allow you to choose from a filter (based on custom security attributes) or individual Entra ID-federated applications.

- **User actions**: You can select the **Register security information** action to assign this policy to the security actions of registering an MFA token or user security questions. The **Register or join devices** option scopes the policy for registering or joining devices to Entra.

- **Authentication context**: This option allows you to choose an authentication context for scoping the policy. An authentication context is essentially a tag that can be applied to scenarios, such as requiring Privileged Identity Management from trusted IP locations.

Under **Network,** select **Not configured** to display the configuration options for networks. You can choose from **Any network or location, All trusted networks and locations, All Compliant Network locations** (devices and networks protected by Global Secure Access), and **Select networks and locations**.

11. Use **Conditions** to define the conditions under which the policy will be applied by clicking **0 conditions selected**, as shown in *Figure 6.4*:

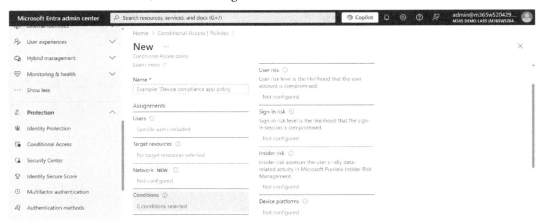

Figure 6.4: Configuring conditions

You can choose from the following conditions:

- **User risk**: Set the level (**High, Medium,** or **Low**) for the policy to be enforced. The user risk level is the likelihood that the user's identity has been compromised.

- **Sign-in risk**: Set the sign-in risk level for the policy to be enforced (**High, Medium, Low,** or **No risk**). The sign-in risk level is the likelihood that a sign-in session has been compromised.

- **Insider risk**: Set the insider risk level for the policy to be enforced (**Elevated, Moderate,** or **Low**). Insider risk is determined through Microsoft Purview Insider Risk Management.

- **Device platforms**: Configure the policy to apply to selected platforms. You can choose to include or exclude the following device types: **Android, iOS, Windows Phone, Windows, macOS,** or **Linux**.

- **Locations**: Configure the policy to apply to selected locations. You can choose to include or exclude the following location options: **Any network or location, All trusted networks and locations, All Compliant Network locations,** or **Selected networks and locations**.

- **Client apps**: Configure the policy to target specific client applications. You can choose from **Modern authentication clients (Browser** and **Mobile apps and desktop clients)** as well as **Legacy authentication clients (Exchange ActiveSync clients** and **Other clients,** which includes legacy email protocol clients).

- **Filter for devices**: Configure a filter to target the policy. Filters can be based on device model and manufacture characteristics, ownership, operating system versions, and many other attributes.

- **Authentication flows**: Configure the policy to target authentication and authorization protocols, such as **Device code flow** (typically used for devices without input capability) and **Authentication transfer** (a mechanism for transferring an authentication state between devices).

Leave the policy as-is, as we'll be continuing to work through it in the next section.

Implementing Conditional Access Policy Controls

Conditional Access policy controls are the decisions that take effect when the conditions specified in a policy are met. They are the *then* in the `if-then` statements that define Conditional Access policies. The controls can be categorized into three main actions: *grant access*, *require additional actions*, or *block access*. The following step-by-step guide shows how to implement a Conditional Access policy:

1. Continuing from where you left off in the previous section, navigate to the **Access controls** area to define the controls enforced when the conditions are met.

2. Under **Grant**, click **0 controls selected** to expose the **Grant** flyout.

Figure 6.5: Configuring grant controls

You can select **Block access** to configure a policy to restrict the user from accessing services specified in the **Assignments** area. You can also select **Grant access** to configure the policy to allow the user to access services specified in the **Assignments** area as long as they can meet one or more of the following requirements:

- **Require multifactor authentication**: The user must fulfill additional requirements, such as verification through a multifactor token app, one-time passcode, phone call, or text message.

- **Require authentication strength**: The user must fulfill additional verification requirements using a specific authentication method or policy.

- **Require device to be marked as compliant**: The device the user is accessing from must be marked as **Compliant** in Intune.

- **Require Microsoft Entra hybrid joined device**: The device the user is accessing from must be hybrid joined to both an on-premises Active Directory environment and Entra ID.

- **Require approved client app**: The application being used for access must *support* Intune app protection policies (which are different from the next grant control).

- **Require app protection policy**: The application being used for access must have an Intune app protection policy applied to it.

- **Require password change**: The policy will be used to force a user to change their password. This option must be used in conjunction with **Require authentication strength** or **Require multifactor authentication**.

- **For multiple controls**: If you have selected multiple grant controls, you can choose to require all of them or just require one of the selected controls.

3. Click **Select** to save the changes to the **Grant** controls section.

Leave the policy as-is as we'll be continuing to work through it in the next section, focusing on continuous access evaluation.

Implementing Session Management

Understanding session management involves configuring policies that control the duration and conditions of user sessions to improve security and compliance. Session management ensures that user access to resources is continuously monitored and evaluated, reducing the risk of unauthorized access due to extended or unattended sessions.

In Azure and Entra ID, session management can be implemented through Conditional Access policies. These policies allow administrators to define session controls that enforce re-authentication requirements, session timeouts, and sign-in frequency. For instance, you might configure a policy that requires users to re-authenticate every four hours when accessing sensitive applications, ensuring that only active and verified users maintain access. This is particularly important when users handle sensitive data, such as financial records or personal information, where prolonged access without re-authentication could pose significant security risks. Follow this step-by-step guide to implement session management in Conditional Access policies:

1. Continuing from where you left off in the previous section, navigate to the **Access controls** area to define the controls enforced when the conditions are met.

2. Under **Session**, click **0 controls selected** to expand the **Session** flyout.

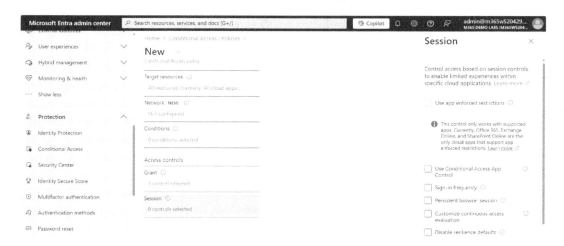

Figure 6.6: Configuring session controls

In the **Session** flyout, select the session controls you want to apply. **Session** options include the following:

- **Use app enforced restrictions**: Configure limited app experiences for Microsoft 365, SharePoint Online, and Exchange Online.

- **Use Conditional Access App Control**: Use a Defender for Cloud Apps access policy.

- **Sign-in frequency**: Specify how often users must re-authenticate. The default setting is 90 days but can be set as low as every hour or every time.

- **Persistent browser session**: Decide whether to keep users signed in on the browser.

- **Customize continuous access evaluation**: Enable customization of the **continuous access evaluation** (CAE) token based on critical events. The options are **Disable** and **Strictly enforce location policies** (which requires configuration of the **Locations** condition).

- **Disable resilience defaults**: Choose whether to extend access sessions during outages.

- **Require token protection for sign-in sessions**: Require a software key binding or hardware security module for long-lived tokens.

- **Use Global Secure Access security profile**: Select a GSA security profile for GSA-targeted or protected resources.

3. Click **Select** to save the options for the session controls.

4. Under **Enable policy**, select **Report-only** to evaluate the policy and report on scenarios where it is triggered, **On** to enable the policy, or **Off** to disable the policy.

A real-world example of session management is in a financial institution where employees access a secure banking application. To mitigate risks, the IAM administrator can set a Conditional Access policy that enforces a session timeout after 30 minutes of inactivity. Additionally, the policy could require users to re-authenticate using MFA if they attempt to access the application from a new device or location. This ensures that even if a session is hijacked, the attacker cannot maintain access without passing the MFA challenge.

To set the session timeout policy for 30 minutes of inactivity, the administrator would select the **Sign-in frequency** control. Setting the sign-in frequency to 30 minutes ensures that users must re-authenticate after 30 minutes of inactivity.

Another scenario involves remote work environments where employees access corporate resources from various locations and devices. In such cases, session management policies can be configured to enforce stricter controls for remote users. For example, a policy might require re-authentication every two hours for users accessing from outside the corporate network while allowing longer session durations for those within the trusted network. This approach balances security with user convenience, ensuring remote access does not compromise the organization's security posture.

Implementing session management also involves understanding the implications of each setting within the Conditional Access policies. For example, setting a short session timeout might enhance security but could also disrupt user productivity, leading to frequent re-authentication prompts. Likewise, longer session durations might improve user experience but could increase the risk of unauthorized access. Therefore, striking a balance that aligns with the organization's security requirements and user needs is essential.

The next section will explore how enforcing device compliance and restrictions can further enhance your organization's security posture by confirming that only secure, managed devices can access sensitive data and applications.

Implementing Device-Enforced Restrictions

As an **identity and access management** (**IAM**) administrator utilizing Azure and Entra ID, you should know how to enforce device compliance to ensure only secure, managed devices can access sensitive organizational resources. This involves configuring Conditional Access policies that evaluate the compliance status of devices before granting access to applications and data. Device compliance policies can include requirements such as installing the latest security updates, being free from malware, and adhering to organizational security configurations.

In Azure and Entra ID, device-enforced restrictions are implemented through Conditional Access policies that integrate with Microsoft Intune or other **mobile device management** (**MDM**) solutions. These policies allow administrators to define conditions under which access is granted or denied based on the device's compliance status. For example, you might configure a policy that requires devices to be marked as compliant in Intune before they can access corporate email. This ensures that only devices meeting your organization's security standards can access sensitive information, reducing the risk of data breaches.

When working with a Conditional Access policy to enforce device restrictions, you'll need to focus on three main areas:

- **Device platforms**, as shown in *Figure 6.7*:

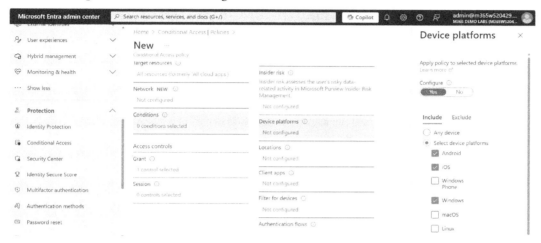

Figure 6.7: Configuring Device platforms restrictions

Here, you'll be able to select which device platforms you want the selected users to access the targeted resources. In the case of enforcing device restrictions, you should select all the device types that are enrolled in your tenant.

- **Grant** access control, as shown in *Figure 6.8*:

Figure 6.8: Configuring Grant controls

In the **Grant** access control, the key options are **Require device to be marked as compliant** (by fulfilling an Intune compliance policy) and **Require Microsoft Entra hybrid joined device**, which will require the device being used to access the resource to be both Active Directory domain-joined as well as Entra ID joined. When creating a policy that includes mobile devices, you'll want to ensure that the multiple-selection option is set to **Require one of the selected controls**.

A scenario illustrating the importance of device-enforced restrictions could involve a healthcare organization where employees access patient records through a secure application. To protect sensitive patient data, the IAM administrator can set a Conditional Access policy requiring all devices accessing the application to comply with the organization's security policies. This might include enabling encryption, a secure password policy, and the latest antivirus definitions. If a device does not meet these criteria, access is denied, and the user is prompted to bring their device into compliance before attempting to access the application.

To work through the preceding scenario, you would take the following steps:

1. To enforce encryption on devices, use Microsoft Intune to create device configuration profiles. For instance, you can configure BitLocker for Windows devices or FileVault for macOS devices.

2. Begin by signing in to the Microsoft Intune admin center and navigating to **Devices** | **Configuration profiles** | **Create profile**.

3. Select the platform, such as Windows 10 and later, and choose a profile type such as **Endpoint protection**.

4. Finally, configure the encryption settings, such as enabling BitLocker for Windows devices. To do this, sign in to the Microsoft Intune admin center with your admin credentials.

5. Navigate to the **Devices** section and click on **Configuration profiles**.

6. Then, select + **Create profile**.

7. In the **Create a Profile** pane, choose **Windows 10 and later** as the platform and select **Endpoint protection** as the profile type.

8. Under **Configuration settings**, expand the **BitLocker** category and configure the necessary settings, such as **Require BitLocker** and any other specific BitLocker policies relevant to your organization.

9. Assign the profile to the appropriate groups or devices by going to the **Assignments** tab.

10. Finally, review your settings and click **Create** to save the profile. This will ensure the profile is successfully applied to the assigned devices and your encryption settings are properly configured through Microsoft Intune.

To enforce a secure password policy, use Microsoft Intune to configure password settings:

1. First, sign in to the Microsoft Intune admin center.

2. Navigate to **Devices** | **Configuration profiles** | **Create profile**, then select the platform, such as Windows 10 and later.

3. Choose a profile type, such as **Device restrictions**, and configure the password settings according to your security requirements. This may include settings such as minimum password length and complexity requirements.

To ensure devices have the latest antivirus definitions, use Microsoft Defender for Endpoint to configure antivirus policies:

1. Begin by signing in to the Microsoft Intune Manager admin center.

2. Navigate to **Endpoint security** and select **Antivirus**.

3. Within the **Endpoint security** section, find and select **Antivirus**.

4. Create or edit a policy:

   - To create a new policy, click on + **Create Policy**

   - To edit an existing policy, select the policy from the list and click on it

5. Configure the settings:

   - Ensure **Real-time protection** is enabled to provide continuous monitoring

   - Configure settings to allow automatic updates for antivirus definitions, ensuring the latest definitions are always applied

6. After configuring the necessary settings, click **Save** to apply the changes.

Integrating your security measures with Conditional Access policies can ensure a robust access control strategy. While specific security settings such as device compliance and encryption are not directly configured within Conditional Access policies, you can use Conditional Access to enforce compliance with these security measures effectively:

1. To do this, sign in to the Azure portal with your admin credentials and navigate to **Azure Active Directory** | **Security** | **Conditional Access**.

2. Choose an existing policy or create a new one by selecting + **New policy** and providing a descriptive name.

3. Under **Assignments**, click **Users and groups** to select the specific users or groups to which this policy will apply, then click on **Cloud apps or actions** to select the applications that need to be protected.

4. Next, configure the conditions under **Sign-in risk** by setting the policy to trigger at a specific risk level, such as **Medium** or **High**, and under **Locations** by selecting **Include | Any location** and **Exclude | All trusted locations** to apply the policy only when sign-ins occur from untrusted locations.

5. Specify the platforms to which the policy should apply under **Device platforms**, for example, **Windows, iOS,** or **Android**. For access controls, click on **Grant** and select options such as **Require multifactor authentication, Require device to be marked as compliant,** or **Require app protection policy**, ensuring that only devices meeting specific criteria can access resources. Consider implementing session controls such as **Sign-in frequency** to prompt for reauthentication, ensuring continuous compliance.

6. Turn on the **Enable policy** toggle to activate the policy, then click **Create** or **Save**. By integrating these conditions and access controls into your Conditional Access policies, you ensure that users accessing your applications and resources comply with your security standards, enhancing overall security.

To ensure that only compliant devices (those meeting encryption, password, and antivirus requirements) can access corporate resources, use Conditional Access policies in the Microsoft Entra admin center:

1. Start by signing in to the Microsoft Entra admin center and navigating to **Protection | Conditional Access | Policies**.

2. Create a new policy and, under **Assignments**, select the users and groups.

3. Next, under **Cloud apps or actions**, select the applications that will be governed by the policy.

4. Under **Conditions**, configure **Device state** to include only compliant devices.

5. Finally, under **Access controls**, select **Grant** and choose **Require device to be marked as compliant**. This ensures that only devices meeting your specified compliance criteria can access your corporate resources.

Another example is in a corporate environment where employees use corporate-owned and personal devices to access company resources. To ensure that only secure devices are used, the IAM administrator can configure a Conditional Access policy that blocks access from non-compliant personal devices while allowing access from compliant corporate-owned devices. This policy can be enforced by integrating with Intune to check the compliance status of each device before granting access. For instance, a policy might require that all devices have installed the latest operating system updates and are free from jailbreaking.

Implementing device-enforced restrictions also involves understanding the implications of each setting within the Conditional Access policies. For example, enforcing strict compliance requirements might enhance security and increase support requests from users whose devices are blocked. Therefore, it is essential to balance security needs with user convenience and provide clear guidance on how users can bring their devices into compliance, such as the considerations we mentioned in the last section regarding user session management.

Next, let's explore CAE to enhance your security measures further.

Implementing CAE

Implementing CAE is a sophisticated feature within Entra that enhances security by providing real-time assessment of user sessions.

Understanding CAE

CAE addresses the limitations of traditional token-based authentication methods. In the past, once a user was granted access, the token would remain valid until its expiration, regardless of any changes in user status or risk profile.

CAE changes this by allowing Entra ID to force a token refresh in response to specific events, ensuring that access rights are always current with the user's situation.

For the SC-300 exam, you'll need to be familiar with the scenarios that trigger CAE, such as the following:

- User account is deleted or disabled
- Multifactor authentication enabled for a user
- User risk changes, including Microsoft Entra Identity Protection and Insider Risk signals
- Password change or reset
- Administrator revokes a session
- Network location change

For exam preparation, focus on understanding CAE mechanisms: specifically, how CAE works and what events can trigger access revaluation. For instance, significant changes in a user's location or if their device becomes non-compliant can prompt CAE to re-evaluate access to ensure the user still meets security requirements.

Several critical events can trigger access reevaluation under CAE. User account changes, such as deletion or disabling, immediately revoke access. When a user's password is changed, their access tokens are invalidated, requiring re-authentication. Enabling or disabling MFA for a user also triggers reevaluation. Administrators can explicitly revoke tokens, forcing users to re-authenticate. Additionally, if an elevated risk, such as suspicious activity, is detected for a user, access is reevaluated to ensure security.

You should be able to configure CAE policies within Entra ID and understand how they interact with other Conditional Access policies. For instance, you might configure a CAE policy that requires re-authentication if a user attempts to access a sensitive application from a new device, ensuring that the user's access is continuously evaluated based on the latest circumstances.

Troubleshooting CAE

In this section, we'll look at how to troubleshoot issues related to CAE, such as unexpected access denials or notification failures. For example, if a user is unexpectedly denied access due to a CAE policy, you should be able to identify whether the denial was triggered by a change in the user's risk level, device compliance status, or location and then take appropriate steps to resolve the issue. This troubleshooting capability is essential for ensuring smooth and secure operations.

To troubleshoot suspected CAE issues, follow these steps:

1. Navigate to the Microsoft Entra admin center (`https://entra.microsoft.com`).

2. Expand **Identity**, expand **Monitoring & health**, and then select **Sign-in logs**.

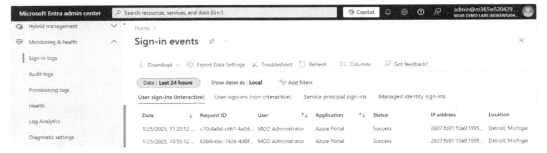

Figure 6.9: Viewing sign-in events

3. Apply the **Is CAE Token** filter to view sign-ins where CAE is applied. To do this, select **Add filters**, select **Is CAE Token**, and apply it.

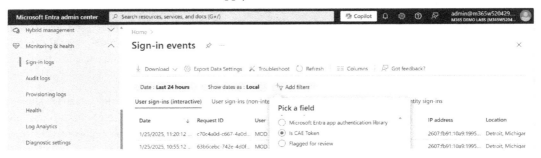

Figure 6.10: The Is CAE Token filter

4. Review the sign-in logs for any anomalies or errors related to CAE.

5. Check for IP address mismatches by comparing the IP address seen by Microsoft Entra ID and the IP address seen by the resource. This can be done using the **Continuous Access Evaluation Insights** workbook in Azure Monitor.

6. Verify that the client applications are correctly configured to support CAE and that they can handle claims challenges and requests.

7. Ensure that network configurations, such as split-tunnel VPNs, are not causing IP address mismatches.

Access issues may be due to a number of factors, such as unexpected network changes that fall outside of the trusted locations or user risk changes.

> **Further Reading**
>
> Microsoft provides additional troubleshooting help for any specific error codes or messages you encounter: `https://learn.microsoft.com/en-us/entra/identity/conditional-access/howto-continuous-access-evaluation-troubleshoot`.

By mastering the implementation of continuous access evaluation, you will ensure that your organization's access policies reflect the most current user statuses, thereby maintaining a robust security posture.

Next, we'll look at managing Conditional Access policies created through templates.

Creating a Conditional Access Policy from a Template

Creating a Conditional Access policy from a template efficiently applies proven security configurations within Azure and Microsoft Entra ID environments. As an IAM administrator, leveraging templates can streamline securing resources while ensuring policies align with best practices.

Templates encapsulate predefined settings that address common access scenarios, such as requiring MFA for external users or blocking access from untrusted locations. By using a template, you can quickly deploy a policy crafted by security experts, reducing the time and potential errors associated with manual configuration. Microsoft provides templates to simplify the deployment of policies. They are designed to align best practices and everyday use cases, making implementing easier for an administrator.

When creating a Conditional Access policy from a template, the process involves the following steps:

1. Navigate to the Entra admin center (`https://entra.microsoft.com`).

2. Expand **Protection** and then select **Conditional Access**, as shown in *Figure 6.11*.

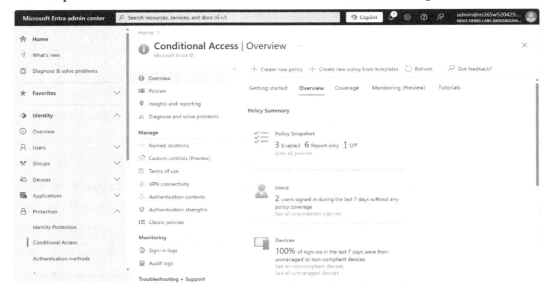

Figure 6.11: Conditional Access Overview page

3. Select **Create new policy from template**.

4. Choose from a list of available templates. These templates are designed to address specific security needs and compliance requirements, ranging from MFA registration to enabling MFA requirements for administration portals. *Figure 6.12* shows some of the policy templates available.

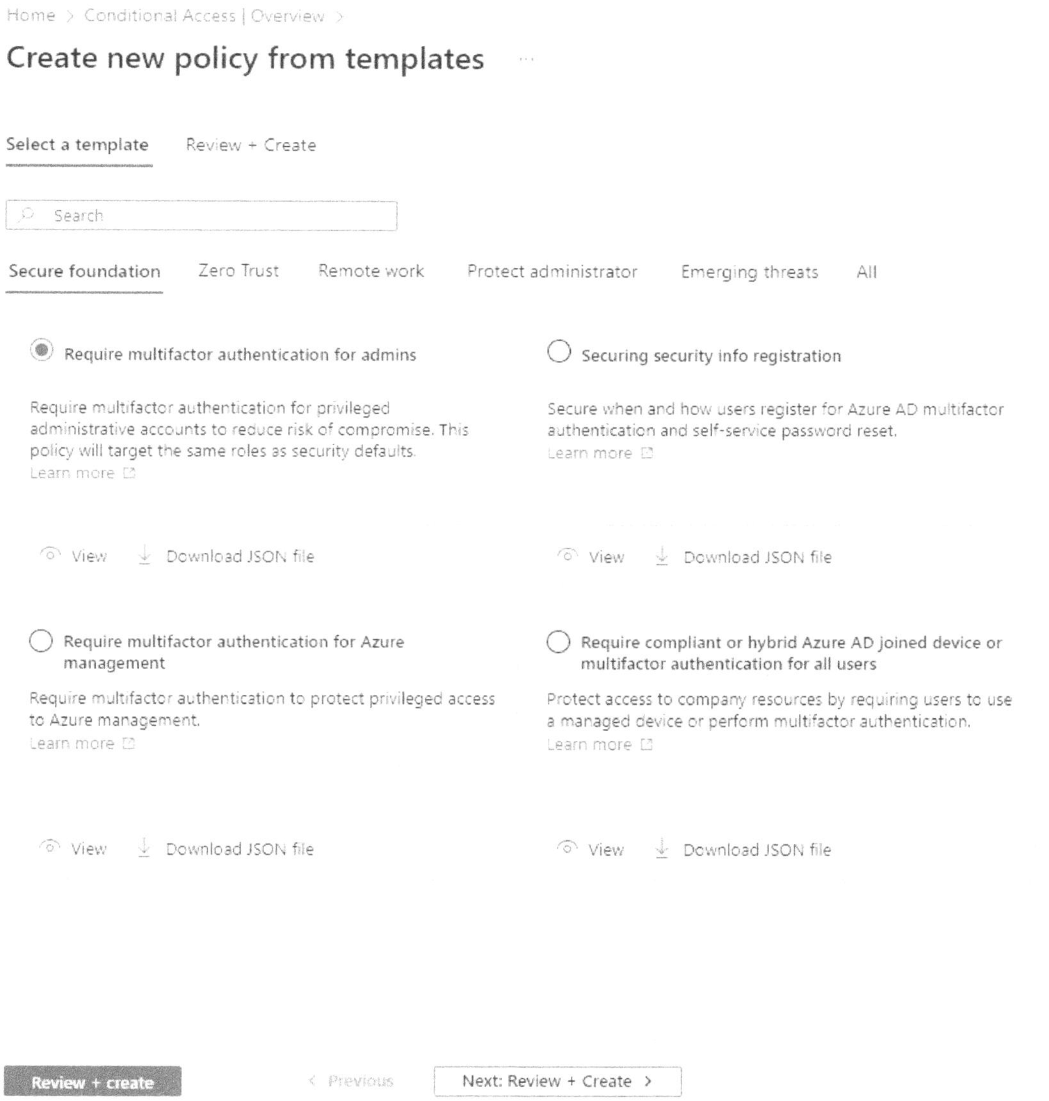

Home > Conditional Access | Overview >

Create new policy from templates ···

Select a template   Review + Create

| ⌕ Search |

Secure foundation   Zero Trust   Remote work   Protect administrator   Emerging threats   All

---

◉ Require multifactor authentication for admins

Require multifactor authentication for privileged administrative accounts to reduce risk of compromise. This policy will target the same roles as security defaults.
Learn more ⌕

◯ View   ↓ Download JSON file

◯ Securing security info registration

Secure when and how users register for Azure AD multifactor authentication and self-service password reset.
Learn more ⌕

◯ View   ↓ Download JSON file

◯ Require multifactor authentication for Azure management

Require multifactor authentication to protect privileged access to Azure management.
Learn more ⌕

◯ View   ↓ Download JSON file

◯ Require compliant or hybrid Azure AD joined device or multifactor authentication for all users

Protect access to company resources by requiring users to use a managed device or perform multifactor authentication.
Learn more ⌕

◯ View   ↓ Download JSON file

Review + create   ‹ Previous   Next: Review + Create ›

Figure 6.12: Conditional Access policy templates

5. Adjust the template's default settings to fit your organization's unique environment. This may include specifying user groups, defining network locations, or selecting targeted applications.

6. Optionally, you can define session controls to refine how access is granted further. This could involve setting sign-in frequency or access lifetimes.

7. Once you have configured the template to your satisfaction, allow the policy and save your changes. You should start with a report-only mode to assess the impact before fully enforcing the policy.

> **Exam Tip**
> By default, the user configuring the policy is excluded from the assignment scope.

Creating policies based on templates can simplify your deployment by allowing you to implement known working configurations.

Summary

In this chapter, you explored Microsoft Entra Conditional Access, a critical component in the modern security infrastructure of any organization. We began by laying the groundwork for planning Conditional Access policies, emphasizing the importance of aligning security measures with business objectives and regulatory requirements.

We then transitioned into the practical application of these plans by implementing Conditional Access policy assignments and controls. Here, you learned how to specify which users and scenarios would trigger the Conditional Access policies, ensuring that only the right individuals have access under the right conditions.

Finally, we looked at creating a Conditional Access policy from a template, learning how to utilize preconfigured templates to deploy policies quickly and effectively that adhere to security best practices.

In the next chapter, we will explore managing risk effectively using Microsoft Entra ID Protection.

Exam Readiness Drill – Chapter Review Questions

Apart from mastering key concepts, strong test-taking skills under time pressure are essential for acing your certification exam. That's why developing these abilities early in your learning journey is critical.

Exam readiness drills, using the free online practice resources provided with this book, help you progressively improve your time management and test-taking skills while reinforcing the key concepts you've learned.

HOW TO GET STARTED

- Open the link or scan the QR code at the bottom of this page

- If you have unlocked the practice resources already, log in to your registered account. If you haven't, follow the instructions in *Chapter 19* and come back to this page.

- Once you log in, click the START button to start a quiz

- We recommend attempting a quiz multiple times till you're able to answer most of the questions correctly and well within the time limit.

- You can use the following practice template to help you plan your attempts:

Working On Accuracy		
Attempt	**Target**	**Time Limit**
Attempt 1	40% or more	Till the timer runs out
Attempt 2	60% or more	Till the timer runs out
Attempt 3	75% or more	Till the timer runs out
Working On Timing		
Attempt 4	75% or more	1 minute before time limit
Attempt 5	75% or more	2 minutes before time limit
Attempt 6	75% or more	3 minutes before time limit

The above drill is just an example. Design your drills based on your own goals and make the most out of the online quizzes accompanying this book.

> First time accessing the online resources? 🔒
>
> You'll need to unlock them through a one-time process. **Head to *Chapter 19* for instructions**.

Open Quiz	
https://packt.link/sc300ch6 OR scan this QR code →	

7

Managing Risk Using Microsoft Entra ID Protection

Managing risk must be an ongoing commitment to protect users and resources. **Microsoft Entra ID Protection** offers a robust set of tools designed to detect, investigate, and remediate potential threats. This chapter discusses Microsoft Entra ID Protection, guiding you through implementing and managing policies that safeguard your organization's identities.

The following topics will be covered in this chapter:

- Implementing and managing user risk policies
- Implementing and managing sign-in risk policies
- Implementing and managing MFA registration policies
- Monitoring, investigating, and remediating risky users
- Monitoring, investigating, and remediating risky workload identities

By the end of this chapter, you'll have the knowledge required to leverage Microsoft Entra ID Protection to its fullest potential, ensuring that your organization's identity landscape is not only secure but also resilient against the threats of tomorrow.

Implementing and Managing User Risk Policies

User risk policies are a cornerstone of proactive identity security within Microsoft Entra ID Protection. Requiring a Microsoft Entra ID P2 license, these policies enable administrators to identify and respond to potential threats before they escalate into full-blown security incidents. Organizations can implement timely and effective mitigation strategies by analyzing user behavior, detecting anomalies, and assessing risk levels. This proactive approach is essential for safeguarding sensitive data and preventing unauthorized access.

Let's explore user risk policies and how to leverage them to bolster your organization's security posture.

Figure 7.1 shows the **User risk policy** blade under **Identity Protection**.

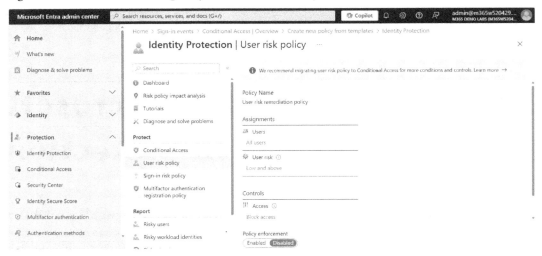

Figure 7.1: User risk policy

User risk levels are determined through a combination of real-time and offline risk decisions by Microsoft Entra ID Protection. The risk levels are calculated using machine learning that analyzes various signal behaviors associated with user accounts:

- **High**: Indicates a high likelihood of compromise.

- **Medium**: Suggests moderate risk of compromise.

- **Low**: Indicates a lower likelihood of compromise.

- **Microsoft's recommendation**: Set the user risk policy threshold to **High**. Setting the user risk policy to **High** reduces the number of times the policy will trigger, minimizing the impact on users while protecting against significant threats.

Within a user risk policy, there are various configuration options, including secure password changes. When a user's risk level is **High**, require a secure password change to force immediate action. The user will not be able to cancel or avoid this action. Additionally, users must perform a password change using Microsoft Entra **multi-factor authentication** (**MFA**) before creating a new password. This ensures that the password change process is secure and verified.

For risk remediation, instead of outright blocking users, it is advisable to encourage self-remediation. Users should be warned to register for MFA *before* encountering a situation that requires remediation and should be forced to register with MFA within a specific number of days, which is a setting within MFA. For hybrid users synced from on-premises environments, it is crucial to enable password writeback in the Entra ID Connect sync. It is also important to note that password changes made outside the risky user policy flow do not meet the secure password change requirements.

Microsoft Entra ID Protection allows granular control over these levels, enabling you to specify conditions that trigger low-, medium-, or high-risk alerts. For example, a sign-in from a new device might be considered low risk, while access requests from a blacklisted IP address would be high risk.

Automated responses are then configured based on these risk levels. High-risk behaviors might trigger an immediate account lockout, requiring administrative intervention, while medium risks could prompt additional verification through MFA. Balancing security with user convenience is essential, ensuring that policies do not hinder productivity.

Managing risk policies involves continuous monitoring and refinement. Microsoft Entra ID Protection offers real-time monitoring capabilities, providing alerts and detailed reports on user risk events. This data is crucial for identifying trends, refining risk definitions, and adjusting automated responses to maintain an optimal security stance.

Microsoft recommends using Conditional Access policies instead of the user risk policy settings, however, as Conditional Access policies can evaluate more signals.

Implementing and Managing Sign-In Risk Policies

Sign-in risk policies in Microsoft Entra ID Protection are designed to assess the risk level of sign-in attempts in real time and apply appropriate controls to mitigate potential threats. Implementing these policies requires a nuanced approach that balances security with user experience. Let's dive into the specifics that an Identity and access management administrator should consider. *Figure 7.2* shows the **Sign-in risk policy** blade under **Identity Protection**.

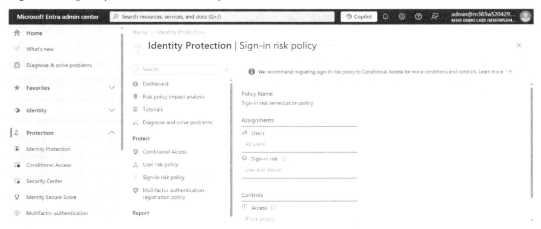

Figure 7.2: Sign-in risk policy

Risk levels in Identity Protection are determined through signals and machine learning algorithms. These levels include *high risk*, which indicates a high probability that the legitimate user may not have performed a sign-in; *medium risk*, suggesting moderate risk associated with the sign-in activity; and *low risk*, indicating a lower likelihood of compromise. Microsoft recommends configuring policies based on these risk levels to balance security and user experience.

> **Note**
>
> It's important to be aware that user and sign-in risk policies will change in 2026. The legacy risk policies configured in Microsoft Entra ID Protection are scheduled to be retired on October 1, 2026. This means that organizations must transition to the updated risk-based Conditional Access policies to ensure continued protection and compliance. Until these legacy risk policies are officially retired, it is necessary to be knowledgeable about them and prepared to address any related questions on the SC-300 exam.

A sign-in risk policy requires MFA for sign-ins when the risk level is **Medium** or **High**. This policy allows users to prove their identity using registered authentication methods, effectively balancing security and user experience.

Within a sign-in risk policy, there are various configuration options for access control. These options include blocking access to stop risky sign-ins, though this may impact legitimate users, and providing self-remediation options that allow users to take action, such as performing a secure password change or using MFA, to mitigate risk. Trusted network locations in Azure are utilized by Identity Protection to minimize false positives in risk detections. By identifying and marking certain networks as trusted, the system can better differentiate between legitimate and suspicious sign-ins, thereby enhancing the accuracy of security measures and reducing unnecessary alerts.

For hybrid users, enabling password writeback allows them to change their passwords securely and ensures compliance with secure password change requirements. It is important to warn users to proactively register for MFA to avoid being blocked during risky sign-ins.

Automated responses are an essential feature of sign-in risk policies. Depending on the risk level detected, the system can prompt Microsoft Entra MFA or even block the sign-in attempt altogether. A secure password change might be enforced for high-risk detections, requiring the user to complete MFA before resetting their password.

As with the legacy user risk policies, Microsoft recommends using Conditional Access policies instead of the sign-in risk policies, as Conditional Access policies can evaluate more signals.

Next, we will examine how MFA registration policies are implemented and managed.

Implementing and Managing MFA Registration Policies

MFA is a critical security feature that adds an additional layer of protection beyond just a username and password. By requiring two or more verification methods, MFA ensures that the risk of unauthorized access to resources is significantly reduced.

Implementing MFA Registration Policies

The implementation process begins with the configuration of the **MFA registration policy**. As an identity and access administrator, ensuring users are protected via MFA is critical to improving the security posture of your organization.

Here are step-by-step instructions for configuring an MFA registration policy:

1. Navigate to the Entra admin center (`https://entra.microsoft.com`).
2. Expand **Protection** and select **Identity Protection**.
3. Select **Multifactor authentication registration policy**, as shown in *Figure 7.3*.

Figure 7.3: MFA registration policy

4. In the **Assignments** section, go to **Users**.
5. In the **Include** section, select either **All users** or **Select individuals and groups** if you want to limit the rollout.
6. In the **Exclude** section, select **Users and groups** and choose your organization's emergency access or break-glass accounts.
7. Set the **Policy enforcement** option to **Enabled**.
8. Click **Save** to apply the policy.

It's important to exclude emergency access or "break-glass" accounts from your policy to ensure they remain accessible in case of an emergency.

The policy must be set to enforce MFA registration for users, which can be done by enabling the policy and saving the changes. Once enforced, users will be prompted to register for MFA during their next interactive sign-in. The default time is to register for MFA within 14 days of a user's first logon to complete the registration. The 14-day period cannot be updated at this time.

Managing MFA Registration Policies

Managing these policies involves monitoring the registration process and ensuring that all users comply with the policy requirements. You should track the progress of user registrations and follow up with users who still need to complete the process within the given time frame.

It's also essential to communicate effectively with users about the importance of MFA and provide guidance on how to register their authentication methods.

Explain to users why MFA is crucial for enhancing security. Highlight how it protects their accounts from unauthorized access and reduces the risk of data breaches.

Offer step-by-step instructions on registering their authentication methods. This can be done through email, internal documentation, or a company intranet page. Include screenshots or a video tutorial to make the process easier to follow. The following steps can be used to help users configure and update their MFA options:

1. Navigate to `https://myaccount.microsoft.com` and log in with a Microsoft 365 identity.

2. Select **Security info**.

3. On the **Security info** page, select **Add sign-in method** to set up a new MFA method. See *Figure 7.4*.

Figure 7.4: Managing MFA methods

4. On the **Add a sign-in method** page, choose which kind of method to configure.

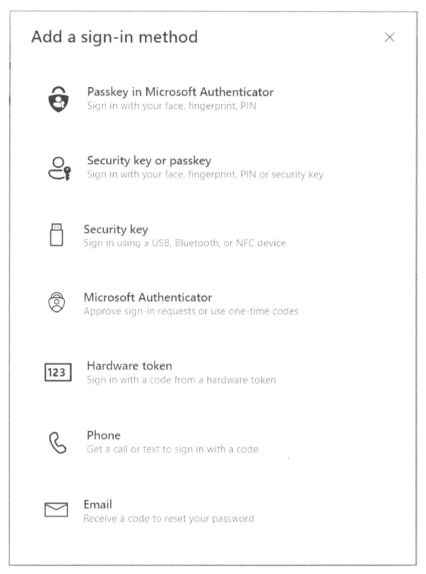

Figure 7.5: Configuring a new sign-in method

5.   Follow the prompts to configure the selected method and validate it.

When instructing users, be sure to send reminders and follow-up communications to ensure all users complete their MFA registration.

Next, we'll look at ways to work with users with risks identified.

Monitoring, Investigating, and Remediating Risky Users

Microsoft Entra ID Protection plays a pivotal role in this context by helping you manage the risk associated with user identities. A risky user is a user account that has exhibited suspicious activity or has a high likelihood of being compromised. This could be due to factors such as multiple failed login attempts from unusual locations, password resets from unfamiliar devices, or the exposure of credentials in a data breach.

These events provide valuable insights into potential threats and help prioritize investigation and remediation efforts. By effectively monitoring these indicators, organizations can enhance their security posture and swiftly address any potential vulnerabilities. To monitor risky sign-ins, follow these steps:

1. Navigate to the Entra admin center (`https://entra.microsoft.com`).

2. Expand **Protection** and select **Risky activities**.

3. Examine the detected risks, as shown in *Figure 7.6*:

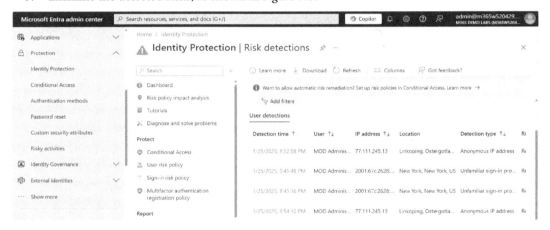

Figure 7.6: Examining the Risk detections report

4. Pay attention to the **Detection type** column, indicating things such as anonymous IP addresses, unfamiliar sign-in properties, atypical or impossible travel scenarios, the use of malware-linked IP addresses, or other factors. These are key indicators of potential account compromise.

5. Look for other anomalous activities, such as multiple failed login attempts or sign-ins from devices not previously associated with the user. These activities can provide additional insights into potential threats.

Use the information gathered from the user risk events to prioritize which incidents to investigate further. Focus on high-risk events that could indicate compromised accounts.

Configuring Risk Alerts

You should also configure alerts to receive notifications when a user's risk level increases significantly. Timely alerts allow you to take swift action. To configure alerts for receiving notifications when a user's risk level increases significantly, follow these steps:

1. Navigate to the Entra admin center (`https://entra.microsoft.com`).

2. Expand **Protection** and select **Identity Protection**.

3. Under **Settings**, select **Users at risk detected alerts**.

4. By default, the initial Global Administrator account is configured to receive alerts. You can add additional email addresses.

5. Click on the alert configuration and specify the user risk level at which you want to receive notifications. By default, this is set to **High**, but you can adjust it to **Medium** or **Low** based on your requirements. See *Figure 7.7*.

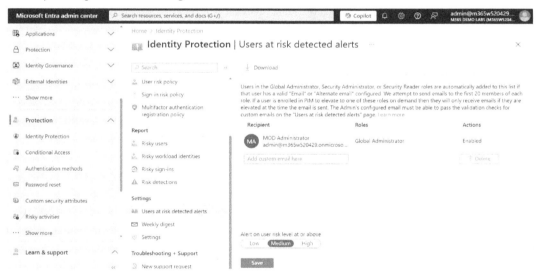

Figure 7.7: Configuring the alert risk level

6. Click **Save** to apply the changes and activate the alert notifications.

Next, we'll look at investigating, managing, and resolving risky sign-ins.

Investigating Risky Sign-Ins

The following investigation tools enable you to drill down into specific user events, view historical data, and identify patterns. Here are some of the essential tools:

- The **Risky users** report: Lists all users whose accounts are currently or were considered at risk of compromise
- The **Risky workload identities** report: Similar to the **Risky users** report, this lists service principals that have been considered at risk of compromise
- The **Risky sign-ins** report: Generated when one or more risk detections are reported for a sign-in
- The **Risk detections** report: Each risk detected is reported as a risk detection
- **Sign-in Diagnostics**: Allows you to review any flagged sign-in events or search for a specific sign-in event

Microsoft intelligence continuously monitors data on cyber threats and risks, keeping you updated on potential dangers. By aggregating data across all of Microsoft's properties, the Microsoft Security Graph allows you to monitor **indicators of compromise** (**IOCs**) and respond proactively to emerging threats. IOCs serve as an early warning system to help you stay ahead in the ever-evolving cybersecurity landscape.

In this scenario, imagine you are an administrator responsible for managing user identities and security. Recently, your organization has noticed an increase in risky sign-in attempts. Your task is to investigate these incidents and take appropriate remediation actions. Let's walk through the steps you will need to take:

1. Navigate to the Entra admin center (`https://entra.microsoft.com`).
2. Expand **Identity**, expand **Monitoring & health**, and then select **Sign-in logs**.
3. Examine the logs for suspicious activity, including uncommon locations or unfamiliar IP addresses.
4. Now, you need to investigate suspicious activity. Under **Identity**, expand **Protection** and select **Risky activities**.
5. Under **Report**, select **Risk detections**. For each risky sign-in, review the details within the **Risky sign-ins** report:
   - **IP address**: Check whether it's from an unusual location. In *Figure 7.8*, you can see the IP addresses and locations from each of the risky sign-in alerts.

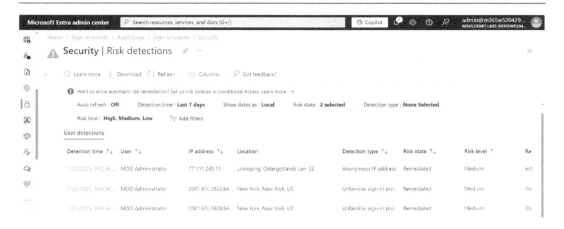

Figure 7.8: Risk detections report

- **Device information**: Look for unfamiliar devices. You can find the device info by selecting a sign-in event and selecting the **Device info** tab.

Figure 7.9: Device info

This information will need to be filled in:

A. Look for device details such as name, operating system, and IP address. Unfamiliar devices might have names or IP addresses that don't match your organization's usual patterns.

B. Identify devices not registered or previously seen in your network. Entra ID can flag new devices that haven't been used before.

C. If an unfamiliar device is detected, verify with the user whether they recognize it. This can help confirm whether the device is legitimate or potentially malicious.

6. To verify the app or service involved, start by accessing the sign-in logs within your security tool to gather detailed information about the sign-in attempts. Identify the specific application or service involved in each attempt, listed under **Application** or **Service principal name**.

7. Next, cross-reference these details with your organization's list of approved and authorized applications, which you can maintain in an internal document or within your security system's settings. Flag any applications or services that are not on this approved list and investigate their legitimacy and potential security risks. If you identify any suspicious or unauthorized applications, take appropriate action, such as blocking access, alerting the security team, or conducting further investigations to prevent potential security breaches.

   Cross-reference with other logs (for example, Entra ID audit logs) to identify patterns. These logs can be found in Entra ID, in the **Monitor** section.

8. For remediation actions, if a sign-in is highly suspicious, such as from a known malicious IP, consider temporarily blocking access for that user using Conditional Access policies. These policies can also be utilized to block access based on the risk level, mitigating potential threats by preventing unauthorized access and ensuring only legitimate sign-ins are allowed, as discussed in *Chapter 6*. If the sign-in risk is moderate, prompt the user for MFA.

9. If the risk is due to compromised credentials in Entra ID, you should take immediate action to mitigate the threat. First, force a password reset for the affected user to prevent any unauthorized access. This can be done through your identity management system, directly in Entra ID, or by configuring Conditional Access policies. You can also perform the steps through the **Risky users** report, as shown in *Figure 7.10*.

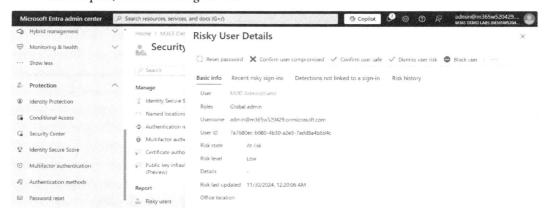

Figure 7.10: Viewing options in the Risky User Details flyout

10. Next, notify the user about the situation and immediately instruct them to change their password. Guide users through creating a strong and unique password to enhance their account security. This proactive approach helps quickly address the risk and safeguard the user's account.

11. Notify affected users about the incident via email, Microsoft Teams message, or phone call, providing clear instructions on the next steps they need to take, such as resetting their passwords or setting up MFA. This ensures that users are informed and can take immediate action to secure their accounts.

Best practices include adhering to the principle of least privilege, conducting regular audits, following security baselines, and educating users about security awareness.

Practical experience and hands-on labs are essential for mastering these concepts. To reinforce your understanding, explore the Azure portal, experiment with policies, and simulate risk scenarios.

The next section will explore how to monitor, investigate, and remediate risky workload identities.

Monitoring, Investigating, and Remediating Risky Workload Identities

Workload identities represent applications, services, and other non-human entities requiring access to your Azure environment's resources. These identities enable secure communication between services and ensure proper access controls. Examples of workload identities are **service principals**, **managed identities**, and **application registrations**. Properly securing these identities prevents unauthorized access and ensures your resources' confidentiality, integrity, and availability.

To effectively monitor workload identities, you should leverage tools such as Azure Monitor and Entra ID Logs. These tools provide invaluable insights into workload identity activities, including sign-ins, role assignments, and resource access. Additionally, monitoring role assignments can help identify unauthorized access or excessive privileges granted to service principals. It's crucial to configure alerts for critical events, such as failed sign-ins or role assignment changes, to enable timely responses to potential security incidents.

Compromised workload identities can grant attackers extensive access to sensitive data and resources, potentially leading to data breaches, system disruptions, or ransomware attacks. Organizations can proactively identify potential threats by evaluating risk factors for workload identities, including excessive permissions, anomalous access patterns, and suspicious authentication attempts. Microsoft Entra ID Protection can be leveraged to detect risky workload identities and assign risk levels (low, medium, or high) based on behavior. For instance, if a service principal with extensive permissions suddenly attempts to access sensitive data from an unusual location, Entra ID Protection can flag this as a high-risk event. The Microsoft Entra ID Protection dashboard provides a centralized view of these alerts, enabling security teams to investigate the incident further and take appropriate actions, such as revoking access or resetting credentials. Staying informed about emerging threats through threat intelligence feeds can also aid in identifying potential attack vectors targeting workload identities.

Several critical steps can be taken to remediate risky workload identities. Implementing granular Conditional Access policies based on risk levels is essential. For instance, requiring MFA for high-risk workload identities accessing sensitive resources can significantly deter unauthorized access. Adhering to the principle of least privilege by granting workload identities only the necessary permissions is crucial. Overly permissive roles can increase the potential damage if a workload identity is compromised.

Promptly resetting passwords or rotating secrets for compromised service principals is imperative. Additionally, carefully reviewing and adjusting role assignments is vital. For example, removing the corresponding role assignment can reduce the attack surface if a service principal no longer requires access to a specific resource.

Transparent communication with application owners and other stakeholders about remediation actions is essential for minimizing disruption. For instance, informing application owners about a service principal password reset and providing guidance on updating application configurations can help prevent service interruption.

Regularly reviewing risk metrics and threat intelligence allows organizations to adapt their security posture. Security teams can identify emerging threats by analyzing trends in risky workload identity behavior and adjust policies accordingly. For example, if an increase in risky workload identities is observed from a specific geographic region, organizations can implement stricter access controls for that region.

Regarding best practices, following consistent naming conventions for service principals is crucial for organization and management. Descriptive and standardized naming schemes, such as incorporating the environment (dev, test, or prod), application name, and purpose, enhance visibility and facilitate troubleshooting.

Secrets associated with service principals, such as certificates and keys, were covered in *Chapter 5*, in the section titled *Implementing Certificate-Based Authentication in Microsoft Entra*. Secrets should be securely managed throughout their lifecycle. This includes generating strong cryptographic keys, storing them in secure vaults such as Azure Key Vault, regularly rotating them, and adhering to strict access controls. Effective lifecycle management involves defining clear expiration dates, implementing automated renewal processes, and conducting regular audits to identify and revoke unused or compromised secrets.

Lastly, enabling auditing for service principal activities can provide valuable insights into potential security threats and operational issues. Organizations can detect anomalies, investigate suspicious behavior, and respond promptly to incidents by monitoring actions such as sign-ins, role assignments, and resource access. Analyzing audit logs can also help identify opportunities for process improvement and compliance adherence.

Summary

This chapter explored the critical aspects of identity risk management. User risk policies were examined to assess and manage risks associated with individual users, enabling organizations to protect user identities from potential threats. Sign-in risk policies were analyzed to evaluate the risk level of each login attempt, allowing for appropriate actions to secure sign-ins and prevent unauthorized access. The implementation and management of MFA registration policies were discussed to enhance security by requiring additional authentication factors during sign-in. Techniques for monitoring and investigating users who exhibited risky behaviors were presented, emphasizing the importance of identifying and addressing potential threats promptly. Finally, the chapter dived into handling risky workload identities within an organization, including the identification of suspicious patterns and corrective measures.

Having explored the strategies for managing risk using Microsoft Entra ID Protection, including identifying and mitigating user and sign-in risks, we now focus on another critical aspect of securing your organization's environment: implementing access management for Azure resources using Azure Roles.

In the next chapter, you will learn about the principles and practices of role-based access control (RBAC) in Azure, providing you with the knowledge and tools to assign and manage permissions across your Azure resources effectively.

Exam Readiness Drill – Chapter Review Questions

Apart from mastering key concepts, strong test-taking skills under time pressure are essential for acing your certification exam. That's why developing these abilities early in your learning journey is critical.

Exam readiness drills, using the free online practice resources provided with this book, help you progressively improve your time management and test-taking skills while reinforcing the key concepts you've learned.

HOW TO GET STARTED

- Open the link or scan the QR code at the bottom of this page

- If you have unlocked the practice resources already, log in to your registered account. If you haven't, follow the instructions in *Chapter 19* and come back to this page.

- Once you log in, click the START button to start a quiz

- We recommend attempting a quiz multiple times till you're able to answer most of the questions correctly and well within the time limit.

- You can use the following practice template to help you plan your attempts:

Working On Accuracy		
Attempt	Target	Time Limit
Attempt 1	40% or more	Till the timer runs out
Attempt 2	60% or more	Till the timer runs out
Attempt 3	75% or more	Till the timer runs out
Working On Timing		
Attempt 4	75% or more	1 minute before time limit
Attempt 5	75% or more	2 minutes before time limit
Attempt 6	75% or more	3 minutes before time limit

The above drill is just an example. Design your drills based on your own goals and make the most out of the online quizzes accompanying this book.

First time accessing the online resources? 🔒

You'll need to unlock them through a one-time process. **Head to *Chapter 19* for instructions**.

Open Quiz	
https://packt.link/sc300ch7	
OR scan this QR code →	

8

Implementing Access Management for Azure Resources by Using Azure Roles

A robust access management strategy is paramount for safeguarding sensitive data and maintaining operational integrity. A compromised Azure environment can result in significant financial losses, reputational damage, and regulatory non-compliance. **Azure roles** provide the cornerstone for establishing granular control over resource access, empowering organizations to implement a security-centric approach. This chapter explores the critical importance of Azure roles, equipping you with the knowledge and skills to design and implement an effective access control system that aligns with your organization's security requirements.

The objectives and skills covered in this chapter include the following:

- Creating custom Azure roles

- Assigning built-in and custom Azure roles

- Evaluating effective permissions for a set of Azure roles

- Assigning Azure roles to enable Microsoft Entra ID login to Azure virtual machines

- Configuring Azure Key Vault role-based access control (RBAC)

By the end of this chapter, you'll have acquired the practical skills you need to implement Azure roles for comprehensive access management. Mastering these skills will allow you to implement a robust access management system for Azure resources, ensuring alignment with both security best practices and organizational requirements.

Creating Custom Azure Roles

While built-in roles such as *Reader*, *Contributor*, and *Owner* offer a basic level of access control, real-world scenarios often demand more granular permissions. For instance, you might need to grant a user the ability to start and stop virtual machines but not modify network security groups. In these cases, custom Azure roles become indispensable. This section dives into the intricacies of creating custom roles, empowering you to define precise permissions aligned with your organization's specific needs.

Passing the SC-300 exam requires a comprehensive understanding of the two pillars of custom roles: permissions and scope.

Permissions define the actions that the users with the assigned role can perform on Azure resources. Permissions are categorized into two main groups, aligned with **Azure Resource Manager (ARM)** operations. ARM uses a declarative, JSON-based language for defining the infrastructure and configuration of your resources, as shown in the following examples:

- **Control plane permissions**: These govern managing Azure resource objects, such as creating, deleting, or modifying them. Examples include `Microsoft.Compute/virtualMachines/write` for creating VMs and `Microsoft.Network/securityGroups/delete` for deleting security groups.

- **Data plane permissions**: These pertain to accessing resource objects' data, such as reading VM logs or retrieving storage blob data. Examples include `Microsoft.Compute/virtualMachines/read` for reading VM information and `Microsoft.Storage/storageAccounts/read` for accessing storage accounts.

Scope determines where the custom role applies. For the SC-300 exam, there are two primary scopes to remember:

- **Management group scope**: This allows the assignment of custom roles to users or groups within a specific management group. They'll then have access to resources contained within that group.

- **Subscription scope**: This grants users or groups assigned to this role access to resources across the entire Azure subscription.

While the SC-300 exam primarily focuses on role assignments at the resource group or individual resource level, understanding the broader context of custom role definition at the management group or subscription level is crucial for effective access management. Although not directly tested in the exam, this knowledge enables you to design a scalable and hierarchical access control strategy. By creating custom roles at higher levels, you can efficiently manage permissions across multiple resources and subscriptions, reducing administrative overhead and minimizing the risk of errors. This holistic understanding of Azure roles will equip you with the ability to implement comprehensive and secure access control solutions within your organization.

Determining Permissions

Before you begin creating a custom role, you'll need to understand both the security and functional requirements of the role you're going to define. Maybe you need to create a very restrictive role for a small subset of users or maybe you need to combine a few permissions from multiple roles. In either case, it's usually best to start by understanding the built-in roles and using them to refine your new custom role.

Built-In Roles

There are dozens (if not hundreds) of built-in roles across the Azure environment. You can view a list of built-in roles and their corresponding permissions here: `https://learn.microsoft.com/en-us/azure/role-based-access-control/built-in-roles`.

Once you've determined your starting roles and the permission set you need, you can start creating a custom role.

Creating a New Custom Role

Creating custom roles is a powerful way to tailor permissions to specific job functions within your Azure environment.

Leveraging built-in roles can significantly streamline the creation of custom roles. Built-in roles provide a solid foundation for many common access scenarios. By starting with a built-in role and then modifying it to add or remove specific permissions, you can create custom roles efficiently. For instance, you might use the *Contributor* role as a starting point for a custom role that allows users to manage virtual machines but restricts access to network resources.

Meaningful naming conventions are essential for maintaining clarity and organization within your Azure environment. Clear and descriptive names for custom roles enhance collaboration and troubleshooting. For example, a role named *VM Contributor - Dev Environment* clearly communicates its purpose and scope.

Principle of Least Privilege

The principle of least privilege is reinforced consistently throughout the SC-300 exam as well as within the Zero Trust model. By granting users only the permissions necessary to perform their tasks, you minimize the potential impact of a security breach. From the custom role perspective, this means selecting the minimum set of permissions required for a specific role.

Let's explore the process of crafting a custom role that precisely aligns with your organization's needs. In this case, we'll create a custom role that allows those assigned the ability to create and manage all aspects of a virtual machine inside a given subscription *except* deleting the disk associated with it:

1. Sign in to the Azure portal (`https://portal.azure.com`) with an account that has either the Global Administrator or a Privileged Role Administrator role.

2. In the search bar, enter `Subscriptions` and then select the Azure subscription where you want to create the custom role.

3. Select **Access control (IAM)**, as shown in *Figure 8.1*:

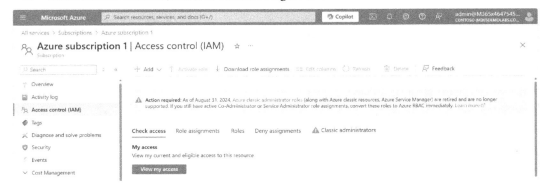

Figure 8.1: Access control page

4. Click **Add** and then select **Add custom role**.

Figure 8.2: New custom role

5. On the **Basics** tab of the **Create a custom role** wizard, provide a name and description for the role.

6. In the **Baseline permissions** area, select **Clone a role**. Under **Role to clone**, select **Virtual Machine Contributor**, as shown in *Figure 8.3*.

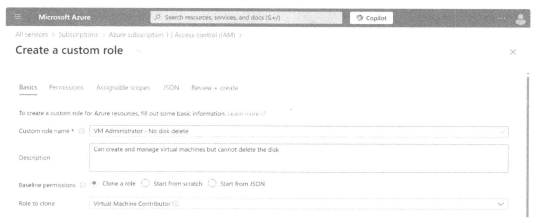

Figure 8.3: Basics tab of Create a custom role

7. Click **Next**.

8. On the **Permissions** tab, locate the **Microsoft.Compute/disks/delete** permission and click the *delete* icon.

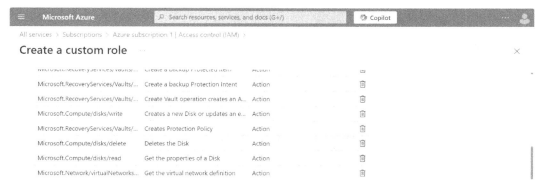

Figure 8.4: Deleting a permission

9. Click **Next**.

10. On the **Assignable scopes** page, select where you want to scope this permission. Since we started at the subscription level, it's scoped there by default. However, if you want this role to apply either more broadly or restrictively, you can select that scope by clicking **Add assignable scopes**. See *Figure 8.5* for an example of selecting an assignable scope.

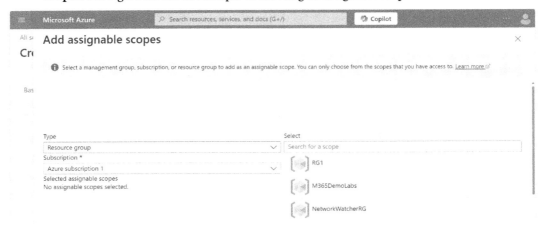

Figure 8.5: Selecting an assignable scope

11. Click **Next**.

12. On the **JSON** tab, you can view the configuration information. You can copy it and save it for future reuse, if desired, or edit it further manually. Click **Next**.

13. Review the permissions on the **Review + create** tab and select **Create**.

Custom roles, whether created from scratch or by using an existing role as a template, can be an important part of managing your organization's security in the context of applying least privilege permissions.

You may have noticed when creating custom roles that you have the ability to exclude permissions. Custom roles allow you to select broad, wildcard permissions, which may be too broad for a given circumstance. You can select a broad permission to include, but then exclude a more restrictive permission inside the set.

When you exclude a permission, it is included as a *NotActions* or *NotDataActions* entry in the role. When the effective permission is calculated, all of the *Actions* and *DataActions* entries are added first, and then all of the *NotActions* and *NotDataActions* entries are subtracted.

> **Creating a New Custom Entra Role**
> You can also use a similar process to create roles within Entra. Entra roles can be created within the Entra admin center (or in the Azure portal under **Entra**). Custom Entra roles can be used to assign permissions to manage Entra objects such as identities.

The following section will review how to assign built-in and custom roles in Azure.

Assigning Built-In and Custom Azure Roles

Built-in roles in Azure are predefined with specific permissions, offering a convenient way to grant common access levels. For example, the *Virtual Machine Contributor* built-in role allows users to manage most aspects of virtual machines, aligning with the permissions defined within the role.

Custom roles offer a more granular approach to managing access control compared to built-in roles. By tailoring permissions to specific job functions, organizations can implement the principle of least privilege effectively. Assigning custom roles involves specifying the security principal (user, group, service principal, or managed identity) to whom the role is assigned, the scope at which the role applies (subscription, management group, resource group, or individual resource), and the precise set of actions permitted within that scope.

When assigning roles, you'll want to take the following into consideration:

- **Security principal selection**: The exam might ask you to identify the appropriate security principal for a specific role assignment. For example, you might need to choose between a user account for an individual administrator, a group for a team of developers, a service principal for an automated application accessing Azure resources, or a managed identity for an Azure virtual machine. Understanding the characteristics and use cases of each security principal type is essential for effective role assignments.

- **Scope selection**: Understanding the different scopes is essential for effective access control. The SC-300 exam might assess your ability to choose the appropriate scope for a role assignment. Assigning a role at the subscription level grants access to all resources within that subscription, while assigning at the resource group level limits access to resources within that group. Selecting the correct scope is crucial for minimizing the attack surface and ensuring that users have only the necessary permissions to perform their duties.

Role assignments exhibit inheritance behavior, meaning roles assigned at a broader scope (such as a subscription) are inherited by child scopes (such as resource groups or individual resources) unless overridden by a more specific role assignment at the child scope. Understanding this concept is crucial for managing access efficiently and avoiding unintended access gaps. For example, a role assigned at the subscription level might grant excessive permissions at the resource group level if not explicitly overridden.

Azure provides multiple methods for assigning roles, including the Azure portal, Azure PowerShell, Azure CLI, SDKs, and REST APIs. Familiarity with these options is essential for flexibility and automation. The *SC-300* exam might present scenarios where you need to use a specific method to assign or modify roles.

To assign either a built-in or custom Azure role, you can follow these steps:

1. Navigate to the Azure portal (`https://portal.azure.com`).

2. Locate the resource to which you wish to assign a role. For example, you can select a resource group, as shown in *Figure 8.6*:

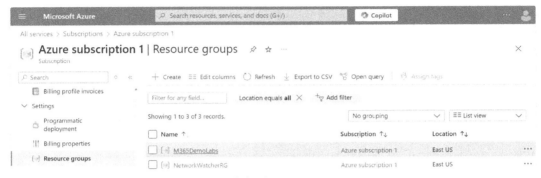

Figure 8.6: Selecting a resource group

3. Select **Access control (IAM)**.

4. Click **Add** and then select **Add role assignment**.

5. Select either a built-in or custom role and click **Next**.

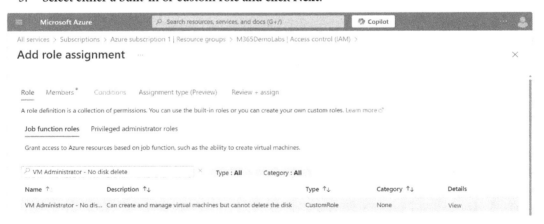

Figure 8.7: Selecting a role

6. On the **Members** tab, select who the role will be assigned to—user, group, service principal, or managed identity—by choosing the appropriate radio button and then clicking **Select members** to add them to the role.

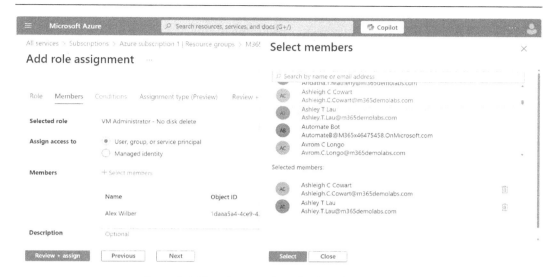

Figure 8.8: Choosing members for the role assignment

7. Click **Next**.

8. On the **Assignment type** tab, select whether the role assignment will be enabled for Privileged Identity Management (just-in-time activation) or permanently assigned and click **Next**.

9. On the **Review + assign** tab, click **Review + assign**.

After the wizard completes, the assignments will be available to use.

Next, we will discuss how to evaluate effective permissions for a set of Azure roles.

Evaluating Effective Permissions for a Set of Azure Roles

Azure RBAC grants access through roles, which define a collection of permissions for specific Azure resources or scopes.

For example, the *Contributor* role grants broad permissions to create and manage objects within a resource group, while the *Reader* role provides read-only access. However, users can be assigned multiple roles, leading to a combined set of permissions. This combined group, known as **effective permissions**, governs what capabilities a user or other identity has inside its delegated scope.

Evaluating these permissions effectively ensures that users have the least privilege principle applied, which aids in minimizing security risks. Consider a scenario where a user is assigned both the *Contributor* and *Reader* roles at the resource group level. While the *Contributor* role grants broad permissions, the *Reader* role might override certain actions, resulting in a net effect of read-only access for the user. Understanding how these roles combine is crucial for effective access management.

Determining effective permissions requires a combination of tools and analysis. The following tools are commonly used for this purpose:

- **Azure portal** (`https://portal.azure.com/`): While the Azure portal doesn't directly display effective permissions, you can analyze them through a combination of the following tools:

  - **Access control (IAM)**: View role assignments for a user, security group, or service principal. This shows the assigned roles at different scopes.

  - **Role definitions**: Investigate the specific permissions included within each assigned role by reviewing the role definition details.

- **Azure PowerShell** or **Azure CLI**: Use cmdlets such as `Get-AzRoleAssignment` to retrieve a user's role assignments across all scopes. Leverage cmdlets such as `Get-AzRoleDefinition` to explore the permissions encompassed within each assigned role.

- **Entra ID Privileged Identity Management** (**PIM**): Focus on evaluating effective permissions for privileged access roles. Use PIM reports to analyze role assignments for privileged roles (for example, Global Administrator and Security Administrator). Identify users with overlapping privileged role assignments, potentially indicating excessive permissions.

Several critical factors influence the evaluation of effective permissions. One key factor is inheritance, where child scopes inherit permissions assigned at a broader scope. For example, a user with a *Contributor* role at the subscription level will inherently have *Reader* access to all resource groups within that subscription, even if not explicitly assigned the *Reader* role on those groups.

Another critical factor is the additive model used by Azure RBAC, where permissions from all assigned roles are combined to determine effective permissions. This means a user with *Reader*, *Contributor*, and any custom roles will have the combined permissions of all roles. Additionally, while RBAC defines access levels, Conditional Access policies and **multi-factor authentication** (**MFA**) can further restrict access based on factors such as device compliance or location.

Implementing effective permission management requires a proactive approach. The principle of least privilege should be followed, granting only the minimum permissions necessary for users to perform their tasks. This involves evaluating roles and removing any unnecessary permissions from assigned roles. Regular reviews of user roles and access are essential, utilizing PIM reports and Azure Monitor logs to identify unused or excessive permissions.

To evaluate the effective permissions on a resource or other object, follow these steps:

1. Navigate to the Azure portal (`https://portal.azure.com`).

2. Using the search, locate any resource.

3. Select **Access control (IAM)**.

4. On the **Check access** tab, select **Check access** to view the effective access for a user, group, service principal, or managed identity.

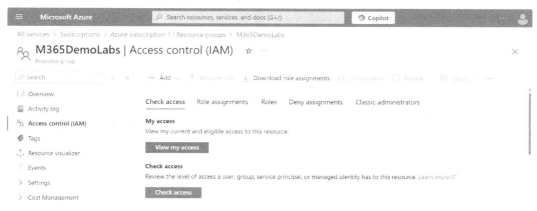

Figure 8.9: Viewing the Access control (IAM) blade of a resource group

5. Enter the identity of a security principal or group to validate.

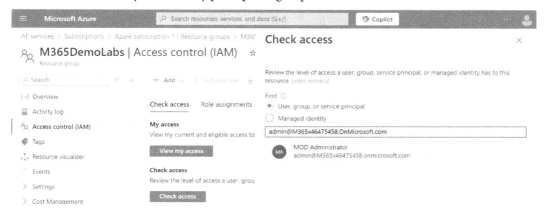

Figure 8.10: Checking access to a resource

6. The **Assignments** page shows both **Current role assignments** and **Eligible assignments**.

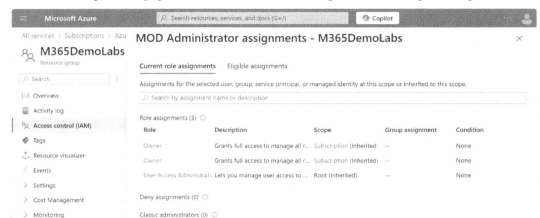

Figure 8.11: Viewing effective permissions

The assignments displayed show the scope where they are delegated, helping you understand how broad a particular assignment might be and where it can be managed.

> **Best Practices for Securing Resources**
>
> In addition to managing permissions to reduce risk, you should utilize **Privileged Access Workstations** (**PAWs**) when performing administration. PAWs are dedicated workstations designed for administering privileged systems and applications, providing an additional layer of security by isolating them from potential threats.
>
> Finally, activating Entra ID Protection features is important for monitoring suspicious activities and detecting potential privilege escalation attempts. For more information about Entra ID Protection, refer to *Chapter 7*.

In addition to using the **Check access** function to evaluate an individual effective assignment, you can also use the **Role assignments** tab to view all of the role assignments for a resource, as shown in *Figure 8.12*:

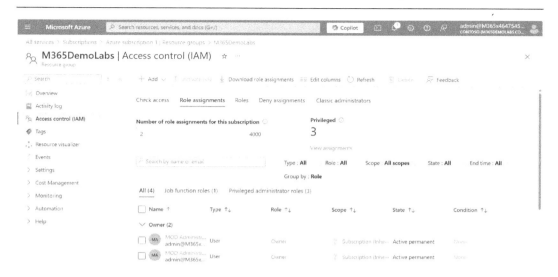

Figure 8.12: Viewing role assignments for a resource

With an understanding of how to evaluate effective permissions for a set of Azure roles and resources, you can ensure users have the right level of access to perform their duties while minimizing security risks. By following these best practices and utilizing the available tools, you can effectively manage Azure RBAC and maintain a secure cloud environment.

We will next look at assigning Azure roles to enable Microsoft Entra ID login to Azure virtual machines.

Assigning Azure Roles to Enable Microsoft Entra ID Login to Azure Virtual Machines

While Entra ID provides user authentication to resources, Azure RBAC controls access permissions on the **virtual machine** (**VM**). Assigning the correct roles ensures users can leverage Entra ID credentials for secure VM access.

Two primary methods exist to enable Entra ID login for VMs. The first method is during VM creation. When creating a new Windows VM in the Azure portal, you can enable the **Login with Microsoft Entra ID** option in the **Microsoft Entra ID** section on the **Management** tab. This option, shown in *Figure 8.13*, automatically configures the new VM for Entra ID login.

Figure 8.13: The Login with Microsoft Entra ID option

The second method allows you to enable Entra ID login for existing VMs. For Windows VMs, you can install the **Azure AD based login for Windows** (AADLogin) VM extension, while Linux-based VMs will require the **Azure AD based SSH Login** extension, which configures the VM for Entra ID authentication. See *Figure 8.14*.

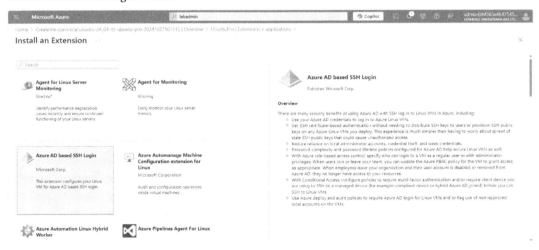

Figure 8.14: Installing the Azure AD based SSH Login extension for a Linux VM

In either case, the extension can be installed on the **Extensions + applications** blade of an existing VM.

Once Entra ID login is enabled, users need appropriate Azure RBAC roles to access the VM with their Entra ID credentials. Here are the relevant roles:

- **Virtual Machine Administrator Login**: This role grants administrative access to the VM. Users with this role can manage users and groups, install software, and configure settings.

- **Virtual Machine User Login**: This role provides basic user access to the VM. Users can perform tasks such as running applications, accessing files, and using remote desktop tools.

Additional considerations are essential for enabling seamless Entra ID integration with Azure VMs. For scenarios requiring finer control, create custom roles with specific permissions for keys, secrets, or certificates, such as read, write, or manage.

Remember that RBAC operates on different scopes. You can assign roles at the vault level for overall access or at the level of individual keys, secrets, or certificates for more granular control. While RBAC defines general access levels, you can further restrict access by implementing MFA and Conditional Access policies based on factors such as device compliance or location.

To enable Entra ID integration with Azure VMs, follow these steps:

1. From inside the Azure portal (`https://portal.azure.com`), navigate to the Azure VM you want to configure.

2. Go to the **Access control (IAM)** blade.

3. Click **+ Add** and then **Add role assignment**.

Figure 8.15: Access control (IAM)

4. Select the appropriate role (**Virtual Machine Administrator Login** or **Virtual Machine User Login**), as shown in *Figure 8.16*.

Figure 8.16: Role selection

5. Under **Assign access to**, choose **Microsoft Entra ID user, group, or service principal**.

6. Select the Entra ID user(s) or group(s) you want to grant access to.

7. Click **Save** to assign the role.

When assigning roles, prioritize the principle of least privilege. Grant only the minimum level of access required for users to perform their tasks.

Be aware of potential security risks associated with granting administrative access (Virtual Machine Administrator Login) to many users. For instance, granting multiple users full administrative privileges on a virtual machine increases the attack surface and the potential for accidental configuration changes. Consider leveraging separate administrative accounts or Entra ID PIM for just-in-time access to these roles. This approach minimizes the risk of unauthorized access and helps maintain a strong security posture.

> **Note**
>
> Keep in mind the difference between Entra ID user authentication and Azure RBAC permissions. While Entra ID verifies user identity, Azure RBAC controls access levels within the VM.

By following the steps and the key considerations, you can effectively assign Azure roles to enable secure and controlled access to Azure VMs using Microsoft Entra ID credentials. Understanding how to leverage Azure roles and Entra ID integration is crucial for demonstrating your ability to secure cloud environments and meet organizational requirements. This knowledge is invaluable for the SC-300 exam, as you may be asked to design and implement RBAC strategies for Azure resources, including virtual machines.

Now that we've covered how to assign Azure roles to enable Microsoft Entra ID login to Azure VMs, let's move on to another important topic: configuring RBAC for Azure Key Vault. This section will guide you through managing access to your organization's keys, secrets, and certificates stored in Azure Key Vault.

Configuring Azure Key Vault Role-Based Access Control (RBAC)

Azure Key Vault (**AKV**) safeguards cryptographic keys and secrets that are vital for your Azure resources, making its security paramount. Since Entra ID provides user authentication, understanding AKV RBAC is essential. AKV RBAC allows you to define who can access specific keys, secrets, and certificates within a key vault, and what actions they can perform. By leveraging RBAC, you can implement the principle of least privilege, ensuring that only authorized Entra ID users have the necessary permissions, thereby minimizing security risks.

Choosing the Right Authorization System

AKV offers two authorization systems:

- **Azure RBAC**: This granular approach leverages Entra ID identities, including Entra ID users, to offer centralized management. Predefined roles within RBAC grant specific access levels to the vault's keys, secrets, and certificates. For instance, you can create a custom role granting a development team read-only access to specific secrets without affecting other vault operations.

- **Legacy access policies**: This method uses vault-specific access policies to control access. While still functional, it offers less centralized management and granular control than RBAC. Legacy access policies are typically used for existing implementations or scenarios where RBAC might not be fully applicable. The SC-300 exam primarily focuses on Azure RBAC due to its enhanced security and management capabilities. However, it may still include some content on legacy policies, though likely a minimum amount. A solid understanding of RBAC is essential as it is a core exam component.

There are several ways to configure AKV to grant access to Entra ID users. Let's walk through them.

Assigning Roles to Your Key Vault in the Azure Portal

The Azure portal provides a user-friendly interface for assigning roles to your key vault. You can manage access control for Entra ID users directly within the portal:

1. Navigate to the Azure portal (`https://portal.azure.com`).

2. In the Azure portal, use the search bar at the top to search for `Key Vaults`. Select **Key Vaults** from the search results and select the key vault you want to manage.

3. In the key vault's navigation pane, select **Access control (IAM)**.

4. Click on the + **Add** button and select **Add role assignment**.

5. In the **Role** dropdown, choose the role you want to assign (such as **Key Vault Contributor**).

6. In the **Assign access to** dropdown, select **User, group, or service principal**.

7. Click on **Select members** and click **Select** to confirm your choices.

8. Review the role assignment details and click **Review + assign** to finalize it.

Once the assignment is complete, you can view the **Role assignments** tab to confirm that the role has been added.

Automating with Azure Tools

You can use Azure PowerShell or Azure CLI cmdlets to automate AKV RBAC configuration. This allows you to manage access at scale and integrate it into your existing **infrastructure as code (IaC)** deployments using ARM templates.

Here is a step-by-step guide to automating AKV RBAC configuration using Azure PowerShell and the Azure CLI:

1. Ensure Azure PowerShell or the Azure CLI is installed on your machine. You can download and install them from the official Azure documentation:

    Azure PowerShell: Launch an elevated PowerShell session and then run the following command:

    ```
    Install-Module -Name Az -Repository PSGallery -Force
    ```

    Azure CLI: Launch an elevated PowerShell prompt session and then run the following command:

    ```
    winget install -e --id Microsoft.AzureCLI
    ```

    For additional installation packages and methods, see https://learn.microsoft.com/en-us/cli/azure/install-azure-cli-windows?tabs=azure-cli.

2. Open your terminal or PowerShell window and sign in to your Azure account using the `az login` command for the Azure CLI or `Connect-AzAccount` for Azure PowerShell.

3. Set the context to the appropriate Azure subscription.

    This is for the Azure CLI:

    ```
    az account set --subscription <subscription-id>
    ```

    This is for Azure PowerShell:

    ```
    Set-AzContext -SubscriptionId <subscription-id
    ```

4. If you don't already have a resource group, create one.

   Use this for the Azure CLI:

   ```
   az group create --name <resource-group-name> --location
   <location>
   ```

   Use this for Azure PowerShell:

   ```
   New-AzResourceGroup -Name <resource-group-name> -Location
   <location>
   ```

5. Next, you will need to create a new key vault or identify an existing one.

   Use this for the Azure CLI:

   ```
   az keyvault create --name <key-vault-name> --resource-group
   <resource-group-name> --location <location>
   ```

   Use this for Azure PowerShell:

   ```
   New-AzKeyVault -ResourceGroupName <resource-group-name>
   -VaultName <key-vault-name> -Location <location>
   ```

6. Use the appropriate command to assign a role.

   This is for the Azure CLI:

   ```
   az role assignment create --assignee <user-or-
   service-principal-id> --role <role-name> --scope /
   subscriptions/<subscription-id>/resourceGroups/<resource-group-
   name>/providers/Microsoft.KeyVault/vaults/<key-vault-name>
   ```

   This is for Azure PowerShell:

   ```
   New-AzRoleAssignment -ObjectId <user-or-service-
   principal-id> -RoleDefinitionName <role-name> -Scope /
   subscriptions/<subscription-id>/resourceGroups/<resource-group-
   name>/providers/Microsoft.KeyVault/vaults/<key-vault-name>
   ```

7. Verify that the role assignment was successful.

   This is for the Azure CLI:

   ```
   az role assignment list --assignee <user-or-service-
   principal-id> --scope /subscriptions/<subscription-id>/
   resourceGroups/<resource-group-name>/providers/Microsoft.
   KeyVault/vaults/<key-vault-name>
   ```

   This is for Azure PowerShell:

   ```
   Get-AzRoleAssignment -ObjectId <user-or-service-
   principal-id> -Scope /subscriptions/<subscription-id>/
   resourceGroups/<resource-group-name>/providers/Microsoft.
   KeyVault/vaults/<key-vault-name>
   ```

To integrate this configuration into your IaC deployments, include the role assignment in your ARM templates. Define the role assignment in the `resources` section of your ARM template, specifying `type`, `apiVersion`, `properties`, and `scope`.

Azure Key Vault offers several options for controlling access to its resources.

Controlling Access to the Key Vault

Microsoft offers pre-built roles for AKV that map to standard access requirements. These include the following:

- **Key Vault Administrator**: Provides complete control over all vault objects (keys, secrets, and certificates)

- **Key Vault Secrets Officer**: Allows managing secrets within the vault (create, read, update, and delete)

- **Key Vault Crypto Officer**: Grants access to manage cryptographic keys for encryption and decryption operations

For scenarios requiring finer control, you can create custom roles with specific permissions for keys, secrets, or certificates, such as read, write, or manage. It's important to understand that RBAC operates on different scopes, allowing you to assign roles at the vault level for overall access or at the level of individual keys, secrets, or certificates for more granular control. Additionally, while RBAC defines access levels, implementing MFA and Conditional Access policies can further restrict access based on factors such as device compliance or location.

This knowledge is essential for the SC-300 exam and demonstrates your expertise as an IAM administrator who can leverage Entra ID for user authentication within a secure Azure environment. Remember, the focus is on using Azure RBAC for centralized and granular access control while adhering to security best practices.

Summary

This chapter discussed Azure RBAC, a crucial aspect of securing your Azure resources. We explored various methods for managing access, including creating custom roles for granular control and leveraging pre-built roles for common tasks.

We discussed the importance of evaluating effective permissions to ensure users have the least privilege necessary for their duties. This minimizes security risks by preventing unauthorized access to critical resources.

The chapter also covered how to assign Azure roles to enable secure login to Azure VMs using Microsoft Entra ID credentials. This integration simplifies user authentication while maintaining strong access controls.

Finally, we explored configuring AKV RBAC, the recommended approach for safeguarding your cryptographic keys and secrets to ensure that only authorized users can access these vital resources.

Having gone through the concepts in this chapter, you are now well-equipped to implement robust access management for your Azure resources using Azure RBAC. You have the knowledge to create custom roles, effectively assign both built-in and custom roles, evaluate user permissions, and configure secure access for Entra ID login to VMs and AKV. This comprehensive understanding will empower you to manage access with precision and maintain a secure Azure environment.

While Azure roles provide a robust framework for managing access to Azure resources, they primarily focus on human users. A more granular approach is necessary to secure and manage the growing number of applications and services within Azure environments. In the following chapter, we will dive into the critical aspects of planning and implementing identities for applications and Azure workloads. We will explore how to assign appropriate identities, secure credentials, and integrate these identities seamlessly with Azure resources, further strengthening your organization's overall security posture.

Exam Readiness Drill – Chapter Review Questions

Apart from mastering key concepts, strong test-taking skills under time pressure are essential for acing your certification exam. That's why developing these abilities early in your learning journey is critical.

Exam readiness drills, using the free online practice resources provided with this book, help you progressively improve your time management and test-taking skills while reinforcing the key concepts you've learned.

HOW TO GET STARTED

- Open the link or scan the QR code at the bottom of this page

- If you have unlocked the practice resources already, log in to your registered account. If you haven't, follow the instructions in *Chapter 19* and come back to this page.

- Once you log in, click the START button to start a quiz

- We recommend attempting a quiz multiple times till you're able to answer most of the questions correctly and well within the time limit.

- You can use the following practice template to help you plan your attempts:

Working On Accuracy		
Attempt	Target	Time Limit
Attempt 1	40% or more	Till the timer runs out
Attempt 2	60% or more	Till the timer runs out
Attempt 3	75% or more	Till the timer runs out
Working On Timing		
Attempt 4	75% or more	1 minute before time limit
Attempt 5	75% or more	2 minutes before time limit
Attempt 6	75% or more	3 minutes before time limit

The above drill is just an example. Design your drills based on your own goals and make the most out of the online quizzes accompanying this book.

First time accessing the online resources? 🔒

You'll need to unlock them through a one-time process. **Head to** *Chapter 19* **for instructions**.

Open Quiz	
https://packt.link/sc300ch8	
OR scan this QR code →	

Implementing Global Secure Access

In today's increasingly distributed and remote work environment, securing access to corporate resources while maintaining productivity can be a challenge. **Global Secure Access (GSA)** is a powerful solution to address this complexity. GSA is Microsoft's **Secure Service Edge (SSE)** solution, securing cloud services and applications regardless of what device is being used or where the end user is located. By providing secure and controlled access to applications and resources, regardless of user location, GSA empowers organizations to operate efficiently while mitigating risks.

This chapter dives into the implementation of GSA, exploring its core components and functionalities. We will examine how GSA can be leveraged to establish secure connections between users and corporate resources, both on-premises and in the cloud. By understanding the fundamental principles of GSA, you will gain the knowledge necessary to design and implement effective access strategies that align with the organization's specific needs. The objectives and skills we'll cover in this chapter include the following:

- Deploying GSA clients
- Deploying Private Access
- Deploying Internet Access
- Deploying Internet Access for Microsoft 365
- Enhancing GSA with Conditional Access

By the end of this chapter, you will be able to deploy GSA clients, configure Private Access for secure connectivity to on-premises resources, and implement Internet Access to protect users from online threats. Additionally, we will explore how to optimize GSA for Microsoft 365 environments, ensuring seamless and secure access to cloud-based productivity tools.

What Is GSA?

Before we get into configuring GSA, let's take a look at what it is and how it works. As mentioned in the introduction, GSA is Microsoft's SSE platform. It wraps several familiar concepts (such as application proxies and split-tunnel virtual private networking) into a new product designed to deliver zero-trust networking for remote users.

Figure 9.1 depicts the interaction of components in the GSA solution:

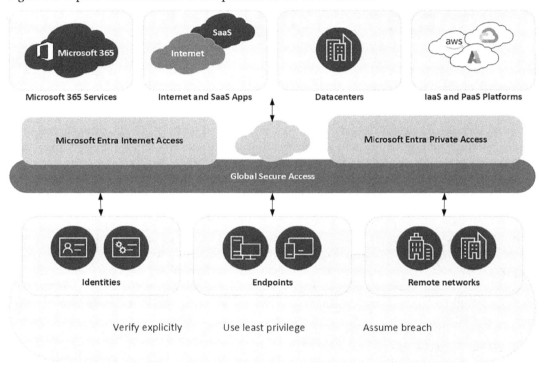

Figure 9.1: Microsoft Global Secure Access

A GSA deployment utilizes several components:

- **Client**: Software installed on endpoints such as laptops and mobile phones to connect to the Microsoft Security service

- **Traffic forwarding profiles**: Policy configuration objects used to send network traffic to the GSA service

- **Connectors**: Agents deployed to data centers and other on-premises networks that allow remote proxy connections into corporate applications

When a device is onboarded with the GSA client, endpoint network traffic is evaluated to see whether it meets one of the configured traffic profiles. Traffic matching a profile is routed to the Microsoft GSA solution, while traffic that doesn't match can be routed out directly, similar to a split-tunnel VPN.

Since GSA is part of the Microsoft 365 ecosystem, you can layer it with other components to create a comprehensive defense-in-depth strategy. GSA-managed connectivity can be integrated with additional Microsoft 365 and Azure security constructs such as the following:

- **Conditional Access policies**: Policies used to manage security conditions (such as group memberships, device health, or network location) to provide access to applications and resources

- **Defender for Cloud Apps**: A **cloud application security broker** (**CASB**) that can be used to control the flow of information through connected apps and services

- **Network security groups**: Configuration objects used to filter traffic to and from Azure resources such as applications and virtual machines

Microsoft is still finalizing the licensing for GSA. While they're currently in preview, Microsoft has announced a few licensing levels for GSA:

- Microsoft Entra Private Access capabilities are included in the Microsoft Entra Suite license and as a standalone subscription. Microsoft Entra Private Access elevates network security as a **zero trust network access** (**ZTNA**) solution.

- Microsoft Entra Internet Access for Microsoft services capabilities are included in a Microsoft Entra ID P1 or Microsoft Entra ID P2 license. Microsoft Entra Internet Access for Microsoft services enhances Microsoft Entra ID capabilities with direct connectivity to supported Microsoft services, improving security, performance, and resilience.

Now that you have a basic understanding of the features, capability, and security of GSA, let's start configuring clients!

Deploying GSA Clients

Understanding the deployment of GSA clients is critical to securing remote access. As an **identity and access management** (**IAM**) administrator preparing for the SC-300 exam, you'll need to understand how to configure both the service and the clients.

Enabling GSA

Before you can begin configuring clients, you will need to ensure that GSA is activated in the tenant. To activate GSA, follow these steps:

1. Navigate to the Entra admin center (`https://entra.microsoft.com`).

2. Expand **Global Secure Access** and select **Dashboard**, as shown in *Figure 9.2*.

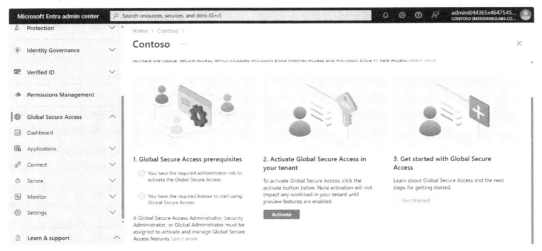

Figure 9.2: Viewing the GSA dashboard

3. Click **Activate**.

4. Wait while the feature is onboarded.

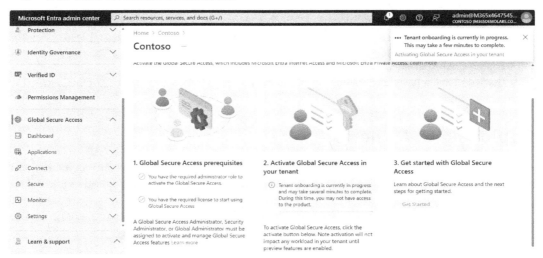

Figure 9.3: GSA feature being enabled

Next, we'll look at onboarding clients.

Client Types and Deployment Methods

GSA clients are typically available for Windows and Android devices, each with its own installation nuances. When deploying GSA clients, you have several options:

- **Manual installation**: This method involves manually installing the client on each device. While time-consuming, it can be helpful for small-scale deployments or specific use cases.

- **Group Policy Objects** (**GPOs**): For Windows environments, GPOs offer a centralized way to deploy and configure clients across multiple machines. This approach is ideal for large-scale deployments and consistently managing configuration settings.

- **Microsoft Intune**: A modern management tool, Intune provides flexibility in deploying GSA clients to Windows and Android devices. You can leverage Intune's capabilities for staged rollouts, compliance checks, and remote troubleshooting.

Depending on the organization's size and what management tools are currently in place, you may use one (or even all three) of these methods for deploying the GSA client. Due to the cloud-first focus of the Microsoft 365 platform, however, Microsoft encourages customers to onboard with Intune.

Onboarding Manually

To onboard a Windows client manually, you can follow these steps:

1. On the device to onboard, open a web browser and navigate to `https://aka.ms/GlobalSecureAccess-windows`. The download should begin automatically.

2. Launch the downloaded `GlobalSecureAccessClient.exe` application.

3. Agree to the license terms and click **Install**, as shown in *Figure 9.4*.

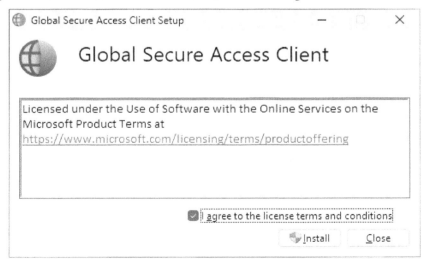

Figure 9.4: Installing the GSA client software package

4. Once the installer has completed, click **Close**.

Next, we'll look at onboarding the client through Intune.

Onboarding through Intune

For many organizations, the preferred method will be to automate onboarding through Intune. To onboard a client through Intune, you'll need to set up a package for deployment using this process:

1. Open a web browser and navigate to `https://aka.ms/GlobalSecureAccess-windows`. The download should begin automatically.

2. Download **Microsoft Win32 Content Prep Tool** from the Microsoft GitHub site (`https://github.com/Microsoft/Microsoft-Win32-Content-Prep-Tool`). This will allow you to package Windows classic apps for an Intune-based deployment.

3. If desired, move the Content Prep tool and the GSA client setup executable to an easy-to-reference folder structure. The executable should be in a folder by itself with no other files, applications, or subfolders.

4. Launch the Content Prep tool with the following parameters:

```
IntuneWinAppUtil.exe -c C:\SourcePath -S
GlobalSecureAccessClient.exe -o C:\DestinationPath
```

This should produce the following output:

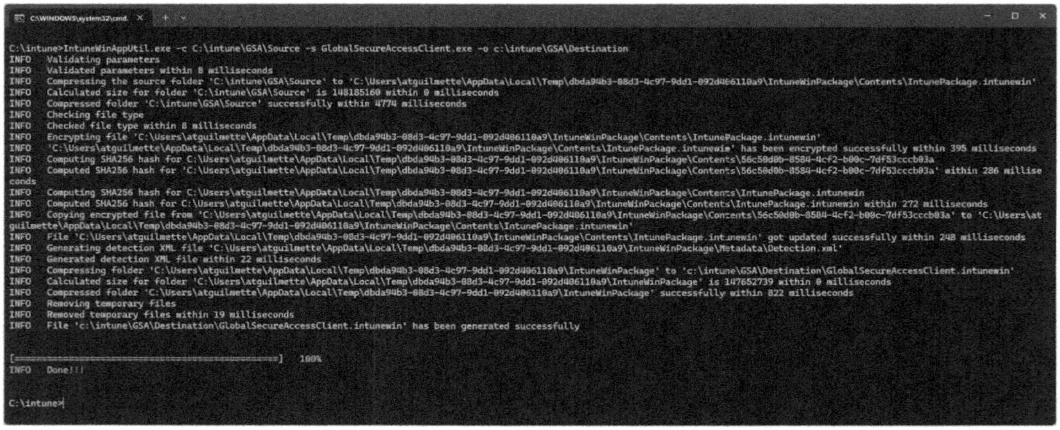

Figure 9.5: Running the Intune Content Prep tool

5. Navigate to the Intune admin center (`https://intune.microsoft.com`).

6. Select **Apps**, expand **By platform**, and select **Windows**, as shown in *Figure 9.6*.

Figure 9.6: Selecting Windows apps

7. Click **Add**.

8. Under **App type**, select **Windows app (Win32)**.

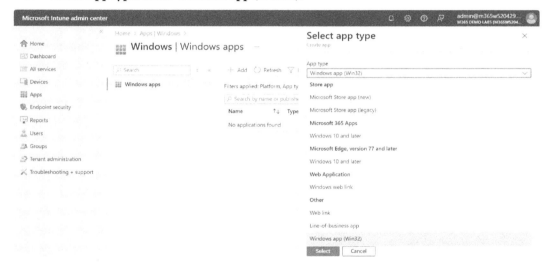

Figure 9.7: Selecting the app type

9. Click **Select**.

10. On the **App information** page, click **Select app package file** to open the **App package file** flyout.

11. Click the folder icon, browse to the `GlobalSecureAccessClient.intunewin` package file that you created using the Microsoft Win32 Content Prep tool earlier, and click **OK**.

Figure 9.8: Selecting the GlobalSecureAccess.intunewin package

12. On the **App information** page, configure as many of the fields as desired. The only required fields are **Name**, **Description**, and **Publisher** (though **Name** and **Description** are populated automatically after importing the package). When it is finished, click **Next**.

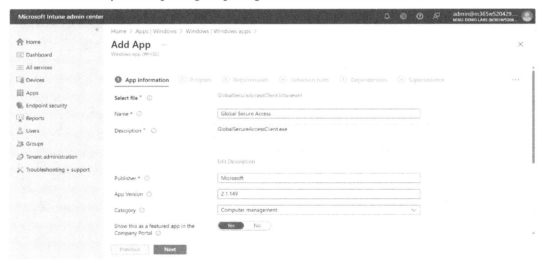

Figure 9.9: Configuring the app details

13. On the **Program** tab, populate the **Install** command using the following value:

```
GlobalSecureAccessClient.exe /install /quiet /norestart
```

14. On the **Program** tab, populate the **Uninstall** command using the following value:

```
GlobalSecureAccessClient.exe /uninstall /quiet /norestart
```

15. Validate the settings and then click **Next**.

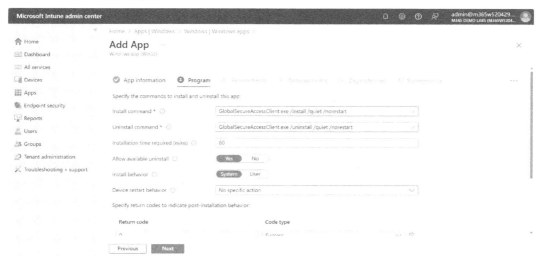

Figure 9.10: Configuring the program details

16. On the **Requirements** tab, choose **64-bit** for **Operating system architecture**.

17. On the **Requirements** tab, choose **Windows 10 1607** for **Minimum operating system**. Click **Next**.

18. On the **Detection rules** tab, select **Manually configure detection rules** for **Rules format**. Click **Add**.

19. On the **Detection rule** flyout, select **MSI** for **Rule type**, enter {DE26D884-6D5D-433E-9F13-B144095F4837} for **MSI product code**, and click **OK**. See *Figure 9.11*.

Figure 9.11: Configuring the app detection rule

> **Note**
>
> The MSI product code may vary based on product versions. This product code is based on version 2.1.149 of the GSA client software package. You may need to install the client on a test machine and extract the product code separately. You can detect the product code for the installed product using the following command:
>
> ```
> Get-WmiObject Win32_Product | ? {$_.Name -eq "Global Secure Access
> Client" } | Format-Table IdentifyingNumber, Name, LocalPackage
> ```

20. Click **Next**.

21. On the **Dependencies** tab, click **Next**.

22. On the **Supersedence** page, click **Next**.

23. On the **Assignments** tab, select devices to deploy the client to and click **Next**.

24. On the **Review + create** page, click **Create**.

After an in-scope devices check, the software package will be deployed.

Let's shift gears and look at some configuration capabilities.

Client Configuration and Policy Management

Once deployed, GSA clients require proper configuration to ensure optimal security and user experience. Client policies allow you to define settings such as the following:

- **Connection preferences**: Specify which network types (Wi-Fi, cellular, and so on) can be used for secure connections

- **Authentication methods**: Determine the authentication factors required, such as **multi-factor authentication (MFA)**

- **Network access controls**: Enforce specific network access rules, such as restricting access to certain IP ranges

For instance, you might create a client policy for sales representatives that only allows access to corporate resources over secure Wi-Fi networks and requires MFA for authentication.

The SC-300 exam focuses on three main configuration scenarios:

- **Private Access**: Connecting to on-premises resources through the GSA client, much like a traditional VPN

- **Internet Access**: Connecting to SaaS applications and routing internet traffic through a secure web gateway

- **Internet Access for Microsoft 365**: Protecting access to Microsoft 365 applications and data through the secure web gateway

In the next section, we'll start by walking through how to deploy Private Access.

Deploying Private Access

Private Access enables remote users to access on-premises resources securely via a connector service. This connector establishes a secure link to the Microsoft cloud, acting as a gateway for remote users. These connectors, based on the Microsoft Azure App Proxy design, can be deployed in a load-balanced configuration to support both high availability and performance requirements.

Private Access, from Microsoft's perspective, is designed to replace VPN connectivity. With Private Access, you can leverage both **fully qualified domain names** (**FQDNs**) and IP ranges when defining your internal resources.

The deployment process involves selecting suitable on-premises servers, installing the connector service, and configuring it to communicate with the cloud service. This includes defining network settings, authentication methods, and other relevant parameters. Careful consideration should be given to the connector's placement within the network infrastructure to optimize performance and security.

Deploying Private Access Infrastructure

As mentioned previously, Private Access requires a connector to provide access to on-premises infrastructure. We'll walk through the steps to onboard the connector:

1. Navigate to the Entra admin center (`https://entra.microsoft.com`).

2. Expand **Global Secure Access** and then select **Connectors**.

3. If you see a message displayed about Private Network being disabled for your tenant, click **Enable Private Network connectors**.

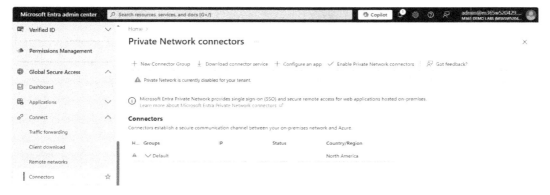

Figure 9.12: Enabling Private Network capability

4. When prompted to enable access to on-premises applications, click **Yes**.

5. Click **Download connector service**.

6. On the **Private Network Connector** flyout, click **Accept terms & Download**.

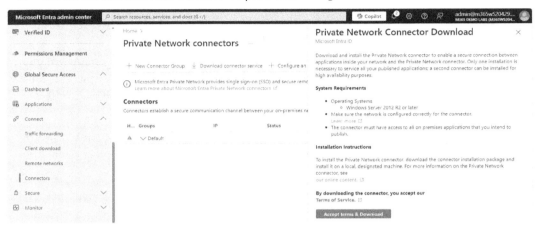

Figure 9.13: Downloading a connector

7. Launch the downloaded file, `MicrosoftEntraPrivateNetworkConnectorInstall.exe`.

8. Agree to the terms and click **Install**.

Figure 9.14: Starting the connector installer

9. When prompted, sign in to the Microsoft 365 tenant to start the connector registration process.

10. Once the connector has been installed, click **Close**.

Figure 9.15: Completing the connector installation

Next, we'll configure Quick Access.

Configuring Quick Access

Quick Access provides the foundational components for accessing applications through Private Access. In this section, we'll configure a Quick Access segment using a network identifier (which can be an IP address, IP address range, or a domain name). This segment will be used to identify which traffic should be routed to the Private Access connector.

To configure Quick Access, follow these steps:

1. Navigate to the Microsoft Entra admin center (`https://entra.microsoft.com`).

2. Expand **Global Secure Access**, expand **Applications**, and select **Quick Access**.

3. Enter a name. Microsoft recommends using **Quick Access**. Select a connector group.

Figure 9.16: Configuring Quick Access

4. Under **Application Segment**, click **Add Quick Access application segment**.

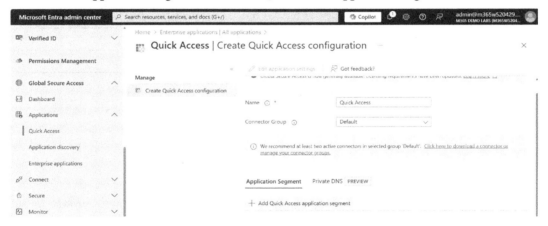

Figure 9.17: Adding an application segment

5. On the **Create application segment** flyout, select a destination type and then fill out the fields accordingly. When finished, click **Apply**.

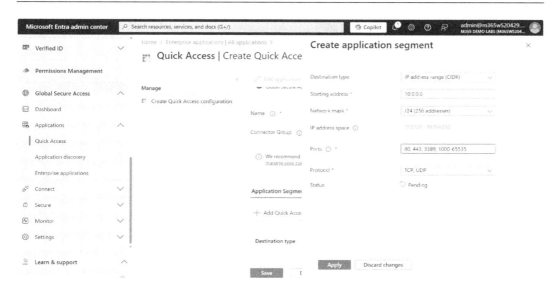

Figure 9.18: Completing the application segment configuration

6. Optionally, select the **Private DNS** tab to add your internal DNS domain as a private DNS suffix. This will allow short names (sites without FQDNs) to be used with Private Access.

Figure 9.19: Configuring Private DNS

7. Select **Enable Private DNS**. Click **Save**.

8. The Quick Access setup will create a new enterprise app on your behalf. You'll need to assign users or groups to it. Refresh the **Quick Access** node to display the configured enterprise app. See *Figure 9.20*.

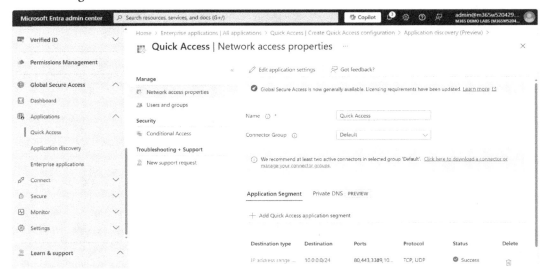

Figure 9.20: Viewing the Quick Access app

9. Under **Manage**, select **Users and groups**.

10. Click **Add user/group** and add the users or groups that will be enabled for **Private Access**.

Figure 9.21: Adding users and groups to Private Access

Users added to the Quick Access enterprise app will be able to access on-premises endpoints through the Private Access profile once it has been enabled.

Next, we'll look at configuring a Private Access application.

Configuring a Private Access Application

Configuring a Private Access application is relatively straightforward, especially if you're already familiar with how Azure App Proxy works. In this section, we'll just configure a sample website as an application in the on-premises environment. The process is similar for other applications:

1. Navigate to the Microsoft Entra admin center (`https://entra.microsoft.com`).

2. Expand **Global Secure Access** and select **Connectors**.

3. Click **Configure an app**.

4. On the **Basic** tab of the **Add your own on-premises application** page, enter a name and an internal URL. The URL you are adding must be protected by an SSL certificate. The **External Url** field will be automatically calculated from the value you add to **Internal Url**, but you can change it. See *Figure 9.22*.

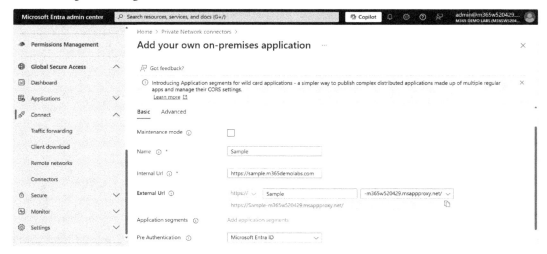

Figure 9.22: Configuring a Private Access app

5. Click **Create**.

After the app has been created and assigned to users, users can see it in the My Apps portal (`https://myapps.microsoft.com`) or on the Microsoft 365 apps page (`https://www.microsoft365.com/apps`) under **Other apps**.

Next, we'll enable the Private Access traffic forwarding profile.

Enabling the Private Access Traffic Forwarding Profile

With all of the configuration for Private Access complete, all you need to do now is enable a traffic forwarding profile. The traffic forwarding profile is what will tell the GSA client to send traffic to on-premises applications:

1. Navigate to the Microsoft Entra admin center (`https://entra.microsoft.com`).

2. Expand **Global Secure Access** and then expand **Connect**.

3. Select **Traffic forwarding**.

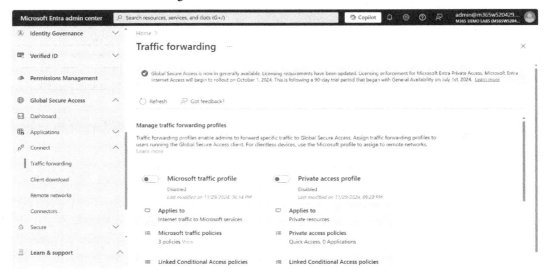

Figure 9.23: Enabling the Private Access profile

4. Slide the **Private access profile** toggle. When prompted, click **OK**.

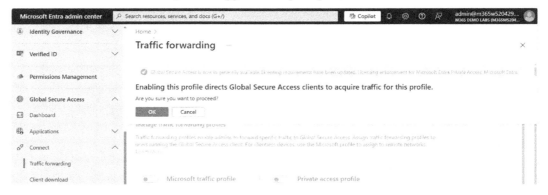

Figure 9.24: Confirming the Private Access profile enablement

5. Scroll to the **User and group assignments** section of the **Private access profile** card. Click **View**.

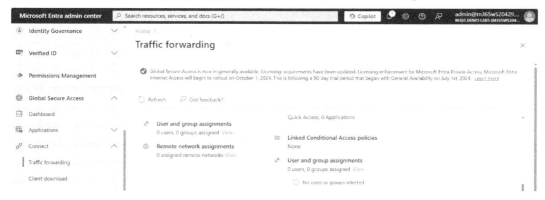

Figure 9.25: Managing user and group assignments

6. On the **User and group assignments** flyout, you can choose **Assign to all users** or assign to select users. When finished, click **Done**.

Figure 9.26: Configuring user and group assignments

Once the policy has been updated and the GSA clients have checked in, the traffic to on-premises apps will be tunneled through the connector to the respective apps.

Establishing secure and reliable private access is crucial for internal operations. Bridging the gap between your internal network and the global digital landscape can optimize business processes and enhance overall productivity.

Deploying Internet Access

Deploying Internet Access within Entra ID primarily revolves around managing and securing access to cloud applications and resources. Unlike Private Access, which focuses on internal network connectivity, Internet Access extends to external users and applications. The core objective is establishing a secure and controlled environment where authorized users can access necessary resources without compromising organizational security.

The Internet Access traffic profile contains three access policies:

- **Custom Bypass**: This customizable policy contains destinations (domains, IP addresses, or IP address ranges) that will be excluded from the GSA Internet Access policy.

- **Default Bypass**: This Microsoft-managed policy contains domains and IP networks matching the Microsoft 365 service endpoints. Traffic destined for these endpoints is excluded from being tunneled to the GSA service.

- **Default Acquire**: This Microsoft-managed policy contains all networks not excluded by one of the bypass rules.

To configure the Internet Access profile, follow these steps:

1. Navigate to the Microsoft Entra admin center (`https://entra.microsoft.com`).

2. Expand **Global Secure Access** and then expand **Connect**.

3. Select **Traffic forwarding**.

4. On the **Internet access profile** card, under **Internet access policies**, select **View**. See *Figure 9.27*.

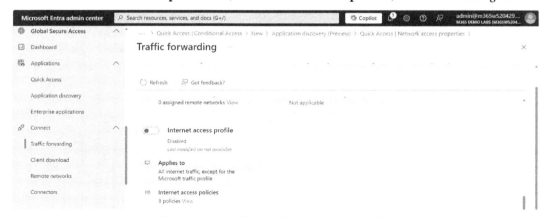

Figure 9.27: Configuring Internet access profile

5. Expand **Custom Bypass** and select **Add rule**.

Figure 9.28: Adding a custom bypass rule

6. Select a destination type (such as **IP address** or **FQDN**), enter a destination, and click **Save**.

7. Click the **X** or **Cancel** button to return to the **Traffic forwarding** page.

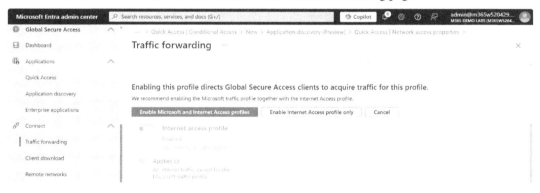

Figure 9.29: Enabling the Internet Access profile

8. Scroll to the **User and group assignments** section of the **Internet access profile** card. Click **View**.

9. On the **User and group assignments** flyout, you can choose **Assign to all users** or assign to select users. When finished, click **Done**.

10. Slide the toggle to enable **Internet access profile**. When prompted, click **Enable Internet Access profile only**.

Once the policy has been updated and the GSA clients have checked in, the traffic destined for the internet will be evaluated against the Internet Access traffic forwarding profile. Traffic on the **Default Bypass** or **Custom Bypass** lists will be routed directly to its destination, while any other traffic will be routed to the GSA service.

Deploying Internet Access for Microsoft 365

When deploying GSA with an Internet Access profile, Microsoft recommends simultaneously deploying the Microsoft traffic profile. This ensures that the traffic for Microsoft 365 services is not acquired by the GSA client.

To enable the Microsoft traffic profile, follow these steps:

1. Navigate to the Microsoft Entra admin center (`https://entra.microsoft.com`).
2. Expand **Global Secure Access** and then expand **Connect**.
3. Select **Traffic forwarding**.
4. On the **Microsoft traffic profile** card, under **User and group assignments**, select **View**.
5. On the **User and group assignments** flyout, you can choose **Assign to all users** or assign to select users. When finished, click **Done**.
6. Slide the toggle to enable **Microsoft traffic profile**. When prompted, click **OK**.

Once the policy has been updated and the GSA clients have checked in, the traffic destined for the internet will be evaluated against the Internet Access and Microsoft traffic forwarding profiles. Traffic on the **Default Bypass** or **Custom Bypass** lists (part of the Internet Access profile) or the Microsoft traffic profile will be routed directly to its destination, while any other traffic will be routed to the GSA service.

Enhancing Global Secure Access with Conditional Access

While traffic to and from most SaaS applications (including Microsoft 365) is protected via SSL/TLS encryption, organizations may want to require endpoints to use the GSA infrastructure to enforce device compliance or other security features.

Conditional Access policies can easily be layered on top of both the Microsoft traffic and Internet Access profiles.

To create a new Conditional Access policy that leverages GSA, follow these steps:

1. Navigate to the Microsoft Entra admin center (`https://entra.microsoft.com`).
2. Expand **Global Secure Access | Settings** and then choose **Session management**.
3. Select the **Adaptive Access** tab and enable the **Enable CA Signaling for Entra ID (covering all cloud apps)** toggle.

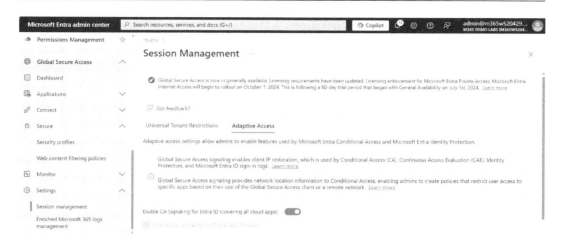

Figure 9.30: Enabling adaptive access signaling

4. Expand **Protection** and then select **Conditional Access**.

5. Select **Policies**.

6. Click **New policy**.

7. Enter a name.

8. Under **Users**, add the users or groups to include or exclude as part of the policy.

9. Under **Target resources**, select **Resources** under the **Select what this policy applies to** dropdown. On the **Include** tab, select **All internet resources with Global Secure Access**.

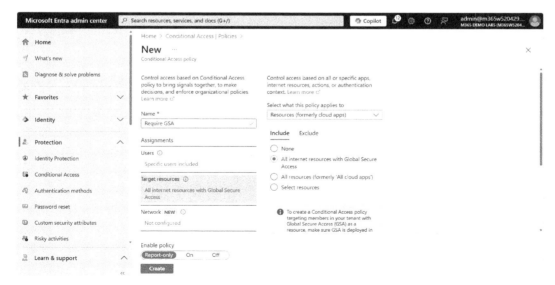

Figure 9.31: Enabling GSA for target resources

10. Under **Grant controls**, select a **Grant** access control (such as **Require multifactor authentication** or **Require Microsoft Entra hybrid joined device**) and then click **Select**.

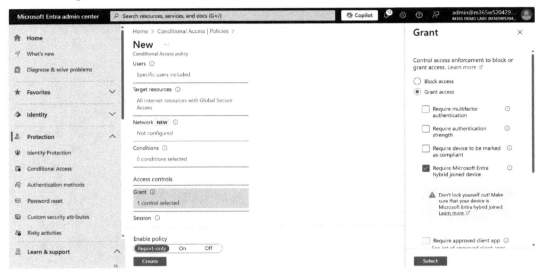

Figure 9.32: Configuring grant controls

11. Click **Create**.

When accessing applications, the in-scope users will be required to be connected to the GSA service.

Summary

This chapter has explored the pivotal role of GSA in safeguarding corporate resources while simultaneously empowering users to work productively in today's hybrid work environment. By exploring the core components of GSA, you've established a strong foundation for securing both on-premises and cloud-based applications.

Having established a robust foundation for secure access with GSA, we can now focus on managing and protecting identities within the application and Azure workload landscape. By effectively planning and implementing IAM strategies, organizations can strengthen their security posture and enable efficient operations. We will explore this topic further in the next chapter, which focuses on planning and implementing identities for applications and Azure workloads.

Exam Readiness Drill – Chapter Review Questions

Apart from mastering key concepts, strong test-taking skills under time pressure are essential for acing your certification exam. That's why developing these abilities early in your learning journey is critical.

Exam readiness drills, using the free online practice resources provided with this book, help you progressively improve your time management and test-taking skills while reinforcing the key concepts you've learned.

HOW TO GET STARTED

- Open the link or scan the QR code at the bottom of this page

- If you have unlocked the practice resources already, log in to your registered account. If you haven't, follow the instructions in *Chapter 19* and come back to this page.

- Once you log in, click the START button to start a quiz

- We recommend attempting a quiz multiple times till you're able to answer most of the questions correctly and well within the time limit.

- You can use the following practice template to help you plan your attempts:

Working On Accuracy		
Attempt	Target	Time Limit
Attempt 1	40% or more	Till the timer runs out
Attempt 2	60% or more	Till the timer runs out
Attempt 3	75% or more	Till the timer runs out
Working On Timing		
Attempt 4	75% or more	1 minute before time limit
Attempt 5	75% or more	2 minutes before time limit
Attempt 6	75% or more	3 minutes before time limit

The above drill is just an example. Design your drills based on your own goals and make the most out of the online quizzes accompanying this book.

First time accessing the online resources? 🔒

You'll need to unlock them through a one-time process. **Head to *Chapter 19* for instructions**.

Open Quiz	
https://packt.link/sc300ch9	
OR scan this QR code →	

10

Planning and Implementing Identities for Applications and Azure Workloads

The proper selection and management of Azure identities are foundational to securing applications and workloads effectively. Azure identities refer to the digital personas or credentials used to authenticate and authorize access to Microsoft Azure resources, such as applications, services, and workloads. Think of them like digital passports that verify who you are and what permissions you have.

If not managed correctly, this key can unlock sensitive information, putting your entire organization at risk. By managing identities effectively, you can control who can perform specific actions on Azure resources, such as reading or writing data, which reduces the risk of unauthorized changes or breaches.

Additionally, proper identity management is critical for organizations that need to comply with regulations such as HIPAA, PCI-DSS, or GDPR, as it ensures that access to sensitive data is restricted to authorized personnel only, demonstrating compliance and reducing the risk of fines or reputational damage. As cloud applications continue to increase in complexity and their reliance on other Azure resources, understanding the nuances of identity becomes critical.

In this chapter, you will be introduced to the different types of identities that can be used for cloud applications and workloads. You will learn how to choose the right identity for your application and how to create, delete, and manage it. Finally, you will work through several examples of selecting and managing these types of identities.

The objectives and skills covered in this chapter include the following:

- Selecting appropriate identities for applications and Azure workloads
- Creating managed identities
- Assigning a managed identity to an Azure resource
- Using a managed identity assigned to an Azure resource

Let's dig in!

Selecting Appropriate Identities for Applications and Azure Workloads

When managing access to resources in Azure, understanding the concept of identities is crucial. Identities, from the Azure perspective, are used to authenticate users, services, and workloads, allowing them to access the resources they need while ensuring that unauthorized access is prevented. So, even though identities are typically associated with user accounts, the term can also refer to those used by applications to access resources and services in a secure manner.

When configuring identity types, you must always consider the principle of least privilege when creating, configuring, or managing any identities.

In the upcoming sections, we'll explore the three primary identities for applications: system-assigned managed identities, user-assigned managed identities, and service principals.

Managed Identities

Managed identities (formerly **MSIs** or **managed service identities**) are identities that are automatically managed and can be used by applications to authenticate to Azure resources that support Microsoft Entra authentication. There are two types of managed identities: **system-assigned** and **user-assigned**.

System-assigned managed identities are automatically generated when you create certain Azure resources, such as in Azure Virtual Machines or Azure App Service, and are tied to those specific resources. In contrast, user-assigned managed identities are standalone Azure identities that can be created independently and associated with one or multiple Azure resources. Each type of managed identity has its own lifecycle and sharing capabilities, which determine its best use case.

> **Support for Managed Identities Is Expanding**
>
> Not all Azure resources currently support system-assigned managed identities. Microsoft maintains a list that is constantly updated and can be accessed here: `https://learn.microsoft.com/en-us/entra/identity/managed-identities-azure-resources/managed-identities-status`.
>
> You should also be aware of some known operational changes that might affect managed identity usage. Microsoft maintains a list that is updated and available here: `https://learn.microsoft.com/en-us/entra/identity/managed-identities-azure-resources/known-issues`.

As you work through the decision to use a managed identity, keep in mind that a user-assigned managed identity can be used when a system-assigned managed identity is not supported. *Table 10.1* describes some common features and use cases to help you decide which type of identity is right for your scenario:

Feature	System-assigned	User-assigned
Creation	Created when the resource is created (such as an Azure VM or App Service).	Created as a standalone object.
Lifecycle	Follows the lifecycle of the correlated Azure resource. When the parent resource is deleted, the corresponding system-managed identity is deleted.	Independent of the lifecycle of any other resources and must be deleted or updated manually.
Sharing across resources	Cannot be shared and is only used with a single Azure resource.	Can be associated with more than one Azure resource.
Common use cases	Workloads contained in a single Azure resource Workloads requiring individual identities Audit logging requiring the specific resource that performed an activity (as opposed to which identity)	Workloads that run on multiple resources Workloads that require pre-authorization or shared access to a resource, or instances where a resource requires access to an identity before provisioning Rapid provisioning of resources Compliance with identity approval processes

Table 10.1: System- and user-assigned managed identities

As an example, one typical use case of a managed identity is when an Azure VM needs to authenticate to Azure Key Vault to access secrets to further enable the application to access Azure resources. Using managed identities frees the application developer from having to hardcode credentials in code, embed secrets in connection strings, or remember to rotate credentials. In this particular use case, either type of managed identity will satisfy the use case, and either type is preferable over a user account assigned for this purpose (sometimes called a **service account**).

You will work through these use cases later when you are creating managed identities.

Service Principals

An Azure service principal is an identity used by applications, services, and automation tools to access specific Azure resources. It is an identity that represents an application or service within Microsoft Entra. It is essentially the local manifestation of a global application object in a specific tenant, allowing the application to be instantiated across multiple tenants while maintaining unique permissions and properties for each instance. When an application is registered with Entra, a corresponding service principal object is automatically created within the tenant where it was registered (the home tenant). This object defines what actions the application can perform in that specific tenant, who can access it, and which resources it can access. *Figure 10.1* shows the relationship between service principals and an app in a multitenant scenario:

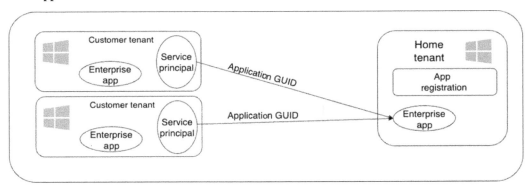

Figure 10.1: Service principal use with single and multitenant applications

The service principal also stores credentials such as client secrets or certificates, allowing it to authenticate securely when accessing other Azure services on behalf of the application. Service principals are often used when configuring automation scenarios or integrating with other applications that need to interact with Azure resources programmatically. By creating a dedicated service principal for these purposes, you can limit the permissions granted to just what is necessary for the specific use case, reducing the risk of accidental data exposure or misuse.

> **Service Principal versus Managed Identity**
>
> Service principals are specific to enterprise applications that are intended to be used within single or multiple tenants. It is important to consider whether you *need* to use a service principal. In general, using managed identities for Azure resources as your application identity is recommended over creating a service principal due to their inherent security benefits and ease of use. Managed identities eliminate the need to manage credentials and provide seamless authentication when accessing supported Azure services, making them an ideal choice for modern cloud applications.

Generally, the scope of an application's deployment (**single tenant**, meaning constrained to a single tenant's boundary, or **multitenant**, meaning it is an application designed to be shared and accessed across tenants) is one of the factors to use when determining what type of identity to use. Next, we'll look at the different types of applications that might require the use of an enterprise identity.

Single-Tenant Applications

These applications are designed for use within a specific tenant only. All user and guest accounts within your organization can potentially sign in to these apps. Single-tenant applications are ideal when you want to limit access to users within your organization.

Authentication is handled through Entra ID. Authorization can be role based, attribute based, or claims based. Typically, an organization might develop single-tenant line of business applications or other internal tools. Applications in this category can generally use managed identities if the resources they are accessing support managed identities.

Multitenant Applications

Unlike single-tenant applications, multitenant applications allow users from any Azure tenant to sign in. This includes employees, partners, or customers who have a work or school account with Microsoft. Multitenant applications are designed to serve multiple organizations or customers and allow resources to be shared across tenants.

Like single-tenant applications, authentication to multitenant applications is handled by Entra ID, with users belonging to different tenants. Authorization can be role-based, attribute-based, or claims-based.

Popular SaaS software packages (think Salesforce or GitHub) or other third-party applications developed and listed in the Azure marketplace are typically enterprise applications. Large organizations with multiple tenants may also develop multitenant applications with the intent to provide access to other business units residing in other tenants.

In either case, these multitenant applications run in the developer's home tenant but are designed to allow users from potentially any tenant to access them. While single-tenant applications can potentially use managed identity, multitenant applications will always require service principals to perform the necessary cross-tenant authentication on behalf of the user.

We will discuss service principals in detail in *Chapter 11, Planning, Implementing, and Monitoring the Integration of Enterprise Applications*, and *Chapter 12, Planning and Implementing App Registrations*.

Logic Apps, Function Apps, and Azure App Service

Logic Apps are serverless workflows that automate tasks and processes across multiple services and systems. Function apps are serverless compute solutions that run code in response to events or triggers. Azure App Service is a platform for building and deploying web applications.

Depending on the use cases, each product or component may use either a managed identity or a service principal. For example, a logic app may use a managed identity to connect to a mailbox but a service principal to connect to an external database. The use cases and capabilities of products you're working with will dictate which type of identities are used.

Choosing an Identity Type

Traditional user account-based authentication systems are widely used due to their familiarity and ease of implementation. However, they pose significant challenges when applied to large-scale applications or complex systems. The requirement for password rotation and management introduces an additional administrative burden, while the limited security features inherent in user accounts render them vulnerable to hacking attempts.

User accounts do not scale effectively with growing system complexity. As a result, they may prove inadequate for applications requiring secure authentication and authorization mechanisms. For instance, e-commerce platforms with thousands of users necessitate an authentication system that can efficiently manage a large volume of traffic while ensuring the security and integrity of user data.

In contrast, managed identities offer a scalable solution for authentication in Azure-based applications. Managed identities eliminate the need for password or credential management, thereby reducing administrative overhead. In some cases (namely, when you are able to use system-assigned managed identities), automatic identity creation and rotation ensure that authentication is both secure and efficient.

For applications requiring fine-grained control over permissions and access, service principals offer a more nuanced approach to authentication. By enabling developers to define explicit permissions and access levels, service principals provide a higher degree of customization and security than user accounts or managed identities.

However, the complexity associated with service principal implementation may deter some organizations from adopting this approach. The manual creation and management of credentials, while offering greater control, also introduce additional administrative burdens (similar to that of creating and managing user-style service accounts).

Now that we have explored the various types of identities available in Azure, let's shift our focus to creating managed identities for Azure resources. Managed identities are a type of identity that is automatically managed by Azure and allows services such as Azure VMs, Azure App Service, or Azure Functions to authenticate without the need for explicit credentials.

Creating Managed Identities

System-assigned managed identities are managed automatically by Microsoft Entra ID. For resources that support system-assigned managed identities, the managed identity is created by going to the **Identity** blade within the resource. The managed identity can be added when the resource is created or added later. In either case, the identity is deleted when the resource is deleted.

> **Note**
>
> You can find a list of resources that support managed identities at the following URL: `https://learn.microsoft.com/en-us/entra/identity/managed-identities-azure-resources/managed-identities-status`.

Creating a System-Assigned Managed Identity

In this exercise, you will create a simple Azure VM and assign a managed identity to it. Follow these steps:

1. In the Azure portal, select **Create a Resource**, then select **Virtual Machine**.

2. On the **Create a virtual machine** page, populate the fields, choosing **Red Hat Enterprise Linux 8.7 (LVM) - x64 Gen2**.

Figure 10.2: Create a virtual machine

3. Choose an SSH public key and the desired name. If you don't already have an existing SSH public key, you can select **Generate new key pair** in the **SSH public key source** field.

4. Select **SSH (22)** in the **Select inbound ports** field and click **Next**.

5. On the **Disks** page, make any changes to the storage resources necessary to support your use case. You can also just accept defaults and click **Next**.

6. On the **Networking** page, ensure the networking selections (including the virtual network or IP address) are accurate for your environment, or continue with the defaults. Click **Next**.

7. On the **Management** page, ensure that **Enable system assigned managed identity** is selected.

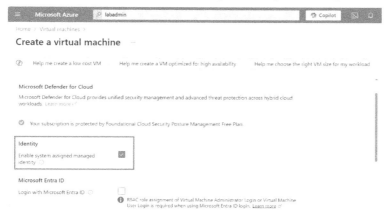

Figure 10.3: Enable system assigned managed identity

8. Scroll to the bottom of the page and click **Review + create**.

9. On the **Review + create** page, confirm your settings selections and then click **Create**. If prompted, download your SSH private key.

To view the system-assigned managed identity that you just created, use the following command in the Azure CLI:

```
az vm identity show --name LabVM --resource-group ManagedIdentityDemo
```

The response should look similar to this:

```
{
  "principalId": "a07ee000-054b-4ff1-b486-a490366c4320",
  "tenantId": "0f172991-846d-4b99-a94d-297ebbbce6bc",
  "type": "SystemAssigned",
  "userAssignedIdentities": null
}
```

You can then use the SSH key you downloaded previously to connect to the VM. When prompted, enter `ssh -I filename.pem azureuser@IP` address.

```
azureuser@abVM:~
C:\demo>ssh -i ssh1.pem azureuser@172.210.20.58
The authenticity of host '172.210.20.58 (172.210.20.58)' can't be established.
ED25519 key fingerprint is SHA256:14XJIpoPm5FoPOC6Hr2Kh7BsZxDcHW9yKYQO9ZZwI0E.
This key is not known by any other names
Are you sure you want to continue connecting (yes/no/[fingerprint])? yes
Warning: Permanently added '172.210.20.58' (ED25519) to the list of known hosts.
Activate the web console with: systemctl enable --now cockpit.socket

Register this system with Red Hat Insights: insights-client --register
Create an account or view all your systems at https://red.ht/insights-dashboard
Last login: Sat Jun 15 18:13:22 2024 from 137.103.217.93
[azureuser@abVM ~]$
```

Figure 10.4: Connecting to a new VM using SSH

Now that you have successfully tested your VM, you can log out. You will use this VM later in the book when we test using managed identities.

Creating a User-Assigned Managed Identity

User-assigned managed identities are Azure resources that are created independently from other Azure resources. They can be used for authentication and authorization to other Azure resources such as Azure Storage accounts. The credentials are managed automatically so there is no need to embed credentials in code or another insecure medium. Unlike system-assigned managed identities, user-assigned identities can be applied to multiple resources. This enables you to share the identity among resources with similar access requirements.

Follow these steps to create a user-assigned managed identity:

1.   In the Azure portal, in the search box, search for `Managed identities`.

Figure 10.5: Searching for managed identities

2. Select **Create a managed identity**.

3. On the **Basics** page, populate the **Subscription**, **Resource group**, **Region**, and **Name** fields.

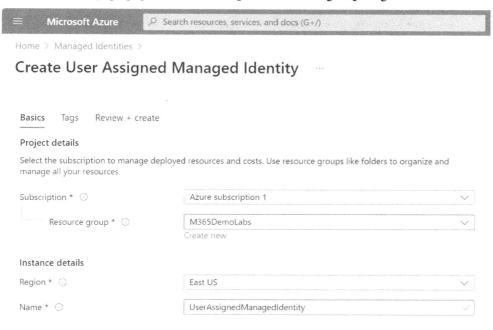

Figure 10.6: Creating a managed identity

4. On the **Tags** page, click **Next**.

5. On the **Review + create** page, click **Create**.

When you complete these steps, you will have a new resource—a user-assigned managed identity—that can be applied to other resources.

In addition to creating user-assigned managed identities in the portal, you can also create them using PowerShell or the Azure CLI and apply them to a resource using PowerShell, the Azure CLI, or a JSON template. You will not need to construct these command lines for the exam, but you might need to recognize them as a possible solution. For reference, the following are the commands that can be used to create the managed identity you created previously using PowerShell and Azure CLI:

- PowerShell:

```
New-AzUserAssignedIdentity -ResourceGroupName "myresourcegroup"
-Name "mymanagedidentity" -Location "East US"


Remove-AzUserAssignedIdentity -ResourceGroupName
ManagedIdentityDemo -Name mymanagedidentity
```

- Azure CLI:

```
az identity create --name myManagedIdentity --resource-group
ManagedIdentityDemo
az identity delete --name myManagedIdentity --resource-group
ManagedIdentityDemo
```

The following are commands to assign a managed identity to an Azure resource using PowerShell, the Azure CLI, and a JSON template:

- PowerShell:

```
Update-AzFunctionApp -Name <app-name> -ResourceGroupName
<group-name> -IdentityType UserAssigned -IdentityId
$userAssignedIdentity.Id
```

- Azure CLI:

```
Az webapp identity assign –resource-group <group-name> --name
<app-name> --identities <identity-name>
```

- JSON template:

```
"identity":{
      "type":        "UserAssigned",
      "userAssignedIdentities": {
           "<RESOURCEID>": {}
      }
}
```

In this section, you created both system-assigned and user-assigned managed identities. The system-assigned managed identity was applied directly to the resource, while we'll use the next section to learn how to apply the user-assigned managed identity to another resource.

Using a Managed Identity Assigned to an Azure Resource

In this section, you will examine how to assign a managed identity to an Azure resource. The first task will be to assign the system-assigned managed identity that you established with a VM and assign it to an Azure key vault. This is a fairly typical scenario for storing your secrets in Azure Key Vault. Once you enable your VM to access the key vault without having to embed sensitive information, you will have access to your secrets in a secure, repeatable manner.

Creating a Key Vault

Before your managed identity can retrieve secrets from the key vault, you'll need to create a vault and assign the identity access to it.

Follow these steps to create a key vault:

1. In the Azure portal, in the search box, enter `Key vaults`. Under **Services**, select the **Key vaults** option.

2. On the **Key vaults** blade, select **Create key vault**.

3. Populate the **Subscription** and **Resource group** fields, leave the remaining default values unchanged, and click **Next**.

4. On the **Access configuration** page, under **Permission model**, ensure that **Azure role-based access control (recommended)** is selected.

5. Under **Resource access**, select all options' defaults to make this key vault public. Click **Next**.

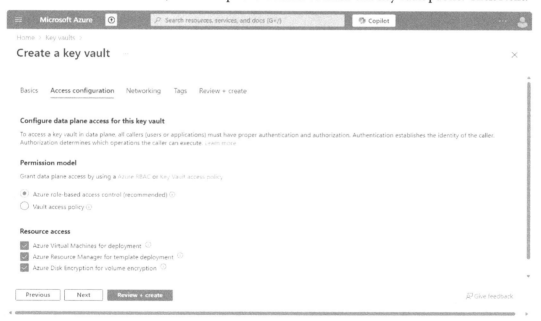

Figure 10.7: Updating the access configuration for the key vault

6. On the **Networking** page, choose what networks can access secrets stored in this key vault. As a best practice for production environments, you should not make your key vault open to the public and should scope it to your organization's trusted networks. For purposes of this demo, however, you can select **All networks** under **Public Access**. If you choose **Selected networks**, you will be given the option to select existing Azure virtual networks or create a new Azure virtual network. See *Figure 10.8*:

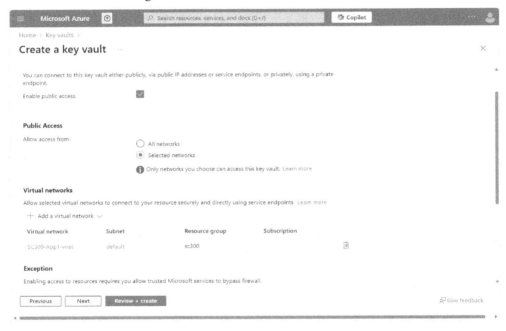

Figure 10.8: Updating the Networking settings

7. Click **Next**.

8. On the **Tags** page, add any tags as desired and select **Review + create**.

9. On the **Review + create** page, click **Create**.

10. Once the deployment of your key vault has succeeded, click **Go to resource**.

Great! Now you've got a key vault that can be used to store and retrieve identities. Next, we'll grant access to the key vault.

Granting Access to a Key Vault

By default, the subscription owner and the creator of the key vault have access to the vault at a high level. If you want identities (either users or managed identities) to be able to retrieve data from the key vault, you'll need to assign them access using the following process:

1. Navigate to the key vault (if you're not already there).

2. Click on **Access control (IAM)**.

3. Click on **Add role assignment**.

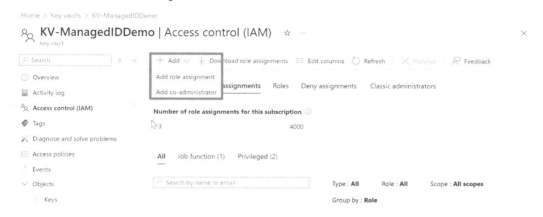

Figure 10.9: Add role assignment to key vault

4. Select **Key Vault Secrets User**, as shown in *Figure 10.10*:

Figure 10.10: Adding the Key Vault Secrets User role assignment

5. On the **Members** page, under **Assign access to**, select **Managed identity**. Click **Select members**.

6. On the **Select managed identities** flyout, choose the subscription, then locate the **Virtual machine** managed identity you created previously.

7. Choose **Review + assign**.

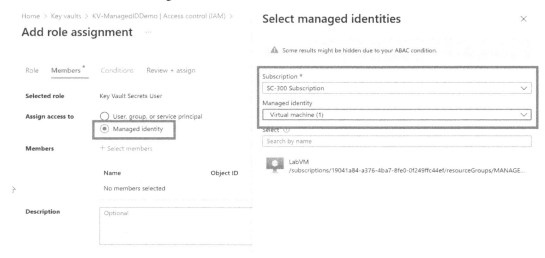

Figure 10.11: Selecting the appropriate managed identities

When considering what permissions should be added to specific resources (including key vaults), remember that the following built-in Azure roles might be available:

- Owner
- Contributor
- Reader
- User Access Administrator
- Other task-specific roles, such as VM Contributor

If the goal is to add access permissions to your resources, only the *Owner* role or *User Access Administrator* role can satisfy that requirement.

Next, we'll look at storing a secret or a credential in the vault.

Storing a Secret in the Key Vault

With all of the access in place, it's time to store a secret!

You can store a secret using the following process:

1. With the key vault open, expand **Objects** and select **Secrets**.

2. Click **Generate/Import** to add a new value pair to the key vault.

3. On the **Create a secret** page, enter the following properties (at a minimum):

    *   **Name**: Secret1

    *   **Secret value**: This is Secret 1

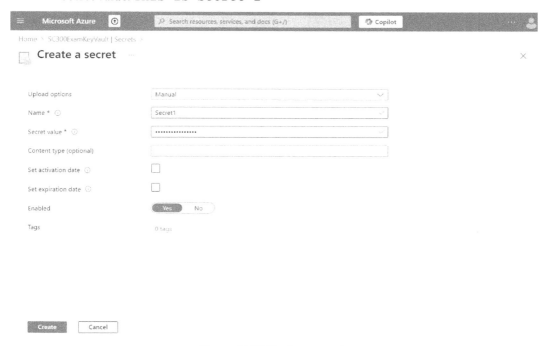

Figure 10.12: Storing a secret

4. Click **Create**.

> **Note**
>
> You may need to add your user account as a key vault administrator before you can store a secret.

Your secret will be added to the vault.

Next, you will validate that we've got everything set up correctly!

Accessing a Stored Secret

Now, you will use the previously provisioned VM to verify access to the secret you just created. Follow these steps but replace the settings with the ones from your own VM:

1. On your own workstation, launch a Command Prompt or Terminal window.

2. In the Command Prompt or Terminal window, enter the following:

   ```
   ssh -i labvm_key.pem azureuser@host
   ```

   Here, key.pem is the key file you downloaded when you created a VM with a managed identity and @host is the IP address or hostname of the VM you created.

3. In the Command Prompt or Terminal window, enter the following:

   ```
   curl 'http://169.254.169.254/metadata/identity/oauth2/token?api-
   version=2018-02-01&resource=https%3A%2F%2Fvault.azure.net' -H
   Metadata:true
   ```

   The output generated after running this command will give you the bearer token that can be used to read a secret using your managed identity. It looks like this:

   ```
   {"access_token":"TRUNCATED.TruncatedAq-pUxM1mmds_
   NcRQQV13gmgWt0ofp3HVAG8Hk6OP9-fEqoP2QypH2jaHZb1hWKJb7PS86
   LiEpneAAf8ItpJ-TruncatedI.  Truncated-Truncated","client_
   id":"01ca6470-f02d-4929-9925-d24dbd197693","expires_
   in":"83857","expires_on":"1718762393","ext_expires_
   in":"86399","not_before":"1718675693","resource":"https://vault.
   azure.net","token_type":"Bearer"}
   ```

4. In the Command Prompt or Terminal window, enter the following:

   ```
   curl 'https://kv-managediddemo.vault.azure.net/secrets/
   Secret?api-version=2016-10-01' -H "Authorization: Bearer
   <Authorization Token>"
   ```

   Here, <Authorization Token> is the value highlighted in the access_token value returned in *step 3*.

In this exercise, you learned about the process of creating and using managed identities as well as configuring managed identities to access a key vault.

Summary

In this chapter, you learned about workload identities. You explored the types of applications that might be present in your Azure tenant, which identity types are available for authentication and authorization to Azure resources, and learned ways to create and manage both system-assigned and user-assigned managed identities.

It's important to remember that, when considering workload identities, system-assigned managed identities are the best option if the resource supports it. This is because they are both more secure and automatically managed, reducing administrative burden. The next best option is a user-assigned managed identity. This must be managed by the administrator but is still more secure than a simple user account.

In the next chapter, you will begin working with Enterprise applications.

Exam Readiness Drill – Chapter Review Questions

Apart from mastering key concepts, strong test-taking skills under time pressure are essential for acing your certification exam. That's why developing these abilities early in your learning journey is critical.

Exam readiness drills, using the free online practice resources provided with this book, help you progressively improve your time management and test-taking skills while reinforcing the key concepts you've learned.

HOW TO GET STARTED

- Open the link or scan the QR code at the bottom of this page

- If you have unlocked the practice resources already, log in to your registered account. If you haven't, follow the instructions in *Chapter 19* and come back to this page.

- Once you log in, click the START button to start a quiz

- We recommend attempting a quiz multiple times till you're able to answer most of the questions correctly and well within the time limit.

- You can use the following practice template to help you plan your attempts:

Working On Accuracy		
Attempt	Target	Time Limit
Attempt 1	40% or more	Till the timer runs out
Attempt 2	60% or more	Till the timer runs out
Attempt 3	75% or more	Till the timer runs out
Working On Timing		
Attempt 4	75% or more	1 minute before time limit
Attempt 5	75% or more	2 minutes before time limit
Attempt 6	75% or more	3 minutes before time limit

The above drill is just an example. Design your drills based on your own goals and make the most out of the online quizzes accompanying this book.

First time accessing the online resources? 🔓

You'll need to unlock them through a one-time process. **Head to *Chapter 19* for instructions**.

Open Quiz

https://packt.link/sc300ch10

OR scan this QR code →

11

Planning, Implementing, and Monitoring the Integration of Enterprise Applications

Organizations rely heavily on a wide range of enterprise applications to drive business operations, streamline processes, and enhance productivity. Some examples include customer relationship management, productivity and collaboration, accounting and financial management, marketing, project management, and identity and cybersecurity. Each industry may have its unique set of required applications, but the need for integration, authentication, and authorization remains a common challenge across all sectors.

Of particular importance when implementing these enterprise applications is the challenge of securely managing access to sensitive resources, users, and data. This is where Entra ID comes into play, providing a robust identity and access management solution to streamline the user authentication and authorization processes. By integrating Entra ID with these enterprise applications, businesses can ensure secure access, simplify user experiences, and maintain compliance with regulatory requirements.

In this chapter, we will explore Microsoft's best practices for enterprise application management. By mastering these concepts, you will be equipped to effectively manage the security and usability of your enterprise applications, ensuring that your organization's digital assets are protected and optimized.

The objectives and skills we'll cover in this chapter include the following:

- Planning and implementing settings for enterprise applications
- Assigning appropriate Microsoft Entra roles to users to manage enterprise applications
- Designing and implementing integration for on-premises apps by using Microsoft Entra application proxy
- Designing and implementing integration for software as a service (SaaS) apps

- Assigning, classifying, and managing users, groups, and app roles for enterprise applications
- Configuring and managing user and admin consent

Planning and Implementing Settings for Enterprise Applications

When it comes to integrating enterprise applications, planning and implementing application-level and tenant-level settings is essential for successful integration. Understanding the required and available settings is crucial to ensuring seamless communication and data exchange between systems. Many of the settings determine how applications interact with each other. Failing to configure them correctly can lead to errors, performance issues, or even security breaches.

Application-level settings are specific to each application and may include configurations such as application type, authentication protocols, scopes, and permissions needed to access Microsoft Graph API or other APIs, **redirect URIs** (locations where users will be redirected after authentication), and any branding elements or customization options. Tenant-level settings, on the other hand, apply across all applications within a tenant and encompass broader settings such as security policies, directory synchronization, and user provisioning.

In this section, you will dive into the application and tenant-level settings, exploring how to plan, implement, and monitor these critical configurations. By understanding how to effectively configure these settings, you'll be able to streamline your organization's enterprise application integration process, reduce complexity, and ensure a seamless user experience across multiple applications.

Tenant-Level Settings

You can access tenant-level settings in several ways. The Azure administration portal can be used to access the tenant-level enterprise application settings. Keep in mind that Microsoft has released a new administration portal along with the Microsoft Entra ID rebranding. Please read up on the Microsoft Entra admin center. This should be the default place you go to perform identity and access management tasks, though not all settings may appear there yet.

Entra.microsoft.com administration portal

The Microsoft Entra admin center is primarily focused on identity and access management, specifically dealing with Entra ID and related identity services. You can learn more about the Entra administration center at `https://learn.microsoft.com/en-us/entra/fundamentals/entra-admin-center`.

The primary tenant-level settings are the following:

- **Conditional Access**: Conditional Access gives you the ability to enforce access requirements when specific conditions occur. In *Chapter 6*, you learned about Conditional Access policies. The access policies and the conditions used to manage access apply to enterprise applications as well. For instance, when you require **multi-factor authentication (MFA)** for users attempting to access sensitive data from outside the company's network.

- **Consent and permissions**: Using tenant-level consent and permissions, you can control when end users and group owners are allowed to grant consent to applications, and when they will be required to request administrator review and approval. Allowing users to grant applications access to data eases the ability to acquire and onboard applications but can represent a risk if it's not monitored and controlled carefully.

Conditional Access Settings

In the following screenshot, you can see the enterprise application configuration page. The configuration settings that are used to apply tenant-wide settings are highlighted.

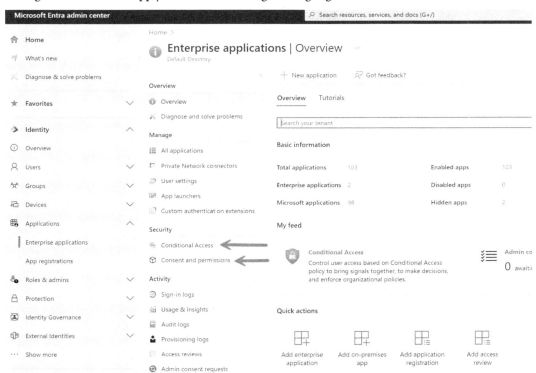

Figure 11.1: Accessing tenant-level enterprise application settings

Selecting **Conditional Access** redirects you to the standard **Conditional Access** blade in the Entra admin center. The **Coverage** tab of the **Overview** page, shown in *Figure 11.2*, highlights the top 10 enterprise applications that have been accessed in the last week with and without protection through Conditional Access.

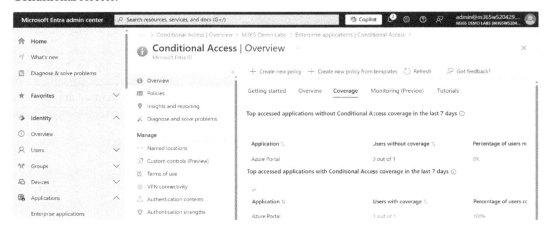

Figure 11.2: Displaying applications with and without Conditional Access protection

This information can help you identify at-risk applications. You can then create additional Conditional Access policies (or modify existing policies) to provide further protections for applications with sensitive data.

Next, you'll learn about consent and permissions.

Consent and Permissions Settings

User consent is a process where users explicitly agree to allow an application to access their data or perform certain actions on their behalf. This is typically required when an application requests access to sensitive information or requires broad permissions beyond what the user initially granted. **Admin consent**, on the other hand, allows administrators to grant permission to applications on behalf of all users in their organization. This is useful for applications that require high-level access to organizational data but do not need individual user consent.

Permissions define the specific actions an application can perform and the level of access it has to user or organizational data. For example, an application may request permission to read a user's email or modify their calendar entries. By granting these permissions, users and administrators allow the application to perform certain tasks on their behalf, which can improve productivity and streamline workflows. However, it is important to carefully review and manage these permissions to ensure that only necessary access is granted and that sensitive data is protected.

It's likely that you will encounter exam questions about enterprise application consent and permissions.

User Consent

User consent covers what permissions users are allowed to grant. By default, users can consent to any application that requests organization data. To configure consent and permissions, use the following process:

1. Navigate to the Entra admin center (`https://entra.microsoft.com`).

2. Expand **Identity**, expand **Applications**, and then select **Enterprise applications**.

3. Under **Security**, select **Consent and permissions**, as shown in *Figure 11.3*:

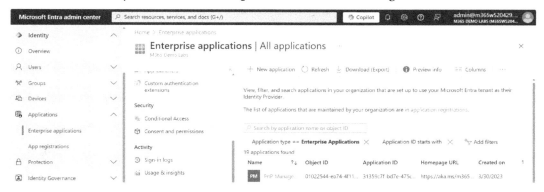

Figure 11.3: Consent

4. Within **Consent and permissions**, we can configure who can grant consent. These options allow only an administrator to grant consent, allow users to grant consent to *low-impact* applications, or allow users to grant consent to all applications. These options are shown in the following screenshot:

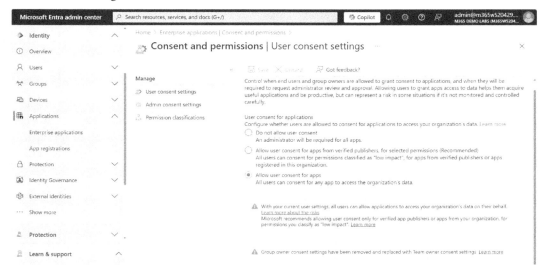

Figure 11.4: User consent options for applications

The default setting allows users to consent to applications (most permissive). Security-conscious organizations will want to change this to either only allow user consent for low-impact applications (moderately restrictive) or only allow admin consent (most restrictive).

Resource-Specific Consent

Previously, you could configure group consent settings in the user consent area.

Group owner consent settings are no longer accessible in the Azure portal or Entra admin center, as of May 14, 2024. They have been replaced with team owner consent or **resource-specific consent** (**RSC**), per https://learn.microsoft.com/en-us/microsoftteams/platform/graph-api/rsc/grant-resource-specific-consent#configure-consent-settings, and can only be configured via PowerShell.

To configure RSC through PowerShell, you'll need to use the Microsoft Graph PowerShell module (available by running `Install-Module Microsoft.Graph` in a PowerShell session). Once you have the required modules, you can use the following steps to configure RSC:

1. Launch a PowerShell console session.

2. Run the following command to connect to Graph with the appropriate permission scopes:

   ```
   Connect-MgGraph -Scopes TeamworkAppSettings.ReadWrite.
   All,Policy.ReadWrite.Authorization,Policy.ReadWrite.
   PermissionGrant,AppCatalog.Read.All
   ```

3. Run the following cmdlet to configure the consent permissions:

   ```
   Set-MgBetaTeamsRscConfiguration -State <Value>
   ```

   In this situation, `<Value>` should be one of the following:

   - `ManagedByMicrosoft`: Default state, allowing chat and team RSC permissions to be consented for all users but can be changed by Microsoft at any time

   - `EnabledForAllApps`: Any app requesting RSC permissions can be consented to by users

   - `DisabledForAllApps`: Users cannot consent to RSC

Configuring these permissions is important if you wish to have the proper levels of security and governance over your applications. If your applications include sensitive information, then allowing users to grant application access could lead to security vulnerabilities. Planning properly and having an appropriate strategy can help you determine the level of governance that's needed for application permissions.

You should always consider relevant industry best practices when configuring security settings, including user and admin consent. Wherever possible, you should implement best practices for securing your environment:

- **Implement least-privilege access**: Ensure that applications are only granted the permissions they strictly need to perform their intended functions. Avoid granting unnecessary permissions, especially for sensitive data or critical resources.

  Periodically audit the permissions that have been granted to applications to ensure they are still necessary. This helps prevent over-permissioning and reduces the attack surface.

- **Use admin consent for sensitive applications**: By default, users can consent to applications requesting access to their data. However, for sensitive or high-risk applications, it's a best practice to require admin consent. This ensures that only administrators can review and approve the permissions being requested.

  Enable the admin consent workflow, which allows users to request consent for an application, but ensures the approval process is routed through an administrator. This ensures better control over which applications can access organizational data.

- **Control user consent settings**: In Microsoft Entra, you can control how users can consent to applications by setting user consent policies. These policies allow you to define what types of permissions users can grant without admin approval. For sensitive data or privileged applications, disable user consent entirely and enforce admin consent for tighter control. Restrict user consent to applications published by verified publishers only. This minimizes the risk of users inadvertently consenting to malicious applications.

- **Monitor and audit consent and permissions**: Turn on audit logs to track when users or administrators grant consent to applications. Monitoring these activities helps identify potential risks and provides a trail for investigation in case of suspicious activity.

  Use Entra ID reports to regularly review the permissions granted to applications. Pay particular attention to apps with high-risk permissions, such as those with access to sensitive user data or administrative functionality.

- **Use Conditional Access for application consent**: Implement Conditional Access policies to control how and when users or administrators can consent to applications. You can enforce policies based on location, device compliance, or risk factors, ensuring that consent is only granted in secure environments.

Next, we'll look at admin consent settings.

Admin Consent

The admin consent settings control escalations for user consent requests and the review process. By default, the review workflow is disabled. You can enable it and then choose reviewers and any timeout or expiration settings, as shown in *Figure 11.5*:

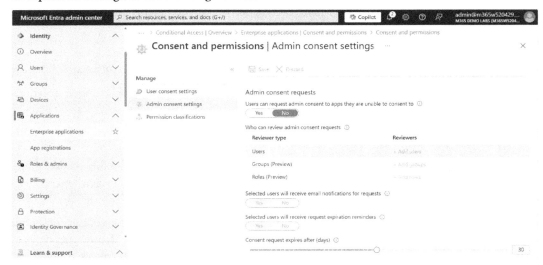

Figure 11.5: Reviewing the admin consent settings

These settings will be invoked if you change the user consent settings to anything but the default.

Permission Classifications

Finally, you can define what qualifies as low, medium, or high-risk permissions on the **Permission classifications** page, based on your organization's policies and risk evaluations.

> **Tip**
>
> The minimum permissions necessary for sign-in are `openid`, `email`, `profile`, and `offline_access`.

To manage permission classifications, follow these steps:

1. Navigate to the Entra admin center (`https://entra.microsoft.com`).

2. Expand **Identity**, expand **Applications**, and then select **Enterprise applications**.

3. Under **Security**, select **Permission classifications**, as shown in *Figure 11.6*.

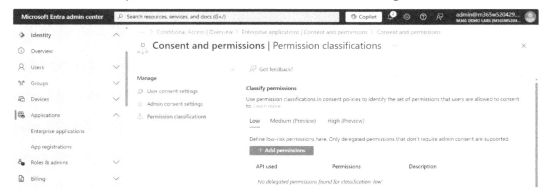

Figure 11.6: Permission classifications page

4. Select a tab (either **Low**, **Medium**, or **High**).

5. Click **Add permissions**.

6. On the **Request API permissions** flyout, choose the API family that you want to use.

7. Choose which permissions to add to the selected classification and click **Add permissions**.

Now that we have looked at tenant-level settings, let's turn our attention to application-level settings.

Application-Level Settings

In addition to configuring permissions and settings at the tenant level, you can also configure them at the application level.

You can configure an application's core settings by choosing the application from the **Enterprise applications** blade in the Entra admin center or the Azure portal. An application's properties are shown in *Figure 11.7*:

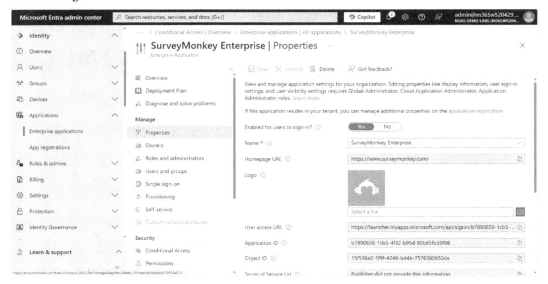

Figure 11.7: Configuring an application's properties

On the **Properties** page, you can configure the following settings that manage the visibility and sign-in properties of an application:

- **Enabled for users to sign-in?**: Select **Yes** to enable assigned users to log in to the application

- **Name**: The name that is displayed in the user's list of applications, as well as the application name that other tenants see.

- **Logo**: An application logo that can be uploaded and displayed on the **My applications** screen.

- **Assignment required**: For federated applications only, this setting defines whether assignment is required for this application. If it is set to **No**, then any user will be able to log in to the app.

- **Visible to users**: Defines whether this application is visible to users who have it assigned to them.

On the **Roles and administrators** page, you can see and manage the roles that grant access for privileged actions. Microsoft recommends using the built-in roles to manage access to application administration features. See *Figure 11.8*.

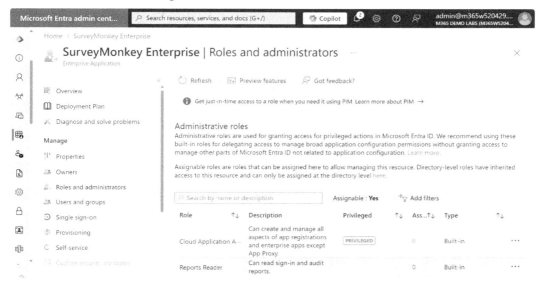

Figure 11.8: Viewing an enterprise application's Roles and administrators page

If you choose to manage or restrict the visibility and access to applications, you can assign users and groups to the application through the **Users and groups** blade.

Some applications may allow you to use your Entra ID as a directory source for authentication (this process is typically referred to as **federation**). Typically, applications are configured to use either **Security Assertion Markup Language** (**SAML**) or **Open Identity Connect** (**OIDC**) authentication.

You can configure this authentication using the **Single sign-on** blade of an application's properties. Alternatively, you can set this up when you add an application from the Microsoft Entra Marketplace or App Gallery.

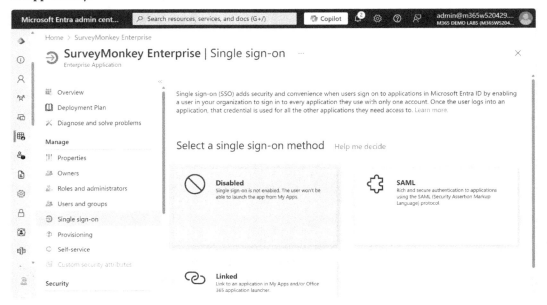

Figure 11.9: Configuring single sign-on properties for an application

In addition to single sign-on capabilities, many third-party enterprise applications also support user provisioning and deprovisioning through **System for Cross-domain Identity Management** (**SCIM**). You can use the **Provisioning** blade to configure applications that support SCIM operations.

If you configure an enterprise application for automated provisioning scenarios using the **Provisioning** blade, you can also configure options on the **Self-service** blade. These settings define whether a user can request access to this application, and what roles are used in that scenario. See *Figure 11.10*.

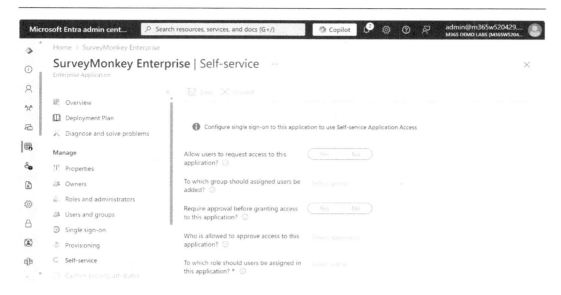

Figure 11.10: Configuring the Self-service blade

The **Security** section of the application's configuration allows you to implement controls to enforce access requirements. Just as you can configure Conditional Access policies at the tenant level, you can also implement policies that are scoped to enterprise applications. The **Conditional access** blade on the application's configuration properties redirects to the main **Conditional access** blade, where you can configure a new policy or edit an existing one.

An application's **Permissions** blade shows the user and admin consent permissions options (similar to a tenant's configuration). You can review and consent to permissions. However, to add permissions, you will need to use the enterprise app's app registration.

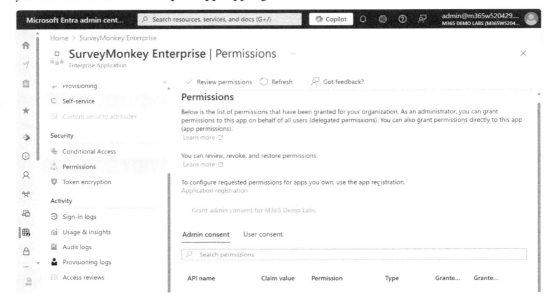

Figure 11.11: Configuring application permissions

Applications can also be configured with encryption that protects the content of SAML communications between Entra ID and the application by uploading a certificate to the application's **Token encryption** blade, as shown in *Figure 11.12*.

Figure 11.12: Configuring token encryption

We have examined general tenant-level settings, as well as application-level settings. Next, we'll explore administration tools for enterprise applications. These include application proxies for on-premises applications, application roles, application collections, and more information about consent.

Assigning Appropriate Microsoft Entra Roles to Users to Manage Enterprise Applications

To protect sensitive data and systems, organizations must implement an effective role-based access control strategy using Microsoft Entra. This approach ensures that only authorized personnel have access to enterprise applications, significantly reducing the risk of unauthorized access or breaches. By implementing this control, organizations can not only maintain compliance with regulatory requirements but also ensure business continuity. The primary roles in the administration of enterprise applications are Application Administrator and Cloud Application Administrator.

Application Administrator Role

The Application Administrator role allows administrators to modify all aspects of an application registration. Due to its high level of access and potential impact, it is essential to assign this role in accordance with the principle of least privilege. To maintain security, grant this role only to highly trusted administrators who require frequent modifications to enterprise roles.

As an additional safeguard, consider using Microsoft Entra's **Privileged Identity Management** (**PIM**) feature to grant this role on a just-in-time basis, further reducing the risk of unauthorized access.

Cloud Application Administrator Role

The Cloud Application Administrator role is similar to the traditional Application Administrator role, but with a critical distinction. Unlike its counterpart, the Cloud Application Administrator has limited capabilities when it comes to managing application proxy settings.

Now that you are familiar with the available roles, we will apply this knowledge to a practical example. In this exercise, we will examine the process of assigning application management roles to Entra users.

Let's say that you have a user that will begin administering a specific SaaS app. They will need to manage SSO, MFA, and application permissions, and monitor user activity. Based on the principle of least privilege, you decide to grant the user the Cloud Application Administrator role for the application.

To configure a user with these settings, use the following steps:

1. Navigate to the Entra admin center (`https://entra.microsoft.com`).
2. Expand **Identity**, expand **Users**, and then select **All users**.

3. Select a user and choose **Assigned roles**.

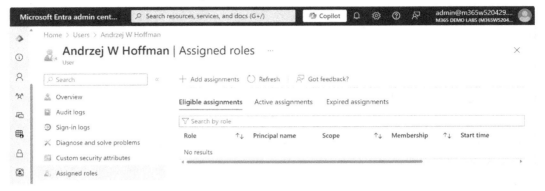

Figure 11.13: Viewing the assigned roles

Deep dive into role assignments

Eligible assignments are the role assignments that a user is eligible to request or accept based on their organizational role, group membership, or other criteria. Eligibility does not guarantee that a user will be granted access to a particular role, but it indicates that they meet the minimum requirements for consideration. **Active assignments** are the current role assignments for a user in Microsoft Entra. Users with active assignments have access to the associated permissions and resources, such as applications, data, and services. **Expired assignments** are role assignments that have expired or been revoked by an administrator. Users who were previously assigned to these roles no longer have access to the associated permissions and resources.

4. Click **Add assignment**.

5. In the **Select role** drop-down box, select the **Cloud Application Administrator** role.

6. Under **Scope type**, select **Application**.

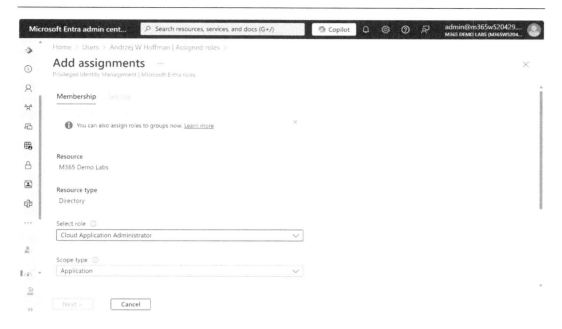

Figure 11.14: Selecting a role and scope

7. Under **Selected scope**, click **No scope selected** and then choose the application to which you wish to assign the user role.

8. Click **Next**.

9. Under **Assignment type**, choose either **Eligible** to require the user to elevate to the role using PIM or **Active** to add the role membership to the user directly. You can also choose whether to make the assignment type permanent (either permanently eligible or permanently active) or enable a start and end time to limit how long the user will have access to the role. See *Figure 11.15*.

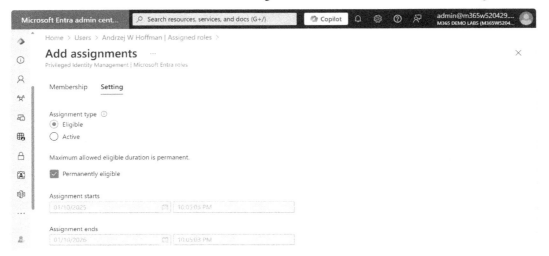

Figure 11.15: Choosing the assignment type

10. Click **Assign**.

From a best practices perspective, it's best to create eligible roles that users can request or activate as needed as opposed to assigning roles permanently. This eliminates the risk associated with the role being active at all times.

Next, we'll shift focus to accessing on-premises applications using Entra application proxy.

Designing and Implementing Integration for On-Premises Apps by Using Microsoft Entra Application Proxy

In today's hybrid IT landscape, organizations often have a mix of cloud-based applications and on-premises applications that require integration for seamless data exchange and user authentication. The Microsoft Entra application proxy enables administrators to integrate on-premises web applications with Entra ID and other cloud-based services. By using the Microsoft Entra application proxy, organizations can provide secure, SSO access to on-premises applications for users, while also enabling these applications to communicate with Entra ID and other cloud-based services.

The Microsoft Entra application proxy provides a range of features that make it easy to integrate on-premises apps, including the following:

- SSO for seamless access to applications

- Reverse-proxy capabilities to publish internal applications to the internet

- Pre-authentication and Conditional Access policies to control who can access which applications

- Support for various authentication protocols, such as Kerberos, NTLM, and OAuth

- Ability to publish web-based applications with different authentication mechanisms, such as claims-based authentication or token-based authentication

- Integration with Azure Active Directory for simplified management and security

- Support for publishing internal APIs for mobile apps, web apps, or other services

In this section, we will dive into the details of designing and implementing integration for on-premises apps using the Microsoft Entra application proxy. We'll cover topics such as the following:

- Planning and designing an integration strategy for on-premises apps

- Installing and configuring the Microsoft Entra application proxy

- Creating application registration and configuration files for on-premises apps

- Troubleshooting common issues with on-premises app integration

Entra application proxy provides integration between Entra and the on-premises application and the authentication settings in Windows Active Directory.

The diagram in *Figure 11.16* shows how Entra application proxy works. Entra application proxy is used to pass credentials for the user to authenticate. The application proxy service makes the on-premises application emulate a cloud application in Entra. This provides SSO authentication through Entra and uses Microsoft 365 and other registered cloud applications as the single identity provider.

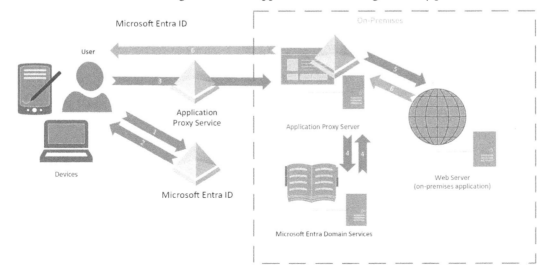

Figure 11.16: Application proxy diagram

Configuring Entra application proxy requires either the Application Administrator or Global Administrator role. Adhering to the principle of least privilege, you should assign users who need to perform this configuration the Application Administrator role.

Configuring App Proxy

Once that role is activated, you can complete the configuration with the following steps:

1. Navigate to the Entra admin center (`https://entra.microsoft.com`).

2. Expand **Identity**, expand **Applications**, and select **Enterprise applications**.

3. Under **Manage**, select **Private Network connectors**.

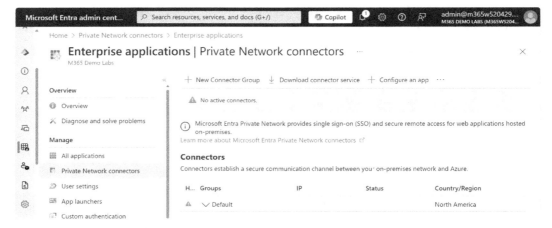

Figure 11.17: Configuring application proxy

4. Select **Download connector service**, as shown in *Figure 11.18*:

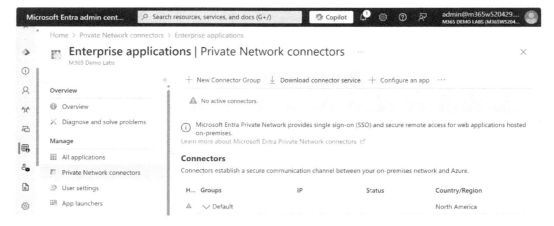

Figure 11.18: Download connector service

5. On the **Private Network Connector Download** flyout, click **Accept terms & download** to get the application proxy connector.

6. Locate the AADApplicationProxyConnectorInstaller.msi file in your Downloads folder and start the installation. Accept the license terms and click **Install**.

> **Note**
>
> Make sure that your network has been configured properly for the application proxy connector to communicate with Entra by following the steps at https://learn.microsoft.com/en-us/entra/identity/app-proxy/conceptual-deployment-plan.

Figure 11.19: Installing the application proxy connector

7. During the installation, sign in to your Azure account (when prompted) with the account that has been assigned the Application Administrator or Global Administrator role.

8. Once you've installed the application proxy connector, refresh the **Private Network connectors** blade in the Entra admin center. Your connector will now be active, as shown in the following screenshot:

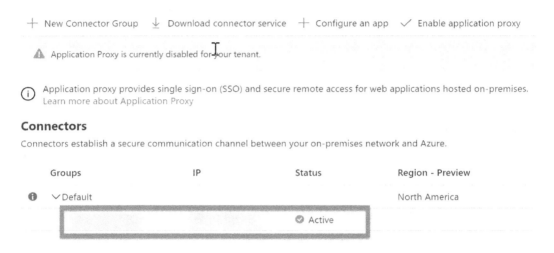

Figure 11.20: Application proxy is active

9. Select **Enable Private Network connectors**.

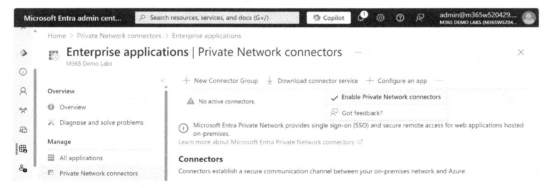

Figure 11.21: Selecting Enable Private Network connectors

10. Select **Yes** when prompted to continue enabling application proxy.

Once application proxy has been enabled, the next step is to configure an app.

Configuring an Application with Application Proxy

To configure an application with application proxy, you'll need an application deployed to a server. Application proxy supports the following types of applications:

- Web apps or APIs

- Applications hosted behind a Remote Desktop Gateway

- Rich client apps that support the **Microsoft Authentication Library (MSAL)**

The application or API you select can use forms-based authentication, SAML authentication, integrated Windows authentication, or header-based authentication. To configure an app, follow these steps:

1. From the Entra admin center (`https://entra.microsoft.com`), navigate to the **Private Network connectors** blade under **Enterprise applications**.

2. Click **Configure an app**.

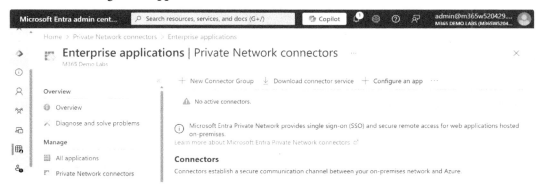

Figure 11.22: Configure an app

3. Enter a name for your application and provide the internal URL for your enterprise application (this URL must be reachable from the application server where the application proxy connector is installed).

4. For the **External Url** value, you can create a name for your app and select a custom URL for your tenant from the dropdown, as shown in *Figure 11.23*. If you do not have a custom URL to use, it will use `tenantname.msapproxy.net` for the external URL value.

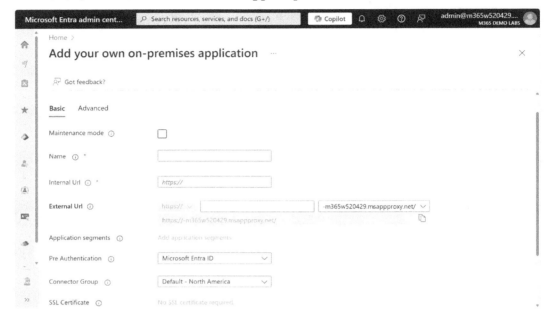

Figure 11.23: Configuring application proxy for an on-premises application

5. Scroll to the bottom of the page and click **Create** to add the on-premises application for SSO.

With that, the application has been added and can be used with Azure AD for SSO.

> **Note**
>
> The application proxy connector should be installed on more than one device in your on-premises network for additional resiliency in case of a device failure. When installing multiple connectors, you can choose to add them to groups for managing failover and redundancy.

Next, we'll discuss integrating third-party or external SaaS applications.

Designing and Implementing Integration for SaaS Apps

SaaS applications have become an integral part of many organizations' IT infrastructure. SaaS apps offer numerous benefits, including scalability, flexibility, and cost-effectiveness. However, integrating these SaaS apps with the rest of the organization's ecosystem can be complex and challenging. Microsoft Entra provides a robust set of features that enable administrators to seamlessly integrate SaaS apps with Entra ID, other cloud-based services, and on-premises applications.

In this section, we will explore how to design and implement integration for SaaS apps using Microsoft Entra. Integration with SaaS apps is critical for ensuring a seamless user experience, while also providing a secure and compliant way to access these applications. SaaS apps can be added to the tenant by using the **Enterprise applications** blade in the Entra admin center or the Azure portal.

When you navigate to the **Enterprise applications** blade, you can browse all of the preconfigured apps in the gallery, as shown in *Figure 11.24*:

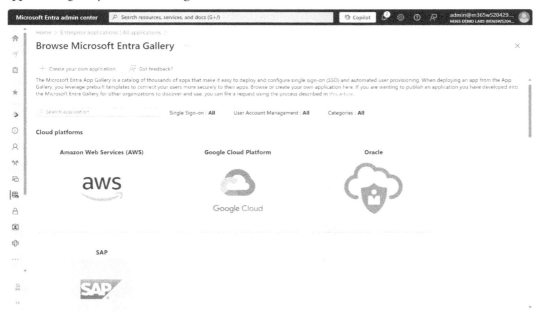

Figure 11.24: Microsoft Entra SaaS App Gallery

The gallery provides you with a list of commercial applications that can be integrated directly with your environment. These app configuration templates have already been validated; frequently, all that you have to do is provide credentials, tokens, or other environment information from the SaaS application vendor to complete the setup.

One item that is clear in the gallery is the preconfigured SaaS apps that support federated SSO and automated provisioning, as indicated by specialized icons (see *Figure 11.25*).

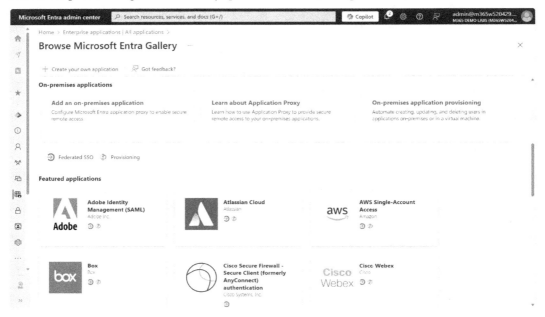

Figure 11.25: Gallery apps that support federated SSO and automated provisioning

Those are among the highest priority when integrating SaaS apps. Microsoft identifies the following primary objectives for the integration of SaaS apps. Provisioning and federation encompass four of the five priorities:

- **Automated user provisioning**: Automatically create new accounts in the external system when a user is granted access to the app

- **Automated user deprovisioning**: Automatically remove the account from the external system when user access to the app is removed

- **Synchronize data between systems**: Ensure that the external system maintains up-to-date user information by continually synchronizing user data

- **Seamlessly deploy in brown-field scenarios**: This means that a provisioning system can begin synchronization with a system that already has user accounts provisioned and can match users as appropriate

- **Govern access**: Provide for logging and auditing of provisioning activity

The SaaS gallery identifies which preconfigured applications support provisioning, but what does that mean?

The two primary ways to enable automated user provisioning and deprovisioning are SAML and SCIM.

SAML is a widely adopted standard used by Microsoft Entra to enable automated user provisioning and SSO for SaaS applications. In the context of automated provisioning, SAML enables Microsoft Entra to automatically create, update, and delete user accounts in SaaS applications based on changes made in Azure AD. When a new user is added to an application's corresponding group in Azure AD, for example, Microsoft Entra can use SAML assertions to automatically create a corresponding account in the SaaS application with the appropriate permissions and attributes. This saves time and reduces errors associated with manual provisioning processes. With SAML-based SSO, users can access their SaaS applications using their Azure AD credentials, eliminating the need for separate usernames and passwords. This improves security and reduces the risk of password fatigue or reuse.

SCIM is an industry-standard protocol that automates user provisioning and deprovisioning between identity providers and applications. By enabling seamless authentication and authorization across multiple systems, SCIM simplifies access to SaaS applications by eliminating the need for manual account creation or updates. The protocol itself is responsible for synchronizing identities and groups.

SSO and automated provisioning using SCIM are two common integration features used with Microsoft Entra. SSO allows users to access multiple applications with a single set of login credentials, making it easier for them to manage their passwords and reducing the risk of forgotten or weak passwords. Automated provisioning using SCIM or SAML, on the other hand, enables the automatic creation, update, and deletion of user accounts in external systems based on changes made within Microsoft Entra. This helps to ensure that user accounts are always up to date and reduces the administrative burden of managing them manually. While SSO provides a more seamless user experience by eliminating the need for multiple login prompts, automated provisioning offers benefits such as improved security and reduced administrative overhead by automating account management tasks. As you explore automated provisioning and deprovisioning, remember that its primary use case involves external SaaS apps that you would otherwise have to administer manually.

For the exam, remember that there are several options to configure automated provisioning and deprovisioning for an external application (SaaS), including SCIM and SAML.

Automated provisioning will satisfy the first four priorities. The last one, governance, can be accomplished by using group membership governance and provisioning logs, which tell you what users have been created across a SCIM or SAML connection.

Now that we have looked at SaaS app integration, we will look at users, groups, and rules used with enterprise applications.

Assigning, Classifying, and Managing Users, Groups, and App Roles for Enterprise Applications

Managing access to enterprise applications is a critical responsibility that demands meticulous planning and execution and ongoing oversight. With its comprehensive set of features, Microsoft Entra allows you to efficiently manage identities by assigning, classifying, and governing users, groups, and app roles for enterprise applications. This section will guide you through the essential skills required to plan, implement, and monitor user, group, and app role assignments in Microsoft Entra ID.

Each of the following activities should be done with at least the Cloud Application Administrator permissions. Using the identity with these permissions, log in to the Microsoft Entra admin center at `https://entra.microsoft.com`. Then, take the following steps:

1. In the left pane, select **Applications**, then **Enterprise applications**.

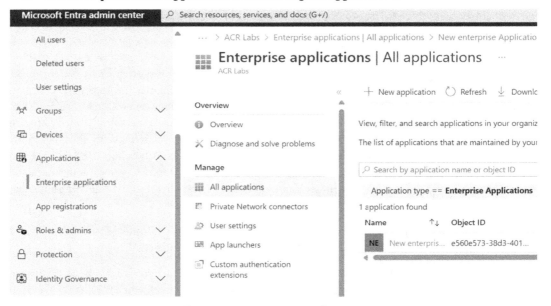

Figure 11.26: Enterprise applications

2. Select the enterprise application to be modified.

3. Select **Users and groups**, then click **Add user/group**.

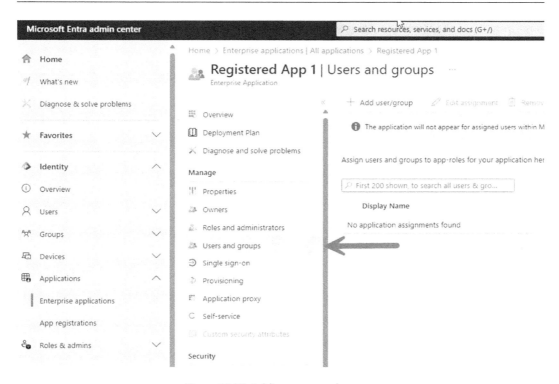

Figure 11.27: Adding users and groups

4. Click **None Selected** under **Users**, then select the appropriate user. Click **Next**.

    Depending on the application, you may be able to select roles or delegate custom roles and permissions.

5. Click **Assign** to complete the user or group assignment operation.

You can also use the **Users and groups** blade to view users and groups that have been assigned to an application, as shown in *Figure 11.28*:

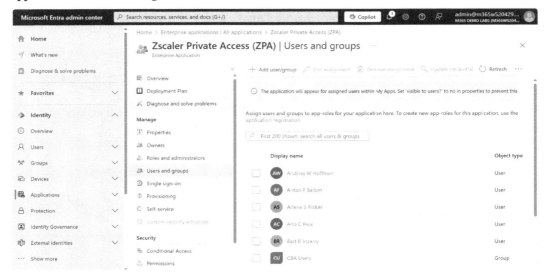

Figure 11.28: Viewing users and groups assigned to an enterprise application

Depending on your organization's security requirements, assigning users and groups to applications may not be necessary (such as in the case of organization-wide applications that everyone needs. A common example might be an HR application employees use to update their benefits).

Summary

In this chapter, you learned about managing settings for enterprise applications—both tenant- and application-level settings around access control, consent, and permissions.

You also learned about assigning the roles necessary to manage enterprise applications and their settings, as well as how to configure Entra ID integration for both SaaS and on-premises applications. Finally, you learned how to assign users to the applications.

In the next chapter, we'll focus on configuring and managing app registrations.

Exam Readiness Drill – Chapter Review Questions

Apart from mastering key concepts, strong test-taking skills under time pressure are essential for acing your certification exam. That's why developing these abilities early in your learning journey is critical.

Exam readiness drills, using the free online practice resources provided with this book, help you progressively improve your time management and test-taking skills while reinforcing the key concepts you've learned.

HOW TO GET STARTED

- Open the link or scan the QR code at the bottom of this page

- If you have unlocked the practice resources already, log in to your registered account. If you haven't, follow the instructions in *Chapter 19* and come back to this page.

- Once you log in, click the START button to start a quiz

- We recommend attempting a quiz multiple times till you're able to answer most of the questions correctly and well within the time limit.

- You can use the following practice template to help you plan your attempts:

Working On Accuracy		
Attempt	**Target**	**Time Limit**
Attempt 1	40% or more	Till the timer runs out
Attempt 2	60% or more	Till the timer runs out
Attempt 3	75% or more	Till the timer runs out
Working On Timing		
Attempt 4	75% or more	1 minute before time limit
Attempt 5	75% or more	2 minutes before time limit
Attempt 6	75% or more	3 minutes before time limit

The above drill is just an example. Design your drills based on your own goals and make the most out of the online quizzes accompanying this book.

> **First time accessing the online resources?** 🔒
> You'll need to unlock them through a one-time process. **Head to** *Chapter 19* **for instructions**.

Open Quiz	
`https://packt.link/sc300ch11` OR scan this QR code →	

12

Planning and Implementing App Registrations

Integrating enterprise applications with Microsoft Entra ID is crucial for securing access to sensitive data and services while streamlining user authentication and authorization processes. As an administrator or developer responsible for implementing Entra ID in your organization, you'll encounter numerous scenarios where app registrations are necessary. An app registration represents a specific application that's registered with Entra ID, allowing users to authenticate and authorize access to the application without having to manually enter their credentials.

App registrations are a fundamental component of Microsoft identity management, enabling organizations to securely integrate various applications with Entra ID while maintaining control over user identities and permissions. However, planning and implementing app registrations can be complex, especially for large-scale enterprise environments with numerous applications and users. To ensure seamless integration, administrators must carefully plan the registration process, considering factors such as application requirements, user access controls, and security protocols.

The objectives and skills that we'll cover in this chapter are as follows:

- Planning for app registrations
- Creating app registrations
- Configuring app authentication
- Configuring API permissions
- Creating app roles

By the end of this chapter, you should be able to define an app registration and identify the benefits of registering an app, understand the authentication and authorization protocols used when registering an app, and define permissions and customize roles for use by a registered app.

Planning for App Registrations

While Microsoft applications will natively support Entra ID authentication and authorization, this is not always the case with non-Microsoft applications or apps that you develop yourself.

> **What Is an App Registration?**
>
> To benefit from Entra's native identity and access capabilities, some apps need to be **registered**. The **app registration** process involves creating an identity for your application within your tenant. This identity allows the application to authenticate users and services as well as access data using its own permissions (**application permissions**) or the logged-in users' permissions (**delegated permissions**).

In the following sections, you will learn how to plan, create, and configure an app registration.

In *Chapter 10*, you learned about the difference between single-tenant applications and multitenant applications. To get the most out of this chapter, you need to have a firm understanding of the application types. Here are a few of the differences:

- **Single-tenant applications**:

  - Only available in the tenant they were registered in (the home tenant)

  - Only users from the home tenant can be authenticated and authorized for access to the application or API

  - Typically use managed identities, provided the required resources

- **Multitenant applications**:

  - Available to any Azure tenant

  - Allow sharing of resources between tenants

  - Managed identities cannot be used

  - Require service principals for cross-tenant authentication and authorization

  - May allow users from any Microsoft Entra directory or personal Microsoft accounts (such as Xbox, Skype, or Outlook)

Registering an application comes with certain benefits related to identity, security, and branding. These benefits include the following:

- Custom branding and logos added to the application login

- Apply restrictions to users based on how they log in to an application and the duration they're allowed to stay logged in

- Use various Microsoft identity resources, such as to store secrets within Microsoft Entra

- Have granular control over permissions

When an application needs access to resources in the same tenant, you can use a managed identity for this purpose. This enables you to simplify the credential management tasks associated with the application. However, for an application to be accessible cross-tenant, you must first register your application to create a service principal.

Figure 12.1 is a diagram of a cross-tenant application. The application is registered in the home tenant with the associated application ID. In the customer tenant, the service principal is used to establish the enterprise app and make it accessible to the tenant.

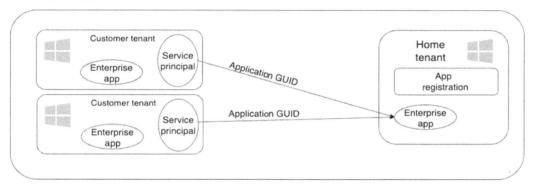

Figure 12.1: Multitenant application

When an application is registered, a unique identifier—called a **globally unique identifier** (GUID)—is generated by Microsoft Entra and assigned to the app. This unique ID is used when the applications require access to resources within the tenant. The unique ID is used to request and generate tokens.

The registration process results in the creation of an application object and a **service principal**. The application object contains the application definition including supported authentication protocols, required identifiers, URLs, secrets, and authentication information. After registration, the application is also issued a client ID that is used for cross-tenant access.

Microsoft Entra ID's role in the app registration process includes handling application authentication and providing the resources necessary for provisioning the application and access across Entra tenants. Entra also handles user consent during token requests and cross-tenant provisioning.

When planning app registrations, you'll want to consider several factors.

Application Type

We've already discussed the decision between single-tenant and multitenant. You'll also want to determine what type of client you'll be using—whether it's a **public** or a **confidential** client. Public clients have no mechanism for key or secret storage, while confidential applications can securely store credential data.

Authentication and Authorization

You'll need to make choices about authentication and authorization.

From an authentication perspective, you need to choose a **grant type** (process to provide credentials and acquire an access token, which allows you to access a resource). When working with OAuth 2.0 and OpenID Connect, your grant types are as follows:

- **Authorization Code Flow**, recommended for web apps and **single-page applications (SPAs)** using **Proof Key for Code Exchange** (PKCE). PKCE is typically used in public client applications to protect against using the application's client secret (compiled in the application) to compromise the application.
- **Client Credentials Flow** for daemon apps or backend-to-backend communication.
- **Implicit Flow**, which is deprecated for SPAs. This stores the access token on the device, making it susceptible to compromise. If you were going to use Implicit Flow, you should use Authorization Code Flow with PKCE instead.

If you're not going to use OAuth or **OpenID Connect** (OIDC), you can also choose certificates or client secrets. **Certificates** are a type of public key cryptography and allow applications to prove their identity. **Client secrets** are essentially passwords and are considered less secure than certificates.

Depending on your organization's requirements, you may need or want to configure the token lifespans. By default, **access tokens** are valid for one hour and **refresh tokens** may be valid for days or weeks.

From an authorization perspective, you can use policies such as Conditional Access, **multi-factor authentication** (MFA), IP restrictions, or device compliance to ensure the right environment is present before allowing application access.

Permissions and Consent

As part of the zero-trust philosophy, you use the principle of least privilege. As we've discussed throughout this book, the concept of least privilege is assigning the lowest level of permissions or other access necessary to perform a task.

From a permissions perspective, you can choose to use either delegated or application permissions. Delegated permissions require user interaction, can only access resources on behalf of the logged-in user, and are limited in scope to whatever the current user has. Application permissions, on the other hand, are typically used with services or background applications. When granting application permissions, you should use the principle of least privilege to ensure the application can only access what is absolutely necessary. When possible, you should choose the delegated permissions model.

Depending on the authentication type, you may need to consider API scopes. With OAuth 2.0, **scopes** are used to determine what permissions are granted.

App roles can be used to help achieve least-privilege access by enforcing **role-based access control** (**RBAC**).

Finally, you should consider what type of consent is necessary. Some permissions may require an administrator to provide consent. Review the permissions you wish to grant and whether they require admin consent and then plan on performing any necessary permissions consent during the configuration or deployment process.

Enterprise Integration

With Entra, you can configure groups to limit access to an application. Consider the role of the application in your organization and whether it should be available to all users or only a select group.

If developing or deploying a multitenant application, consider how **single sign-on** (**SSO**) and **federation** might be used, including what tenants or directories you want to allow access from.

Operational Best Practices

Configuring and deploying apps is only part of the story. Once you're up and running, you want to have processes in place to manage the ongoing security and governance of your apps. Things you should consider include the following:

- Use Entra ID logs and Microsoft Defender to track anomalies
- Automate secret and certificate rotation using Azure Key Vault
- Implement automation to ensure standardized deployments
- Enforce naming conventions
- Perform regular audits

Now that we have looked at the elements to consider when planning your enterprise application, we are ready to walk through the registration process. In the next section, we will create an enterprise app registration.

Creating App Registrations

Now that you understand what an app registration is and how you should be planning for an application, you will register an application in your Entra tenant. In the example, you will register a single-page web app:

1.   Sign in to the Microsoft Entra admin center (`https://entra.microsoft.com`), expand **Identity**, expand **Applications**, and select **App registrations**.

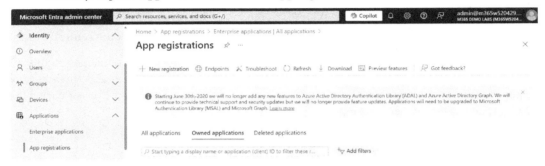

Figure 12.2: Creating a new registration

2.   Click **New registration**.

3.   Enter a name and select the supported account types. For this example, select **Accounts in any organizational directory (Any Microsoft Entra ID Tenant - Multitenant)** and click **Register**.

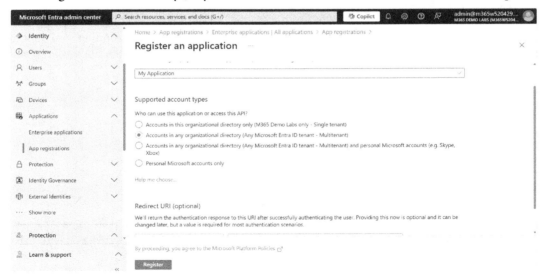

Figure 12.3: Registering an application

4. On the newly created application's **Overview** page, note the **Application (client) ID** value.
    Also, note that it's available for multiple organizations.

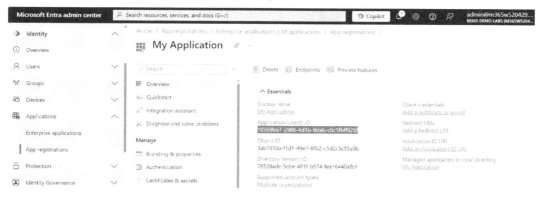

Figure 12.4: Creating the new application

At this point, a new application has been registered and a service principal has been created. However, there is no application currently associated with this registration, and there are no authentication parameters associated with this app. In the next section, we will examine the authentication options available and complete the registration of our application.

Configuring App Authentication

Configuring app authentication in Entra ID enables secure access control for applications by managing how users and services authenticate and obtain authorization. It defines authentication methods (OAuth, OpenID Connect, or SAML), assigns permissions, and enforces security policies such as MFA or Conditional Access. Proper configuration ensures secure access to resources while minimizing risk through least-privilege access and credential management.

In the following sections, we'll explore the authentication options associated with an app registration.

Understanding Authentication Protocols and the Redirect URI

Authentication and authorization are two fundamental concepts that govern access control to systems and resources. While they are closely related, they serve distinct purposes. In order to continue examining the app registrations, let's take a brief look at this process.

Authentication is the process of verifying the identity of a user, system, or entity attempting to access an application. It ensures that the person or system attempting to access resources is who they claim to be. This is typically done through mechanisms such as passwords, biometrics, or MFA, which combines something the user knows (password), something they have (a phone or token), or something they are (biometric data). Authentication is crucial because it prevents unauthorized access, helping to protect sensitive data and systems.

Authorization, on the other hand, is the process that determines what an authenticated user is allowed to do. Once a user's identity is verified through authentication, authorization controls what resources or operations they have permission to access. For example, a user might be authenticated to a system but only authorized to view certain files, not modify or delete them. This distinction between verifying identity (authentication) and controlling access (authorization) is key to maintaining a secure information environment.

When managing app registration in Azure, you will be primarily concerned with OAuth 2.0. OAuth is an authentication and authorization standard that is very popular and geared toward third-party access. This makes it perfect for today's ever-increasing number of social media applications.

> **Further reading**
>
> The OAuth standard is documented in RFC 6749 and RFC 6750. See `https://www.rfc-editor.org/rfc/rfc6749` and `https://www.rfc-editor.org/rfc/rfc6750` for more information.

A **redirect URI**, or **uniform resource identifier**, is a publicly accessible endpoint (such as a web page) where the user is redirected after being authenticated. It is sometimes also referred to as a redirection endpoint or a callback URL.

> **Cross-Origin Resource Sharing (CORS)**
>
> OAuth 2.0 supports many different flows. The particular flow type used in Entra SPAs is Authorization Code Flow with PKCE. PKCE is a modification to the out-of-the-box flow to mitigate the risk of a threat actor hijacking the authorization code and requesting an access token.

When creating a redirect URI, you should adhere to the following best practices and constraints:

- Use a secure protocol such as HTTPS
- Unlike URLs, URIs are case-sensitive
- The URI must match the path URL of the application
- Do not include the following special characters: !, $, ', (,), ;, and ,

At this point, we can continue with our application registration.

Finishing the Application Registration

In this section, we'll update the application registration with additional configuration parameters:

1. In the Entra admin center, navigate to the new app registration.
2. Under **Manage**, select **Authentication**.

Figure 12.5: Configuring authentication

3. Click **Add a platform**.

4. On the **Configure platforms** flyout, click the **Single-page application** tile, as shown in *Figure 12.6*.

Figure 12.6: Single-page application

5. Under **Redirect URIs**, enter a valid path. You can use a path such as `http://localhost` or `http://localhost/MyApp` for calls that never leave the local device or a public endpoint. See *Figure 12.7*.

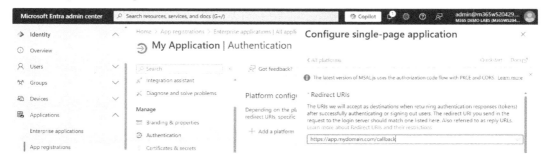

Figure 12.7: Configuring a redirect URI

6. Optionally, configure a front-channel logout URL. When an application is no longer being used, it is necessary to cancel any existing refresh tokens so that there is no chance a refresh token can be used to hijack the session.

Figure 12.8: Configuring the front-channel logout URL

7. When finished, scroll to the bottom of the flyout and click **Configure**.

By registering an application in Microsoft Entra and configuring authentication settings, you have ensured that only authorized users can access your application's resources.

In the following section, we will explore how to configure permissions for APIs, allowing you to control access to sensitive data and resources. This will enable you to further understand the app registration process.

Configuring API Permissions

When you register an application in Microsoft Entra, API permissions define what resources your application can access and what actions it can perform. These permissions are critical to controlling access to Microsoft APIs (such as Microsoft Graph) or non-Microsoft APIs (custom APIs), ensuring that applications only access the data they are authorized to use.

API permissions can be broken down into two broad categories: delegated permissions and application permissions.

Delegated Permissions

These permissions are used when an application is accessing resources on behalf of a signed-in user:

- The user's identity is delegated to the application via OAuth 2.0 authorization flows, meaning the app can act with the same permissions as the user.

- Consent is typically required from the user before the application can use these permissions. The scope of these permissions is limited to what the user can do themselves.

Next, we'll look at application permissions.

Application Permissions

These are used when the application needs to access resources without a signed-in user, acting by itself:

- Typically granted to service accounts or background services, these permissions allow apps to access data at a much broader level, often requiring admin consent because the app is not limited by the user's permissions. For example, a background process that runs reports on all users in an organization might require application-level permissions.

- Admin consent is mandatory because of the potential access level to all organizational resources.

API permissions in Microsoft Entra are governed by OAuth 2.0 scopes, which define the level of access the application is requesting. OAuth 2.0 is the authorization framework that manages the process of obtaining consent and issuing tokens (such as access tokens) that allow applications to interact with APIs securely.

Each API has its own set of scopes, representing granular levels of access. For example, a scope might be User.Read, indicating that an application can read user data, or Mail.Send, allowing it to send email on behalf of the user. Scopes are typically defined by the API provider and are used to limit the access an application has to sensitive resources. During the OAuth 2.0 authorization flow, an application requests specific scopes, and if the user or admin grants consent, the access token issued to the app will include those scopes.

When a user delegates permissions to an application, OAuth 2.0 ensures that the app can only act within the scope granted by the user. This means that the application is limited to accessing only the resources it has been explicitly authorized for.

In contrast, when an application requires broad access to resources, OAuth 2.0 scopes are used to secure this level of access. However, these scopes often require admin consent, which provides an additional layer of security and control over who can access sensitive data.

Overall, OAuth 2.0 scopes provide a flexible and secure way for applications to interact with APIs while also giving users and admins fine-grained control over the level of access they grant. This brings us to our next topic: scopes and grants. These concepts are important to understand and are necessary to truly understand how app permissions work.

A scope defines the level of access an application requests to a resource or API. It represents the specific permissions or actions that an application wants to perform on behalf of the user. Think of a scope as a permission that an application needs to request from the user or admin.

A grant is a way for an application to obtain an access token, which allows it to access a protected resource or API. A grant represents the actual permission granted to an application by the user or admin, based on the scopes requested.

When an application requests a scope (e.g., User.Read), the authorization server issues a grant in response, which includes an access token that grants access to the requested scope. The grant can be thought of as a ticket that allows the application to access the protected resource or API. The following steps provide an overview of the authorization process:

1. An application requests the User.Read scope from a user.

2. The authorization server grants the request and issues a grant that includes an access token with the User.Read scope.

3. The application uses the access token (grant) to read user data, within the limits of the User. Read scope.

Microsoft Entra uses OIDC for authentication while authorization is performed using OAuth 2.0. OAuth uses grants to enable an application to access data and other resources on behalf of a user. An application is designed to require (and request) access to certain data, and the user and/or administrator must consent to this access.

You can use the following process to add either delegated or API permissions scopes to an app registration:

1. Navigate to the app registration you created earlier.

2. Under **Manage**, select **API permissions**.

3. Click **Add a permission**, as shown in *Figure 12.9*.

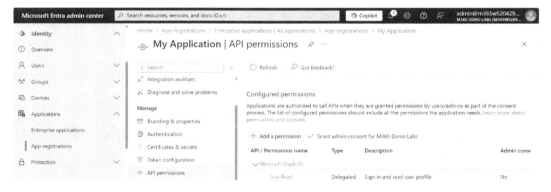

Figure 12.9: API permissions

4. Choose an API whose permissions you wish to grant.

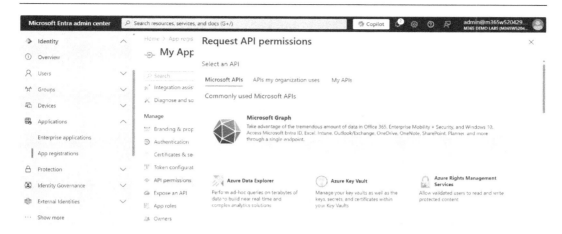

Figure 12.10: Selecting an API

5. Choose either **Delegated permissions** or **Application permissions**.

Figure 12.11: Choosing application or delegated permissions

6. Select the permissions to add and then click **Add permissions.**

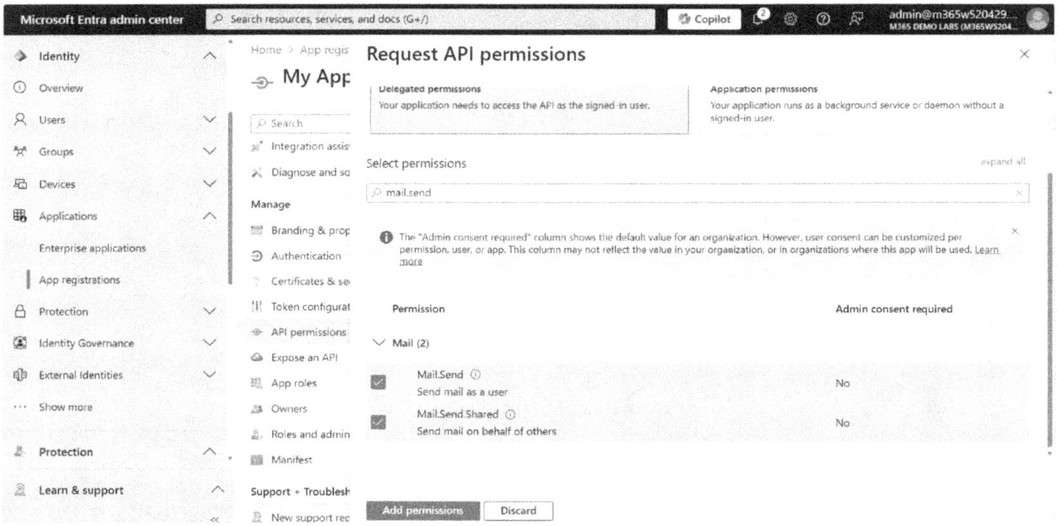

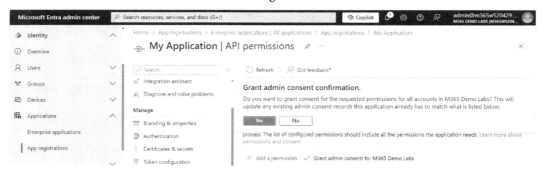

Figure 12.12: Selecting permissions

7. If necessary, click **Grant admin consent** to consent to the added permissions. Click **Yes** in the **Grant admin consent confirmation** dialog box.

Figure 12.13: Granting consent

Next, we'll look at creating app roles for managing the permissions that users can access.

Creating App Roles

In Microsoft Entra ID, an **app role** is a custom attribute associated with an application registration that defines a set of permissions or capabilities for users to access.

Think of an app role as a label that identifies a user's permission level within an application. App roles are used to grant specific privileges to users, such as the following:

- **Admin**: A high-level admin role with full access to the application
- **Editor**: A lower-level role with editing permissions but not full administrative rights
- **Viewer**: A read-only role that allows users to view data but not modify it

When a user is assigned to an app role, their permissions are automatically updated in the application.

To create an app role, use the following steps:

1. Navigate to an app registration.
2. Under **Manage**, select **App roles**.

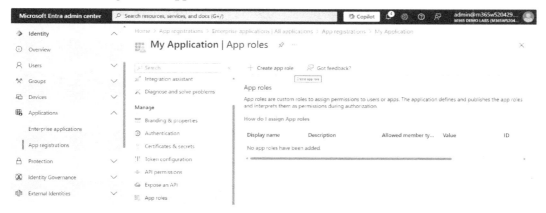

Figure 12.14: The App roles page

3. On the **Create app role** flyout, enter a display name.
4. Under **Allowed member types**, select what types of objects this role will impact.

Figure 12.15: Creating an app role

5. In the **Value** field, enter a value that will appear in the role's authorization data. For example, if your application code has authorization for a role called `SurveyData_Read`, you would enter `SurveyData_Read` in the **Value** field.

6. Enter a value in the **Description** field.

7. Click **Apply**.

This enables organizations to manage access and permissions across multiple applications using a single, unified identity system.

Once app roles are created, you can also assign them. To assign app roles to users, groups, or applications, follow these steps:

1. Navigate to the Entra admin center (`https://entra.microsoft.com`).

2. Expand **Identity**, expand **Applications**, and select **Enterprise applications**.

3. Under **Manage**, select **All applications**.

4. On the **Applications** page, locate the application that you created roles for and click it.

Figure 12.16: Editing an application

5. Under **Manage** for the application, select **Users and groups**.

6. Click **Add user/group** to add a user or group with a role assignment.

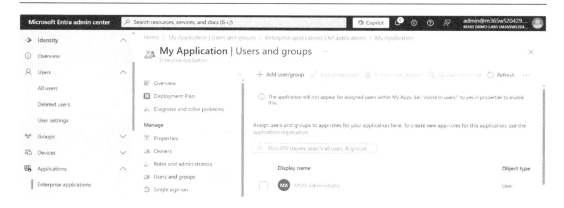

Figure 12.17: Adding a user or group assignment to the application

7. On the **Add Assignment** page, under **Users and groups**, click **None Selected** and choose one or more users or groups (depending on the scope of the role that you created).

8. On the **Add Assignment** page, under **Select a role**, click **None Selected**.

9. On the **Select a role** flyout, click the role to assign. You can only select one role per assignment.

Figure 12.18: Selecting roles

10. Scroll to the bottom of the flyout and click **Select**.

11. On the **Add Assignment** page, click **Assign**.

When working with a registration, using app roles can help with managing access and permissions. App roles provide you with a centralized approach to manage permissions across multiple apps at once. Additionally, they help prevent over-privileging. You can ensure that each user only has the necessary permissions for their specific role. Finally, app roles make it very easy to manage large numbers of users and applications. Without app roles, granting permissions across applications would require you to manually set up permissions for each app.

Summary

In this chapter, you have explored the process of registering your application in Microsoft Entra. You began by planning for a successful app registration, considering factors such as authentication protocols and API permissions.

Next, you walked through the process of creating an app registration in Microsoft Entra using the Microsoft Entra admin center. You then learned how to configure authentication protocols, including OAuth, OpenID Connect, and SAML, to secure your application's login experience.

You also explored how to configure API permissions, ensuring that your application has access to the necessary resources. Finally, you learned about app roles, which enable you to define custom roles within your application for granular permission control.

In the next chapter, you'll begin examining how Defender for Cloud Apps can be used to secure other SaaS apps.

Exam Readiness Drill – Chapter Review Questions

Apart from mastering key concepts, strong test-taking skills under time pressure are essential for acing your certification exam. That's why developing these abilities early in your learning journey is critical.

Exam readiness drills, using the free online practice resources provided with this book, help you progressively improve your time management and test-taking skills while reinforcing the key concepts you've learned.

HOW TO GET STARTED

- Open the link or scan the QR code at the bottom of this page

- If you have unlocked the practice resources already, log in to your registered account. If you haven't, follow the instructions in *Chapter 19* and come back to this page.

- Once you log in, click the START button to start a quiz

- We recommend attempting a quiz multiple times till you're able to answer most of the questions correctly and well within the time limit.

- You can use the following practice template to help you plan your attempts:

Working On Accuracy		
Attempt	Target	Time Limit
Attempt 1	40% or more	Till the timer runs out
Attempt 2	60% or more	Till the timer runs out
Attempt 3	75% or more	Till the timer runs out
Working On Timing		
Attempt 4	75% or more	1 minute before time limit
Attempt 5	75% or more	2 minutes before time limit
Attempt 6	75% or more	3 minutes before time limit

The above drill is just an example. Design your drills based on your own goals and make the most out of the online quizzes accompanying this book.

First time accessing the online resources? 🔒

You'll need to unlock them through a one-time process. **Head to** *Chapter 19* **for instructions**.

Open Quiz	
`https://packt.link/sc300ch12` OR scan this QR code →	

13

Managing and Monitoring App Access Using Microsoft Defender for Cloud Apps

Cloud apps offer many benefits such as scalability, flexibility, and cost savings. However, this shift toward the cloud also brings new security challenges that require careful consideration. One of the most significant challenges is managing and monitoring app access to prevent unauthorized use, data breaches, and compliance violations. This requires implementing strong security controls and policies that govern who can access what resources and under what conditions. Traditional network- or perimeter-based security models are no longer sufficient in the cloud era since users can access apps from anywhere, anytime, and on any device.

In this chapter, we will explore how **Microsoft Defender for Cloud Apps** (**MDCA**; formerly known as Cloud App Security) can help organizations address these challenges and secure their cloud environment. MDCA is a **cloud access security broker** (**CASB**) solution that provides visibility, control, and protection for cloud applications. It enables organizations to discover shadow IT, monitor user activity, and enforce policies across multiple cloud services. By using MDCA, organizations can gain insights into their cloud usage, identify potential threats, and take action to mitigate risks.

The objectives and skills we'll cover in this chapter include the following:

- Configuring and analyzing cloud discovery results by using Defender for Cloud Apps
- Configuring connected apps
- Implementing application-enforced restrictions
- Configuring Conditional Access app control
- Creating access and session policies in Defender for Cloud Apps
- Implementing and managing policies for OAuth apps
- Managing the cloud app catalog

By the end of this chapter, you should be able to effectively manage your organization's cloud application security posture using the power of Defender for Cloud Apps. You will do this by configuring connected apps to ensure deep visibility into application activity that might affect your security posture.

Configuring and Analyzing Cloud Discovery Results by Using Defender for Cloud Apps

Cloud discovery is an important step in securing an organization's cloud environment as it helps identify shadow IT and unsanctioned apps that may pose a security risk. By discovering these apps, organizations can gain visibility into their cloud usage and take appropriate actions to manage and secure them.

Let's talk about a common misconception in the world of technology management. When IT administrators are asked to estimate how many cloud applications their employees use, they often provide a number that is far lower than the actual number. Research shows that the actual number of separate cloud apps being used by employees is much higher than expected. This phenomenon is known as **shadow IT**.

So, what exactly is shadow IT? It refers to the use of unauthorized software applications and services by employees without the knowledge or approval of the IT department. These apps may be installed on personal devices or accessed through web browsers, allowing employees to bypass traditional security measures. The issue with shadow IT is that it can pose significant risks to an organization's security and compliance. When employees access cloud apps from outside the corporate network, they can bypass traditional security measures such as firewalls.

> **Data Points**
>
> A Gartner industry survey (`https://www.cio.com/article/234745/how-to-eliminate-enterprise-shadow-it.html`) found that shadow IT accounts for up to 40% of IT spending in enterprise organizations. In 2024, IBM published a report showing that shadow IT services (and shadow data repositories) amounted to 35% of data breaches (`https://www.ibm.com/think/insights/hidden-risk-shadow-data-ai-higher-costs`).

The bottom line is that organizations need to adapt their security strategies to account for shadow IT by using tools such as cloud discovery reports.

The first step to discovering shadow IT in your organization is identifying the cloud apps that are being used by employees, often without the knowledge or approval of the IT department.

Setting Up Cloud Discovery

Microsoft Defender for Cloud Apps (MDCA) is a cloud service within Microsoft 365 that provides cloud access security broker (CASB) services. A CASB is used as a policy enforcement point between the consumers and the providers so that applications adhere to the baseline security requirements of the company. MDCA provides these capabilities for Microsoft, third-party cloud, and registered on-premises applications.

MDCA uses machine learning algorithms and behavioral analytics to detect anomalous activities that could indicate a security breach. It also provides real-time alerts and reporting capabilities that enable organizations to respond quickly to security incidents. Additionally, it integrates with other Microsoft security solutions, such as Microsoft Entra ID and Microsoft Intune, to provide a comprehensive security posture for the organization's cloud environment.

MDCA will identify all applications that are being accessed using a combination of Entra ID and Microsoft Defender for Endpoint. These applications are then reported on the **Discovery** dashboard. Knowing the applications that are being used and accessed allows us to plan for authorized applications and block unauthorized applications.

For companies that have users in office locations with a firewall or a device that is able to log network traffic, MDCA has discovery capabilities that allow you to connect these logs, and it will create a discovery report that lists the applications that are being accessed.

The workflow and architecture of MDCA is shown in *Figure 13.1*:

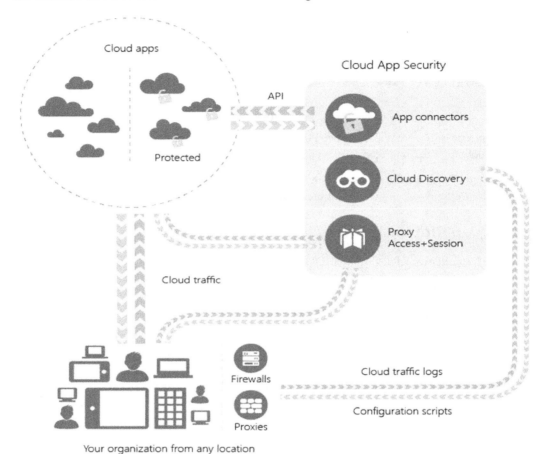

Figure 13.1: MDCA architecture

Traffic logs from on-premises firewalls will provide a snapshot report on the most common applications and the users that are accessing these apps. Traffic from managed devices with Microsoft Intune will be fed into the MDCA **Cloud Discovery** overview dashboard, as shown in *Figure 13.2*:

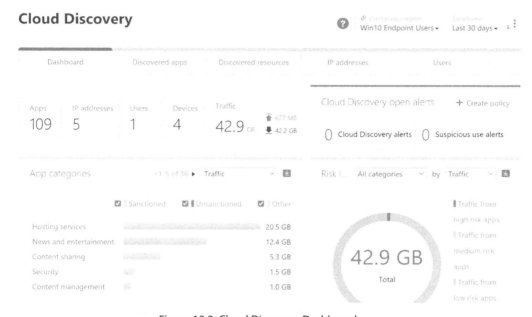

Figure 13.2: Cloud Discovery Dashboard

This information is the basis for planning and monitoring your application use within your organization. You can build documentation on user usage and habits, and then evaluate the legitimate need for various applications to be used on company resources. For additional information on cloud app discovery, please visit this link: `https://docs.microsoft.com/en-us/cloud-app-security/set-up-cloud-discovery`.

Once we have determined the applications that we are going to allow as a company, we can create a plan and enforce policies utilizing MDCA. *Figure 13.3* shows the architecture of monitoring applications within Entra ID. This architecture simplifies the management and monitoring of applications through MDCA:

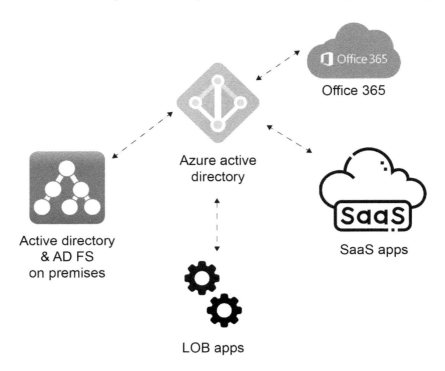

Figure 13.3: AD/Entra ID and cloud apps managed by MCAS

When this architecture is used, the company can utilize the various security, compliance, and governance tools available within Entra ID, Microsoft 365, and Azure for the cloud and on-premises **line-of-business (LOB)** applications, such as MDCA, Conditional Access policies, and Entra ID Protection.

MDCA provides various discovery methods, including log collection from firewalls and proxies, API connectors, and agent-based discovery. In this section, we will discuss how to configure these discovery methods and analyze the results to identify potential security risks. We will also cover how to categorize discovered apps based on their risk level and create policies to manage them effectively.

Configuring Cloud Discovery

In this lab walk-through, we will demonstrate how to configure and analyze cloud discovery results using MDCA. This lab assumes that you have access to a Microsoft Azure subscription with the necessary permissions to create and manage resources.

1. Navigate to the Microsoft 365 Defender portal (`https://security.microsoft.com`).

2. Expand **Cloud apps** and then select **Cloud discovery**.

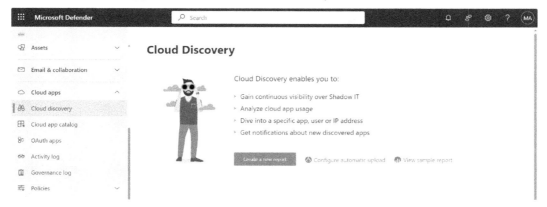

Figure 13.4: Cloud apps portal

3. Click on **Create a new report**.

4. On the **Create new Cloud Discovery snapshot report** page, click **Next**.

5. Enter a report name. Optionally, you can enter a description.

6. Select **Squid (Native)**.

7. Click **View log format** and then click **Download sample log** to obtain a sample log file that can be used for this exercise.

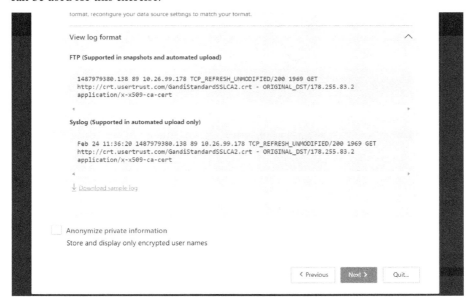

Figure 13.5: Downloading a sample log file

> **Note**
>
> Cloud discovery is limited to processing logs that are 90 days old. Some of the samples downloaded may contain data with timestamps older than 90 days, which will prevent cloud discovery from processing the files. You may need to update the timestamp's value column in your log file for it to be processed correctly.

8. Extract the downloaded ZIP file.

9. Click **Next**.

10. Navigate to and upload the sample log files provided. Click **Upload Logs**.

11. Wait for the snapshot report to be generated.

Figure 13.6: Report being generated

It will likely take some time before the report is completed. It could be up to 24 hours, but it's usually less. If you used the demo report log data provided in *step 7*, it will likely take about 30-40 minutes.

After the snapshot report has been generated, you can view it in the Microsoft Defender portal under **Settings | Cloud apps | Cloud Discovery | Snapshot reports**, shown in *Figure 13.7*.

Figure 13.7: Processing a snapshot report

Once the **Status** column has changed from **Processing** to **Ready**, you can review the results. Click on the report to display the results. The report's dashboard is displayed in *Figure 13.8*:

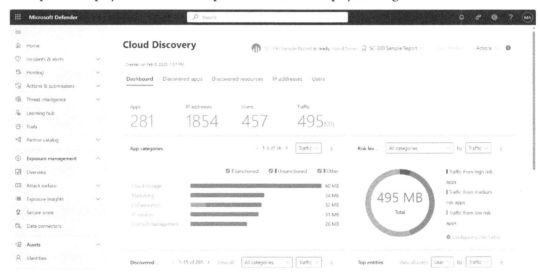

Figure 13.8: Discovering apps

The **Dashboard** tab displays high-level information about the ingested log file, including app traffic categories and the data volume transmitted, segregated by the apps' corresponding risk categories. Scrolling to the bottom of the dashboard reveals additional summary tiles, such as the discovered apps and users involved in transactions.

Figure 13.9: Viewing additional data on the dashboard

Each of the tabs along the top of the page provides detailed data for the tile snapshots.

In the next section, we'll look at ways you can parse the information.

Analyzing Cloud Discovery Results

Now that you have set up cloud discovery and have uploaded your first set of log files, let's explore how to analyze the results. You'll want to do the following:

- **Create custom reports**: Since policies vary across different user groups, regions, and business units, you might need to create separate shadow IT reports for each of these entities.

- **Review cloud app usage**: Look at the reports generated by cloud discovery and examine the dashboard for a comprehensive view of app usage. It's a good idea to categorize apps to identify non-sanctioned apps being used for legitimate work-related purposes.

To better understand the risks associated with each discovered app, you can use the Defender for Cloud Apps catalog, which assesses over 31,000 apps based on 90 risk factors. These factors range from general information about the app to security measures and controls.

When reviewing and analyzing the results of cloud discovery, you'll want to begin by understanding your organization's security policies, your organization's level of risk tolerance, regulatory or compliance requirements, and what applications or third-party services have been authorized for use. Then, you can use the filtering capabilities of the Cloud Discovery snapshot to identify apps that meet or fall outside of your organization's written policies and craft a technology policy through **sanctioning** or **unsanctioning** apps to align accordingly. Follow these steps:

1. Navigate to the Microsoft Defender portal (`https://security.microsoft.com`).

2. Expand **Cloud apps** and then select **Cloud discovery**.

3. Select the **Discovered apps** tab, as shown in *Figure 13.10*.

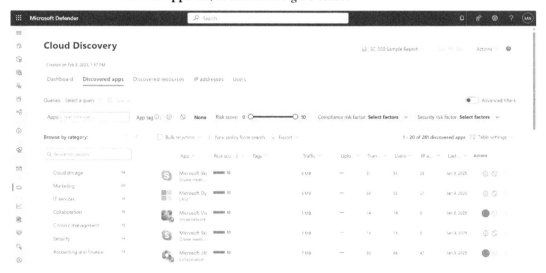

Figure 13.10: Discovered apps

4. Filter for apps.

    You can use the pre-canned **Queries** to identify applications using a variety of criteria. See *Figure 13.11*.

Figure 13.11: Viewing default queries

You can also use the **Advanced filters** toggle to identify apps with specific risk scores or characteristics.

Figure 13.12: Viewing advanced filters

5. Select an app name and then click on the **Info** tab to view details about its security risk factors.

Figure 13.13: Viewing data about risk factors

6. Expand the ellipsis (**...**) at the top of the page and view the actions that you can take, including tagging an app as **Sanctioned** or **Unsanctioned**.

Figure 13.14: Viewing actions

By following these steps, you can gain a better understanding of the cloud apps being used within your organization, assess their potential risks, and prepare to take actions to prohibit them from being used.

> **Further Reading**
>
> After tagging apps as **Unsanctioned**, you can download scripts from Defender for Cloud Apps to use with network appliances. These scripts can be used to configure a variety of proxy and firewall devices to prevent clients from accessing unsanctioned apps. For more information, see `https://learn.microsoft.com/en-us/defender-cloud-apps/governance-discovery`.

If you deploy Microsoft 365 Defender for Endpoint, you can also enable the blocking of unsanctioned apps by selecting **Settings | Cloud apps | Microsoft Defender for Endpoint** and then selecting **Enforce app access**, as shown in *Figure 13.15*:

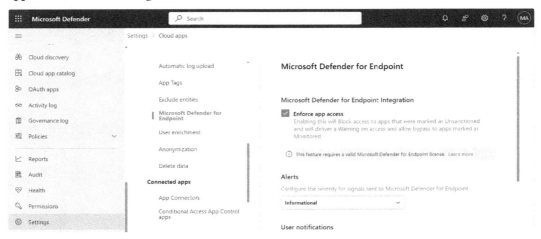

Figure 13.15: Enforcing Microsoft Defender for Endpoint app access integration

Now that you have analyzed a discovery report, you can see the wealth of information available as to app usage within your enterprise. Let's move on to controlling the apps and new ways to limit and control them.

Configuring Connected Apps

In order to keep an eye on cloud apps, MDCA uses special helpers called **app connectors**. These connectors use APIs that are provided by the cloud app and service vendors.

App connectors use these APIs to give MDCA more visibility into what's happening with the connected apps. This means that MDCA can get a better understanding of how the apps are being used, which helps it make decisions about security and policy.

Each service has its own set of rules about how APIs can be used. These rules are called **limitations**. For example, some services might limit how many requests an app can receive in a certain amount of time (**rate limiting**) or how much data can be transferred over a period of time (**throttling**). Others might have different windows of time when they're available to communicate.

MDCA works with each service to make sure it's using the APIs efficiently. This means that it takes into account all the rules and limitations while using only the allowed capacity. Some tasks, such as scanning all files in a tenant, might need to use many APIs over a long period of time.

You can use the following process to connect to apps.

1. Navigate to the Microsoft Defender portal (`https://security.microsoft.com`).

2. Click **Settings** and then select **Cloud apps**.

3. Under **Connected apps**, select **App Connectors**.

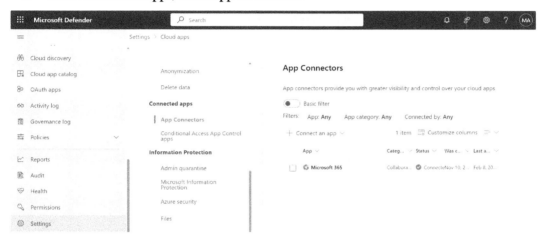

Figure 13.16: Managing OAuth apps

4. Click **Connect an app** and then select an app from the list.

Figure 13.17: Selecting a new app to configure

5. Walk through the wizard to connect the app. You'll typically need to provide some sort of credential, certificate, or API key. In this case, we'll connect a Google Workspace instance as an example. Type a value in the **Enter instance name** field that describes the connector and click **Next**.

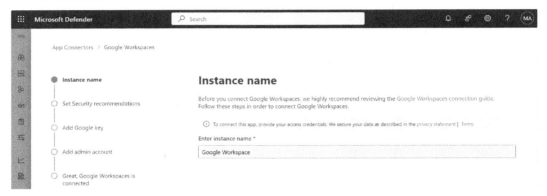

Figure 13.18: Connecting a new cloud app

6. On the **Set Security recommendations** page, select the **Security recommendations** checkbox to allow MDCA to send data to Microsoft Security Exposure Management to get security updates and recommendations from the app connector. Click **Next**.

7. Add the security details required and click **Next**.

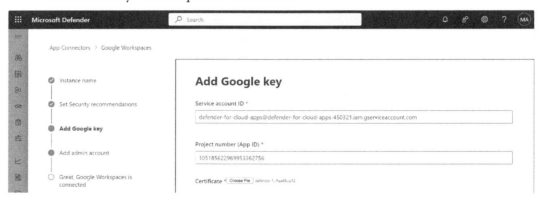

Figure 13.19: Adding authentication data

8. On the **Add admin account** page, enter the admin account for your Google Workspace environment and click **Next**.

9. Click **Done**.

Once the app has been successfully connected, the **Status** column will be updated. See *Figure 13.20*.

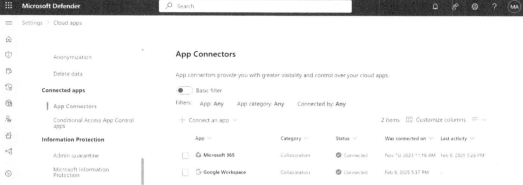

Figure 13.20: Reviewing a newly connected app

Connected apps are automatically sanctioned (since you've added them to your environment).

Next, we'll shift focus to implementing restrictions on cloud apps.

Implementing Application-Enforced Restrictions

As discussed previously, enterprises are increasingly reliant on cloud applications to enhance productivity and collaboration. However, this reliance also introduces new security challenges, such as data breaches and unauthorized access. MDCA offers a solution through application-enforced restrictions, which act as customizable security policies within cloud applications. These restrictions enable organizations to control user activities, safeguard sensitive information, and ensure compliance with regulatory standards.

You can use application-enforced restrictions to block or limit access to Exchange, SharePoint, or OneDrive from unmanaged devices. Application-enforced restrictions require two separate configurations:

- A Conditional Access policy configured with the app-enforced restrictions control
- App configurations in SharePoint and Exchange to control experiences for unmanaged devices

Let's take a look at both of those areas next.

Creating an Application-Enforced Restrictions Policy

The simplest way to configure this feature is by using a Conditional Access policy template. To configure the template, follow these steps:

1. Navigate to the Entra admin center (`https://entra.microsoft.com`).

2. Expand **Identity**, expand **Protection**, and then select **Conditional Access**.

3. On the **Overview** blade, click **Create new policy from templates**.

4. On the **Select a template** tab, enter O365 in the search box.

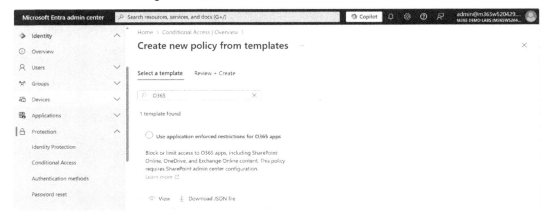

Figure 13.21: Creating a new policy from a template

5. Select the radio button in the **Use application enforced restrictions for O365 apps** card. Click **Review + create**.

6. Click **Create**.

The template will automatically exclude the account you use to create the policy from the application scope and will configure the policy in report-only mode.

Next, you'll need to configure the services for the behaviors you want to enforce. First, we'll configure SharePoint.

Configuring App-Enforced Restrictions in SharePoint

SharePoint Online provides three options for access control. To select an option, follow these steps:

1. Navigate to the SharePoint admin center. Go to `https://admin.microsoft.com`, expand **Admin centers**, and select **SharePoint**.

2. Expand **Policies** and select **Access control**.

3. Click **Unmanaged devices**.

Figure 13.22: Selecting Unmanaged devices

4. Select one of the available options, as described after the following figure:

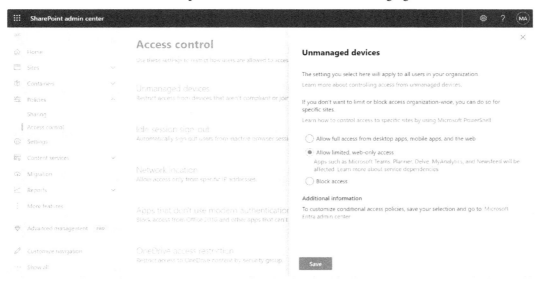

Figure 13.23: Selecting an app experience

- The default option, **Allow full access from desktop apps, mobile apps, and the web**, enables any device (whether managed or not) to have the full experience

- The **Allow limited, web-only access** option means that users can only access data through a web browser and are restricted from downloading, printing, or syncing files to their devices

- **Block access**, the final option, prevents unmanaged devices from accessing the service

5. Click **Save**.

After the policy has been updated, unmanaged devices will adhere to the combined policy enforcement actions of SharePoint and the Conditional Access app-enforced restriction policy.

In addition to managing access tenant-wide, you can also update the settings of individual sites or OneDrive using the SharePoint Online Management Shell module. For example, to enable limited access to the Marketing SharePoint site, use the following command:

```
Set-SPOSite -Identity https://tenant.sharepoint.com/sites/Marketing
-ConditionalAccessPolicy AllowLimitedAccess
```

> **Further Reading**
>
> For more examples and options, see https://learn.microsoft.com/en-us/sharepoint/control-access-from-unmanaged-devices.

Next, we'll look at the steps necessary for configuring Exchange Online to support the app-enforced restrictions.

Configuring App-Enforced Restrictions in Exchange

You can prevent users on unmanaged devices from downloading email attachments in Outlook on the web (formerly known as **Outlook Web App**, or **OWA**) and the new Outlook for Windows. However, they can still view and edit these files using Office Online without saving them locally. Additionally, you can block users from accessing attachments entirely in these Outlook versions on unmanaged devices.

These restrictions are managed through mailbox policies in Exchange Online. Every Microsoft 365 organization with Exchange Online mailboxes includes a default policy called **OwaMailboxPolicy-Default**, which is automatically applied to all users. You can also create custom policies to enforce different restrictions for specific user groups.

To configure these restrictions, you must use the Exchange Online PowerShell module. In the following example, the default mailbox policy will be updated to block both the reading and downloading of attachments:

```
Set-OwaMailboxPolicy -Identity "OwaMailboxPolicy-Default"
-ConditionalAccessPolicy ReadOnlyPlusAttachmentsBlocked
```

> **Further Reading**
>
> For more examples, see https://learn.microsoft.com/en-us/security/zero-trust/zero-trust-identity-device-access-policies-exchange.

After the policy has been updated, unmanaged devices will adhere to the combined policy enforcement actions of Exchange and the Conditional Access app-enforced restriction policy.

Configuring Conditional Access App Control

Configuring Conditional Access app control within MDCA plays an important role in an enterprise security strategy. It allows organizations to monitor and control user activities in real time, ensuring that only authorized users can access sensitive data under defined conditions.

This aligns with the zero-trust security model, which operates on the principle of "never trust, always verify." By enforcing real-time monitoring and control, organizations can prevent unauthorized access and data exfiltration.

For example, let's say you want to configure a policy to prevent or block downloads from devices outside your network. To create this Conditional Access app control policy, follow these steps:

1. Navigate to the Entra admin center (https://entra.microsoft.com).

2. Expand **Identity**, expand **Protection**, and then select **Conditional Access**.

3. On the **Overview** blade, click **New policy**.

4. Enter a name.

5. Under **Assignments**, select the users (or groups) to whom the new policy will apply.

6. Under **Assignments**, select the target resources to which the new policy will apply. You can select **All resources (formerly 'All cloud apps')** as a starting point but may want to exclude the admin portals using the **Microsoft Admin Portals** resources object.

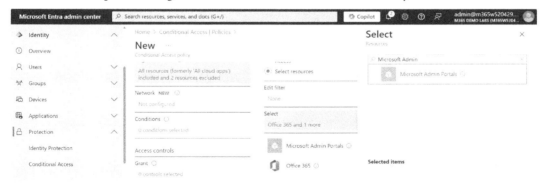

Figure 13.24: Selecting cloud resources

7. Under **Assignments**, specify the conditions under which this policy will apply. In the case of blocking downloads on devices outside the network, select the **Locations** condition and then click the option to exclude all trusted networks and locations, as shown in *Figure 13.25*:

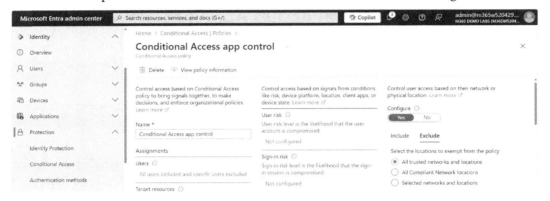

Figure 13.25: Excluding locations from the policy

8. Under **Access controls**, add the **Use Conditional Access App Control** option. Select either **Monitor only (Preview)**, **Block downloads (Preview)**, or **Use custom policy….** The **Use custom policy…** option requires a supported SAML application.

Figure 13.26: Configuring the Session control

9. Click **Select**.

10. Choose a policy enablement mode (**Report-only**, **On**, or **Off**).

11. Click **Create**.

Session policies such as Conditional Access app control take effect immediately.

Creating Access and Session Policies in Defender for Cloud Apps

MDCA provides Conditional Access control policies. You can create session or access policies to manage any application or device that supports Conditional Access policies. Access policies are configured on a per-application basis and enable you to define what devices can gain access to applications, and how they can access the app.

For example, you can create an access policy that requires only managed devices to access a specific application, so if you attempt to authenticate using an unmanaged device, the attempt will fail. Similarly, session policies will define what actions can be performed under various circumstances. For example, you can require that data that is classified as sensitive cannot be downloaded on unmanaged devices. We will now look at the details for each policy type individually.

A **session policy** is essentially a set of rules that govern how users interact with cloud applications while they're logged in. It's like setting up a security checkpoint that monitors and controls what users can do within the app. For example, you might want to prevent users from downloading sensitive files or uploading files that contain malware.

To create an access policy, you start by defining who the policy applies to—this can be specific users, groups, or even entire departments. Then, you choose what conditions must be met for the user to gain access to the app.

One common condition is the **Client app** filter, which specifies whether a user can access the app from a browser session or a mobile and desktop client app. For example, you might want to block users from accessing sensitive data in a cloud application from their personal devices or unmanaged browsers.

MDCA policies are based on Conditional Access policies—and take into consideration the conditions that are specified. That means that access to an app is only granted if certain conditions are met—for instance, if a user has multi-factor authentication enabled or if they're accessing the app from a specific IP address.

When you set up an access policy, you can also specify what actions should be taken when a user's session doesn't meet the required conditions. For example, you might want to block the user's access entirely or redirect them to a different page with more information about why they were blocked.

One key benefit of access policies is that they allow for fine-grained control over who can access cloud applications and under what conditions. This helps organizations ensure that sensitive data is only accessible by authorized users and devices.

Access policies also work in conjunction with session policies, which we discussed earlier. When a user tries to perform an action within a cloud app (such as uploading a file), the access policy checks whether they're allowed to access the app at all, while the session policy monitors their activities once they're inside.

In summary, access policies are like a gatekeeper that controls who can enter a cloud application and under what conditions. By setting up access policies, organizations can ensure that sensitive data is only accessible by authorized users and devices, adding an extra layer of security to their cloud applications.

Here's an example to illustrate how access policies work: Let's say you're the administrator for a company that uses Microsoft 365. You want to make sure that employees from the finance department can only download sensitive financial documents in SharePoint Online when they're using a compliant device. To do this, you create an access policy that grants access to SharePoint Online only to users who are part of the finance department and are either accessing from a hybrid-joined device or are on a device marked as compliant.

If someone from the marketing department tries to access SharePoint Online without meeting these conditions, they'll be blocked by the access policy. However, if a user from the finance department meets all the requirements, they'll be granted access to SharePoint Online and can download documents.

To create an access policy using your organization's process, follow these steps:

1. Navigate to the Entra admin center (`https://entra.microsoft.com`).

2. Expand **Identity**, expand **Protection**, and then select **Conditional Access**.

3. Under **Manage**, select **Authentication contexts**.

4. Click **New authentication context**.

Figure 13.27: Creating an authentication context

5. On the **Add authentication context** flyout, enter a name. Leave the rest of the defaults and click **Save**.

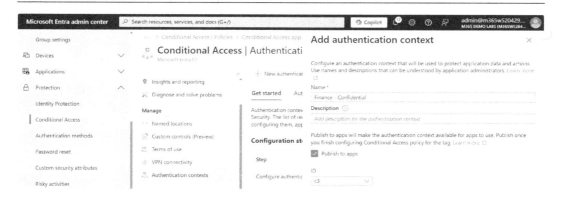

Figure 13.28: Adding the authentication context

6. Connect to SharePoint Online using the SharePoint Management shell.

7. Run the following command to apply the newly created authentication context to a SharePoint site, replacing the values for the site and authentication context with your own:

```
Set-SPOSite -Identity https://m365w520429.sharepoint.com/
sites/finance -ConditionalAccessPolicy AuthenticationContext
-AuthenticationContextName "Finance - Confidential"
```

8. Navigate to the Microsoft Defender portal (`https://security.microsoft.com`).

9. Expand **Cloud apps**, expand **Policies**, and select **Policy management**.

10. Click **Create policy** and then select **Access policy**.

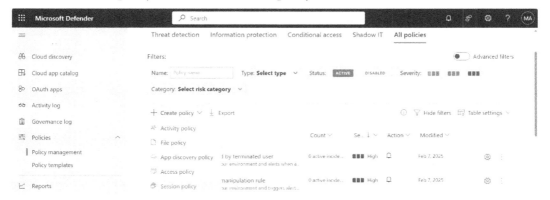

Figure 13.29: Creating a new session policy

11. Enter a policy name.

12. Under **Session control type**, select **Control file download (with inspection)**.

13. Under **Activities matching all of the following**, configure the following filters:

- **Device | Tag | does not equal | Intune compliant, Microsoft Entra Hybrid joined**
- **App | equals | Microsoft SharePoint Online**

Figure 13.30: Configuring an activity filter

14. Under **Actions**, ensure the **Block** action is selected.

15. Click **Create**.

To test the policy, log in as a user who has access to the restricted site but on a device that does not meet the requirements to download.

Notice, as displayed in *Figure 13.31*, that you receive a notification that SharePoint is protected through Defender for Cloud Apps. You can also examine the URL bar to see that the request has been approved.

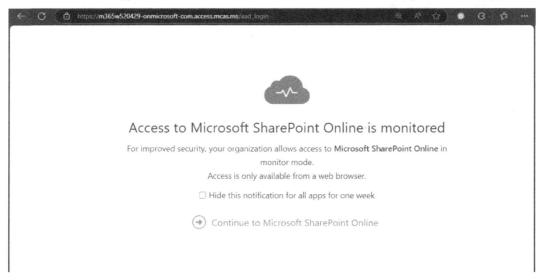

Figure 13.31: Defender for Cloud Apps splash page

After continuing to the site, you'll also notice a banner is displayed, indicating that certain activities are prohibited. See *Figure 13.32*.

Figure 13.32: Warning banner

Access and session policies in MDCA are important tools that help organizations secure their cloud applications. Since it is yet another tool you can use to secure your cloud applications, it's essential to learn how to effectively use policies to augment your company's overall security policy.

Next, we'll look at managing OAuth policies using Defender.

Implementing and Managing Policies for OAuth Apps

OAuth app policies let you review requested permissions and user authorizations for Microsoft 365, Google Workspace, and Salesforce. You can approve or ban these permissions, with banned ones disabling the associated enterprise application.

You might use an OAuth policy to track a new, uncommon app that attempts to use permissions on behalf of a user or requests risky permissions. Upon detection, you can generate alerts or disable the application altogether.

To create an OAuth app policy to disable an uncommon OAuth app that requests a high level of permissions, follow these steps:

1. Navigate to the Microsoft Defender portal (`https://security.microsoft.com`).

2. Expand **Cloud apps**, expand **Policies**, and select **Policy management**.

3. Click **Create policy** and select **OAuth app policy**.

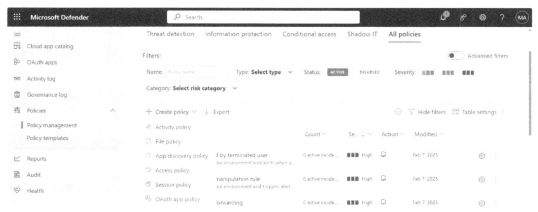

Figure 13.33: Creating a new OAuth app policy

4. Give the policy a meaningful name, such as `Risky App Policy`.

5. Under **App selection**, select a service to which to apply the OAuth policy.

6. Under **Create filters for the policy**, click **Add a filter** and select **Permission level | equals | Medium severity** and **High severity**.

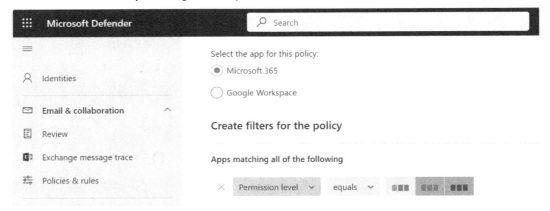

Figure 13.34: Configuring app filter parameters

7. Under **Governance actions**, expand **Revoke Office OAuth app** and select the **Revoke Office OAuth app** checkbox.

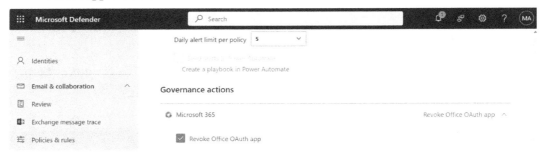

Figure 13.35: Configuring Governance actions

8. Click **Create.**

In addition to detecting a variety of trigger conditions, you can also use an OAuth policy to execute a playbook in Power Automate. Power Automate playbooks can be used to further automate actions, such as disabling user accounts, posting messages to Teams channels, or working with any other connectors and apps.

Managing the Cloud App Catalog

Microsoft's catalog of cloud apps provides you with a wealth of information on some of the most commonly used applications. As with other components of a security strategy, the risk introduced by various apps will differ from enterprise to enterprise. In order to accommodate this variability, it's important to understand how you can manage the cloud app catalog to more accurately reflect your company's particular requirements. This involves regularly reviewing the catalog to ensure that all authorized apps are properly tracked and monitored. You'll also need to keep an eye on the catalog's roadmap to stay informed about upcoming additions and updates.

If you notice that an app isn't currently included in the catalog, don't worry—there are options available. By default, Defender for Cloud Apps cannot discover apps that aren't in the catalog. However, you can check the roadmap to see whether the app is planned for future inclusion, suggest a new app for addition to the catalog, or even create a custom app to meet your organization's unique needs.

You can use the cloud app catalog to sanction or unsanction apps, as well as create new policies based on your selections.

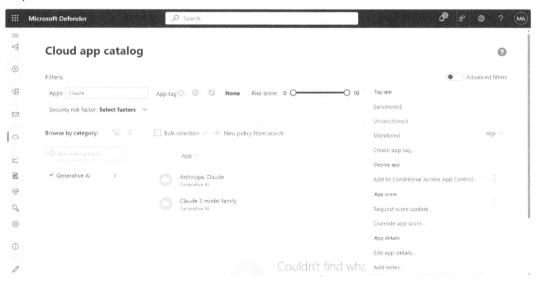

Figure 13.36: Viewing the cloud app catalog

Next, we'll explore the Microsoft Defender risk score.

Calculating the Risk Score

When calculating risk scores, Defender for Cloud Apps takes a comprehensive approach that combines regulatory certifications, industry standards, and best practices. This ensures that each app's score accurately reflects its maturity level in terms of enterprise use. The total score is essentially a snapshot of an app's reliability, compiled from various subscores that fall into distinct risk categories. These subscores are weighted to provide a comprehensive view of an app's strengths and weaknesses. An example of a high-risk app is depicted in *Figure 13.37*.

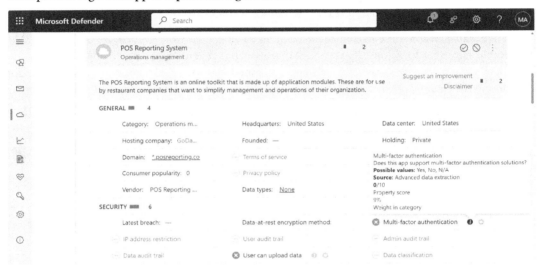

Figure 13.37: Viewing an app's risk score

The risk categories used to assess an app's reliability are carefully chosen to cover critical areas that impact security and data protection. For instance, the **Data Loss Prevention** category evaluates how effectively an app safeguards sensitive information. Another important area is **User and Account Management**, which examines the security measures in place for user accounts and authentication processes. Additionally, the **Device Security** and **Session and Authentication** categories scrutinize the app's ability to ensure device security and maintain a secure login process, respectively. By evaluating these key areas, Defender for Cloud Apps provides a thorough risk assessment that helps organizations make informed decisions about which apps to use and how to protect their cloud environment.

When evaluating an app's risk score, multiple properties within each risk category are considered. These properties can be either simple true or false values, such as whether the app has a certain compliance certification, or continuous properties such as the age of the domain. Each property is assigned a preliminary score based on its value, and these scores are then weighted against all other existing fields in the category to create a subscore.

To keep risk scores current, Defender for Cloud Apps uses a combination of automated data extraction, advanced algorithms, and continuous analysis by an expert team. For example, the app's attributes such as compliance certifications and terms of service are extracted automatically, while algorithms analyze HTTP security headers and other attributes. Additionally, customers can request updates to the cloud app catalog, which are reviewed by the analysis team. This ensures that risk scores stay up to date and accurately reflect an app's reliability.

Occasionally, you may encounter an unscored app, which usually means that its properties have not been evaluated yet or are unknown. However, with Defender for Cloud Apps' comprehensive approach to risk scoring, you can make informed decisions about which apps to use and how to secure your organization's cloud environment.

Customizing the Risk Score

By default, all the different parameters evaluated by Defender are given equal weight. But what if some factors matter more to your company than others? For instance, maybe you're particularly concerned about data protection or user authentication.

To adjust these weights, you can go into the settings of Microsoft Defender and make changes to how important each parameter is. Think of it like setting a priority list for your company's online security needs. You can slide a slider from **Ignored** all the way up to **Very High**, depending on how critical that particular factor is to you.

One other thing to keep in mind is what happens when there's no information available for a certain parameter—maybe it's not applicable or it's just plain missing. By default, these unknown values can actually hurt your score, but you have the option to check a box and tell Defender to ignore them if they're not relevant or not available. This way, you get a more accurate picture of each app's risk level based on the information that does matter to your organization. Let's look at that next.

Overriding the Risk Score

There may be instances where you need to override the risk score of a specific app without changing its weighted scoring methodology. This is particularly useful when you want to immediately reflect your organization's stance on a particular app, even if its risk score doesn't accurately represent that.

For instance, consider a scenario where an application has been assigned a risk score of 7, yet it is actually a vital tool for your organization, crucial for meeting business objectives or addressing specific security concerns. In this situation, you may wish to adjust the risk assessment solely for that particular app, assigning it a rating of 10 to reflect your company's standpoint.

Let's walk through how to do that:

1. Navigate to the Microsoft Defender portal (`https://security.microsoft.com`).

2. Expand **Cloud apps** and then select **Cloud app catalog**.

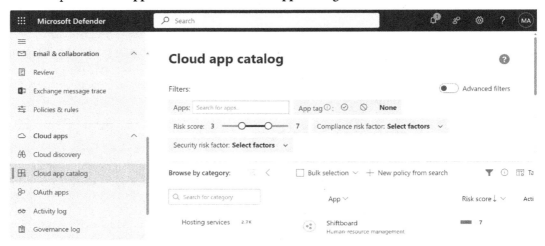

Figure 13.38: Cloud app catalog

3. Locate an app whose risk score you wish to modify.

4. Click the ellipsis to expand the app's context menu and select **Override app score…** from the list of options.

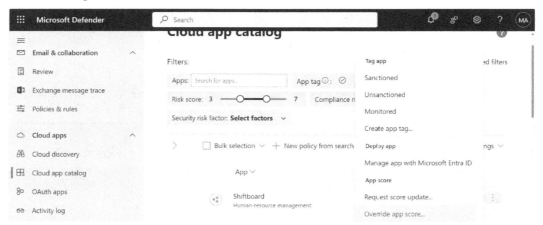

Figure 13.39: Selecting the app's properties

5.   Assign a new risk score by clicking the **Select score...** dropdown and choosing a new numeric value.

Figure 13.40: Overriding the app risk score

6.   If desired, expand the app notes and write a brief update to describe the justification for the change.

7.   Click **Save**.

With the updated risk score, the app now reflects your organization's policy position.

Summary

In this chapter, you learned about MDCA. We reviewed cloud discovery results, both continuous reports and snapshot reports. We then looked at how to analyze the results of the discovery reports. We looked at connected apps and learned how to connect an app to Defender for Cloud Apps. We then reviewed how to improve your app security posture by using application-enforced restrictions as well as using Conditional Access app controls. We then discussed access and session policies and studied how they can be used to secure your apps. We looked at how to monitor OAuth apps for logging and control. Finally, we discussed the cloud app catalog, and how to manage it to customize your risk profile according to your enterprise's unique requirements.

In the next chapter, we'll look at entitlement management with Microsoft Entra.

Exam Readiness Drill – Chapter Review Questions

Apart from mastering key concepts, strong test-taking skills under time pressure are essential for acing your certification exam. That's why developing these abilities early in your learning journey is critical.

Exam readiness drills, using the free online practice resources provided with this book, help you progressively improve your time management and test-taking skills while reinforcing the key concepts you've learned.

HOW TO GET STARTED

- Open the link or scan the QR code at the bottom of this page

- If you have unlocked the practice resources already, log in to your registered account. If you haven't, follow the instructions in *Chapter 19* and come back to this page.

- Once you log in, click the START button to start a quiz

- We recommend attempting a quiz multiple times till you're able to answer most of the questions correctly and well within the time limit.

- You can use the following practice template to help you plan your attempts:

Working On Accuracy		
Attempt	Target	Time Limit
Attempt 1	40% or more	Till the timer runs out
Attempt 2	60% or more	Till the timer runs out
Attempt 3	75% or more	Till the timer runs out
Working On Timing		
Attempt 4	75% or more	1 minute before time limit
Attempt 5	75% or more	2 minutes before time limit
Attempt 6	75% or more	3 minutes before time limit

The above drill is just an example. Design your drills based on your own goals and make the most out of the online quizzes accompanying this book.

First time accessing the online resources? 🔓

You'll need to unlock them through a one-time process. **Head to *Chapter 19* for instructions.**

Open Quiz	
https://packt.link/sc300ch13	
OR scan this QR code →	

14

Planning and Implementing Entitlement Management

Whether you're working with people at an external business partner or your own internal teams, automating the process of granting and removing access to things such as teams, sites, groups, and apps is important to ensure you maintain a least-privilege access control stance while also enabling people to have access to the resources they need to do their jobs.

The objectives and skills we'll cover in this chapter include the following:

- Planning entitlements
- Creating and configuring catalogs
- Creating and configuring access packages
- Managing access requests
- Implementing and managing terms of use
- Managing the lifecycle of external users
- Configuring and managing connected organizations

By the end of this chapter, you should understand the Entra entitlement management solution and be able to plan and implement an entitlement design.

Planning Entitlements

So far, you've been involved in planning and implementing security, identity, and access across an organization's tenants. This has included managing access for both internal members and external users or guests. When adding a user, whether they're internal or external, it's essential to ensure they have the appropriate permissions to access resources. Entitlement management is both a technical and business process that can be used to help ensure users have access to the resources they need to complete their job duties.

Before we start configuring things, let's take a look at some of the terminology that we'll be using throughout this chapter:

- **Entitlement management**: The process of determining and managing who has access to what resources in your tenant or organization.

- **Resources**: Individual objects in the environment used to identify groups, applications, teams, or sites. Any group, application, team, or site made available through entitlement management is referred to as a resource.

- **Access package**: A specific set of settings, resources, and policies used to grant and administer access for a particular requirement. For example, an access package might have a Team, a configuration specifying which users can request access, approval settings, and lifecycle settings that govern how long users will retain access to the resources once access is granted.

- **Policy**: A set of rules used to govern an access package.

- **Catalog**: A container of resources and access packages used to logically group things together.

- **Assignment**: The process by which a user is granted the ability to consume the resources in an access package.

- **Requestor**: The individual requesting access to a resource.

- **Approver**: The individual granting access to a resource.

- **Connected organization**: Another Microsoft 365 or Entra-based organization that you have established a relationship with. Users from connected organizations will be able to request access to resources.

- **My Access portal**: A web portal that lists both the resources you have access to as well as an interface for requesting access to additional resources.

The access packages created define which resources a user or group is authorized to use, providing clear governance over their access once authenticated to the tenant. Entitlement management facilitates this governance by allowing you to create catalogs and access packages tailored to specific groups of users.

> **Tip**
> Entitlement management is a premium feature and requires Entra ID Premium P2 licenses for users and groups.

Figure 14.1 depicts a high-level overview of the Entra entitlement management process.

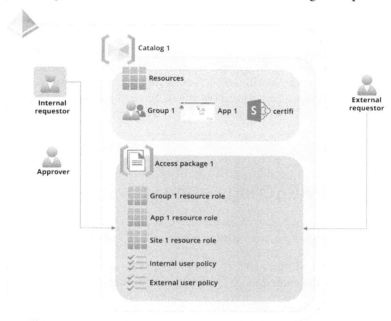

Figure 14.1: Entitlement management

The entitlement management process allows users (either internal or external) to request access to resources. Depending on the configuration, it can be sent to an approver for review and become part of the entitlement management lifecycle, where either the requestor or approver can attest to the requestor's continued access to the resources.

As previously mentioned, catalogs are collections of users, groups, SaaS applications (such as Salesforce, Workday, and ServiceNow), enterprise applications, and SharePoint sites.

Users and groups are added to catalogs to access the applications and sites within them. These users and groups can include both internal and external members. The flexibility to include both types of users in catalogs supports entitlements for project-based access or for specific uses, such as branch offices or departmental assignments.

SharePoint sites can also be included in a catalog. These might be internally created project-based sites, file share sites on SharePoint, or any SharePoint URL that needs to be assigned to the catalog. Multiple sites can be added to a single catalog.

> **Exam tip**
>
> While Global Administrators and Identity Governance Administrators manage all aspects of entitlement management, there may be scenarios (especially in large organizations) where they may be too far removed from the business requirements to understand what resources need to be managed together or how long a user might need access to a resource. To help in these scenarios, business users can be delegated the **Catalog creators** role to create and manage catalogs and access packages. For more information, visit `https://learn.microsoft.com/en-us/entra/id-governance/entitlement-management-delegate`.

Next, we'll look at managing catalogs.

Creating and Configuring Catalogs

After planning the necessary groups, apps, and resources for the entitlements, it's time to start building catalogs and access packages. First, we'll create a catalog:

1. Navigate to the Entra admin center (`https://entra.microsoft.com`). Under **Identity**, expand **Identity Governance** and select **Entitlement management**.

2. Under **Entitlement management**, select **Catalogs**.

Figure 14.2: Preparing to create a new catalog

3. Select **New catalog**.

4. On the **New catalog** flyout, enter a name and description. Choose whether to set the catalog to **Enabled** and whether to set it as **Enabled for external users**. In this example, select **Yes** for both toggles and select **Create**.

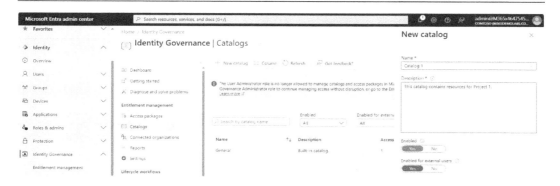

Figure 14.3: Creating a new catalog

5. Select the newly created catalog from the list.

6. Under **Manage**, select **Resources** and then select **Add resources**.

Figure 14.4: Managing resources for a catalog

7. Add the roles, sites, applications, groups, and teams that you want to be part of this particular catalog.

Figure 14.5: Adding resources to the catalog

While access packages can't be moved between catalogs, the same resources can be added to more than one catalog if necessary.

> **Note**
>
> Adding roles to an access package requires a Microsoft Entra ID Governance license.

8. Once you have finished adding resources, select **Add**.

Figure 14.6: Saving the catalog

In addition to resources and roles, a catalog can also contain **custom extensions**. Custom extensions are Logic Apps workflows (similar to Power Automate flows) that can be triggered during different phases of the access package lifecycle.

Now that you've created a catalog that contains the resources you want to manage, you can move on to creating an access package!

Creating and Configuring Access Packages

As discussed in the previous section, catalogs define the groups, teams, applications, SharePoint sites, and roles available to be used within the Identity Governance framework. Creating a catalog does not establish access for end users to these catalogs. You must go through the creation of **access packages** to provide a mechanism for users to request and obtain access to the resources.

To create an access package, follow these steps:

1. Navigate to the Entra admin center (`https://entra.microsoft.com`). Under **Identity**, expand **Identity Governance** and select **Entitlement management**.

2. Under **Entitlement management**, select **Access packages**.

3. Select **New access package**.

Figure 14.7: Creating a new access package

4. On the **Basics** tab, enter a name and description. Select the catalog that you created with your resources and select **Next: Resource roles**.

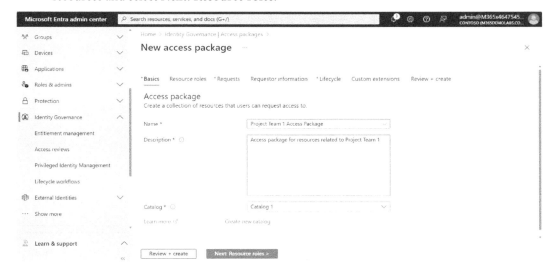

Figure 14.8: Configuring the Basics tab

5. On the **Resource roles** tab, click the **Groups and Teams, Applications, SharePoint sites,** and **Microsoft Entra role** buttons to add the resources that you want to include in this access package. Select **Next**.

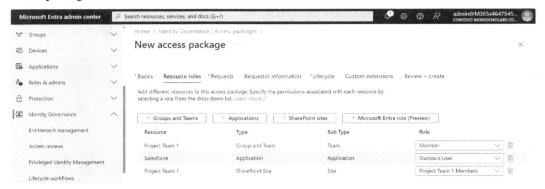

Figure 14.9: Adding resources to the access package

6. On the **Requests** tab, select the scope of users who can work with this access package.

Under **Users who can request access**, you can select **For users in your directory, For users not in your directory**, or **None (administrator direct assignments only)**, which controls the scope of the users:

- **For users in your directory** includes sub-selections. From least restrictive to most restrictive, they are **All users (including guests), All members (excluding guests),** and **Specific users and groups**. The **All users (including guests)** option includes people who already have been invited to your tenant via Entra B2B.

- **For users not in your directory** is used to manage external users that are part of **connected organizations**. You can select **Specified connected organizations** (if you have already set some up), **All configured connected organizations** (includes all current and future connected organizations with the status of **Configured** but does not include connected organizations with the **Proposed** status), and **All users (All connected organizations + any new external users)**.

> **Note**
> Connected organizations are separate from the cross-tenant access settings. They are not interchangeable.

- **None (administrator direct assignment only)** is used to configure an access package that the administrator must assign. End users cannot request access to it.

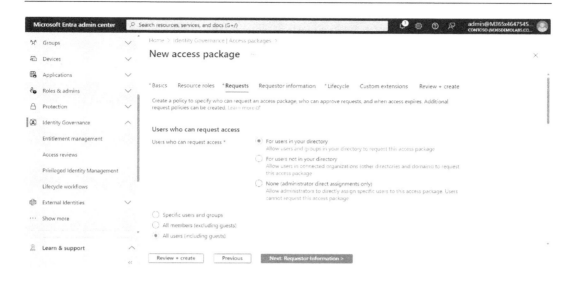

Figure 14.10: Configuring Users who can request access

7. Scroll down to the **Approval** settings section.

8. Select the **Approval** options, such as **Require approval** and the number of stages of approval necessary. You can also specify requirements for both requestor and approver justification and backup approvers if the primary approver doesn't respond within a certain time period.

Figure 14.11: Configuring the Approval settings

9. Scroll down to the **Enable** section.

10. Manage the toggle for **Enable new requests** to allow or prevent this policy from accepting new requests. If you have configured the Entra Verified ID service (which requires a Microsoft Entra ID Governance subscription), you can also enforce further verification of requestors. Select **Next: Requestor Information**.

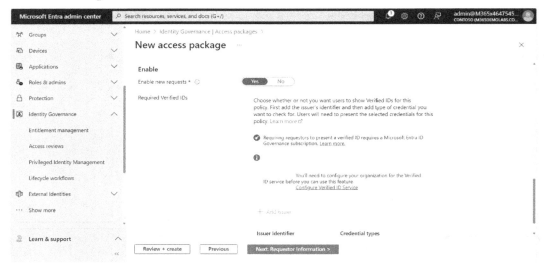

Figure 14.12: Configuring the Enable settings

11. The **Requestor information** page is used to gather any additional information from the requestor that needs to be provided during the process. You can create free-form questions as well as select attributes from the source tenant to be provided. Select **Next** when finished.

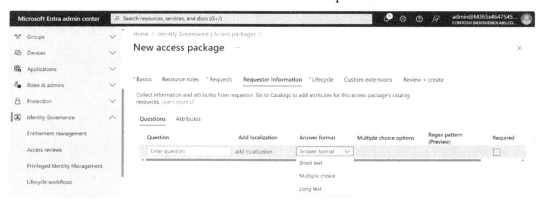

Figure 14.13: Configuring Requestor information

12. On the **Lifecycle** tab, you can select how long the access package assignment will stay active and how the review cycle will run. You can choose options for **Review frequency** as well as who will be responsible for conducting the review (**Self-review** or self-attestation, **Specific reviewers**, or **Manager** (with fallback reviewers). Select **Next: Rules**.

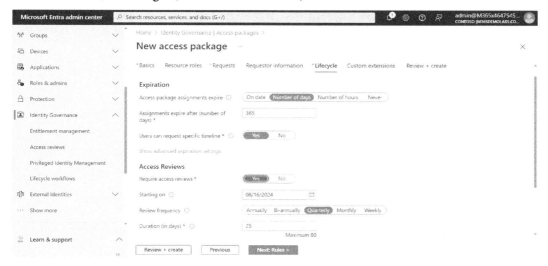

Figure 14.14: Configuring the Lifecycle options

13. If you have a Microsoft Entra ID Governance subscription, you can also include a **Custom extension**—essentially, a Logic App (Azure version of a Power Automate workflow) to execute. Extensions are triggered based on the stage of the access package workflow. You can have rules or extensions configured for each of the access package workflow stages shown in *Figure 14.15*:

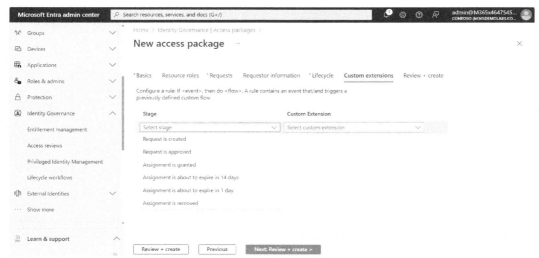

Figure 14.15: Configuring Custom extensions

14. Click **Next: Review + create**.

15. On the **Review + create** page, confirm your settings and select **Create**.

After the package has been created, you can view its settings, including the link to the package, which can be shared with individuals who want to request access.

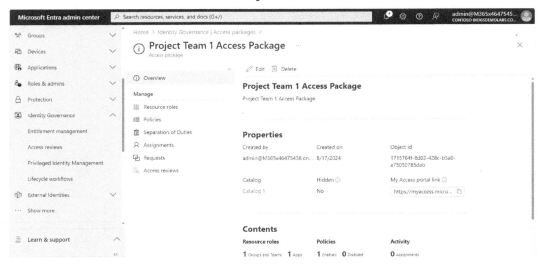

Figure 14.16: Viewing the configured access package

Next, we'll look at working with access requests.

Managing Access Requests

From the user's perspective, working with access packages is fairly straightforward. The user requesting access can select the link provided by the owner of the access package, provide their own Microsoft 365 identity, and request access, as shown in *Figure 14.17*:

Figure 14.17: Requesting an access package as an external user

When an access request is made, the person who was configured to grant the approval will receive an email similar to the one depicted in *Figure 14.18*:

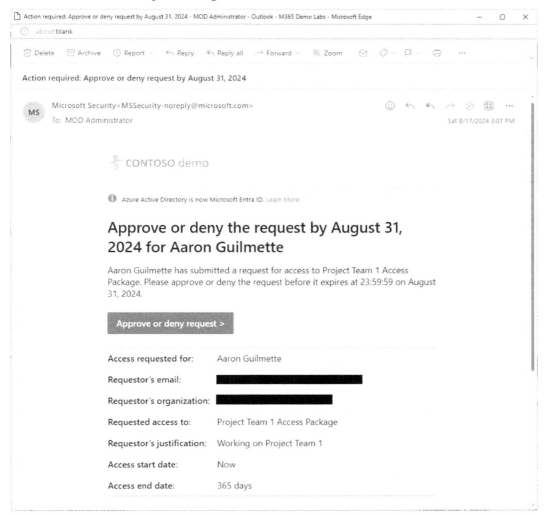

Figure 14.18: Viewing an access package approval request

The approver can click the **Approve or deny request** link, which will redirect them to the **My Access** portal for final approval or denial. See *Figure 14.19*.

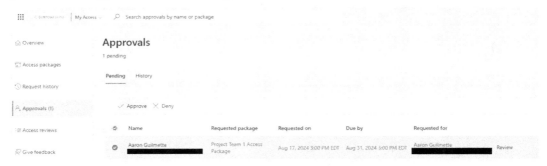

Figure 14.19: Viewing an access package request in My Access

Access requests can also be viewed and processed in the Entra admin center, under **Identity Governance | Access Packages | <Access package name> | Requests**.

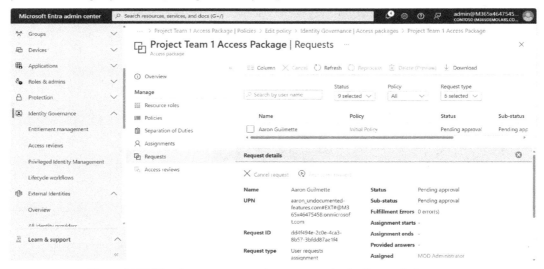

Figure 14.20: Viewing an access package request in the Entra admin center

You cannot accept an access request in the Entra admin center (though you can cancel one). Access requests, at this time, must be approved in the My Access portal by the persons specified in the access package configuration.

Implementing and Managing Terms of Use

If your company requires that users acknowledge terms of use (such as consenting to the monitoring of access or warnings about privileged information) for the applications or sites that are being accessed by members and guest users, Entra Identity Governance allows companies to assign these terms of use and tie them to a Conditional Access policy to allow access to the application.

The terms of use configuration page can be accessed through the Entra admin center under **Identity Governance | Terms of use**.

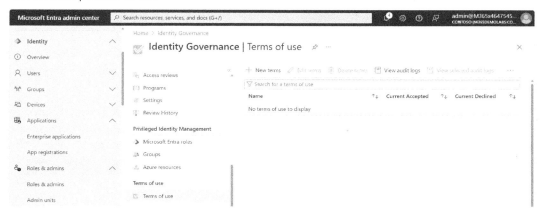

Figure 14.21: Managing the terms of use

When creating new terms of use, you must upload a PDF containing the verbiage that you wish to display to users, as shown in *Figure 14.22*.

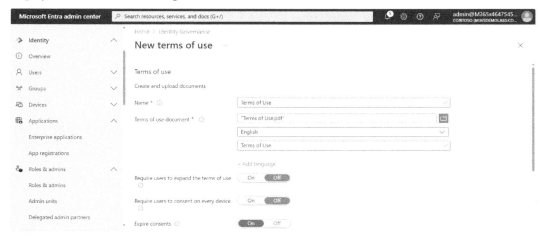

Figure 14.22: Configuring terms of use

The terms of use configuration supports multiple languages. For each language, you'll need to upload a separate document.

You can also require that users open the terms prior to granting consent, requiring consent individually on each device they use, and the frequency for how often they'll need to consent. You can also link the terms of use object to a Conditional Access policy.

Next, we'll switch gears to the lifecycle governance components.

Managing the Lifecycle of External Users

Access packages govern the lifecycle of access to resources for both internal and external users. Disused accounts can be a security risk for an organization since the account may have weak or reused passwords or could potentially be exploited without the original owner being aware. Consequently, managing the lifecycle of those external users is important.

The **Identity Governance** platform includes an option to manage external users. Located on the **Identity Governance | Settings** page, you can use it to control what happens to guest users who had accounts provisioned through an access package once their assignments are removed, as shown in *Figure 14.23*:

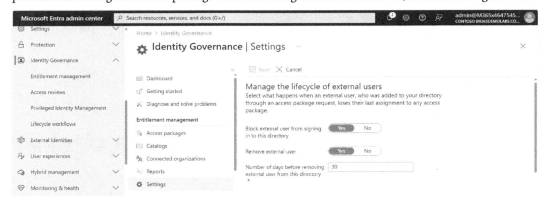

Figure 14.23: Managing the external user lifecycle

Next, we'll look at how you can use connected organizations with entitlement management.

Configuring and Managing Connected Organizations

The term *connected organization* describes a relationship between two organizations that share resources such as applications, teams, or groups.

Connected organization configurations are located in the Entra admin center under **Home | Identity Governance | Connected organizations**.

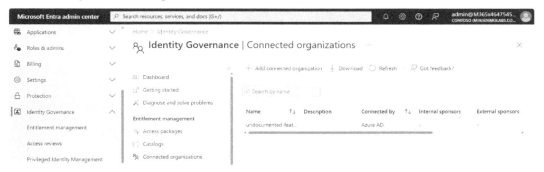

Figure 14.24: Reviewing connected organizations

To create a new connected organization, follow these steps:

1. Navigate to the Entra admin center (`https://entra.microsoft.com`). Expand **Identity**, select **Identity Governance**, and then select **Connected organizations**.

2. On the **Basics** tab, enter a name and a description. For **State**, select either **Configured** or **Proposed**. **Configured** organizations have been either explicitly defined or approved by an administrator, while the **Proposed** status is designed to apply to entries automatically created when an access package is requested by a user in an unconfigured organization.

Figure 14.25: Configuring the Basics tab

3. On the **Directory + domain** tab, enter one of the domains for the remote tenant and select **Add**. Click **Select** when finished.

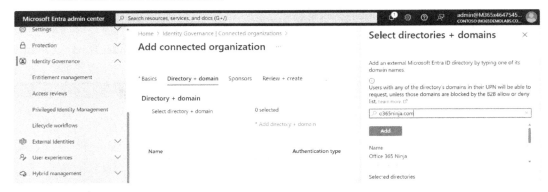

Figure 14.26: Adding a domain

4. Select **Next**.

5. On the **Sponsors** tab, you can select the people who will be the organizational contacts for guest users from this connected organization. Select **Next**.

Figure 14.27: Configuring sponsors

6. On the **Review + create** tab, review the settings and then select **Create**.

After the configuration is complete, the new organization will show up on the **Connected organizations** page. You can select the organization entry to edit the parameters, such as the sponsors (which internal person is ultimately responsible for authorizing access), name, or description.

Summary

This chapter covered the planning and implementation of entitlements, including the concepts of catalogs and access packages. Entitlement management can be used to govern the lifecycle of resource access for both internal and external users. Entitlement management supports the identity pillar of the zero trust architecture by helping ensure individuals have the appropriate access to resources.

In the next chapter, you'll build on the foundation of entitlement management by learning how to implement access reviews.

Exam Readiness Drill – Chapter Review Questions

Apart from mastering key concepts, strong test-taking skills under time pressure are essential for acing your certification exam. That's why developing these abilities early in your learning journey is critical.

Exam readiness drills, using the free online practice resources provided with this book, help you progressively improve your time management and test-taking skills while reinforcing the key concepts you've learned.

HOW TO GET STARTED

- Open the link or scan the QR code at the bottom of this page
- If you have unlocked the practice resources already, log in to your registered account. If you haven't, follow the instructions in *Chapter 19* and come back to this page.
- Once you log in, click the START button to start a quiz
- We recommend attempting a quiz multiple times till you're able to answer most of the questions correctly and well within the time limit.
- You can use the following practice template to help you plan your attempts:

Working On Accuracy		
Attempt	Target	Time Limit
Attempt 1	40% or more	Till the timer runs out
Attempt 2	60% or more	Till the timer runs out
Attempt 3	75% or more	Till the timer runs out
Working On Timing		
Attempt 4	75% or more	1 minute before time limit
Attempt 5	75% or more	2 minutes before time limit
Attempt 6	75% or more	3 minutes before time limit

The above drill is just an example. Design your drills based on your own goals and make the most out of the online quizzes accompanying this book.

First time accessing the online resources? 🔒
You'll need to unlock them through a one-time process. **Head to** *Chapter 19* **for instructions**.

Open Quiz	
https://packt.link/sc300ch14	
OR scan this QR code →	

15

Planning, Implementing, and Managing Access Reviews in Microsoft Entra

In *Chapter 14, Planning and Implementing Entitlement Management,* we covered the process of implementing the entitlement management portion of the Microsoft Identity Governance product. Entitlement management implements a lifecycle component responsible for auditing, maintaining, and updating access grants through a process called **access reviews**.

So far, we've addressed configuring catalogs and resources to cover the lifecycle of accessing resources. A critical part of entitlement management and access governance is ensuring that identities are subject to the same security controls as the resources through the access review process. Access reviews can also stand independently, used to govern identity processes outside the entitlement management system. In this chapter, we'll look at the standalone capabilities of access reviews as they relate to the following SC-300 exam objectives:

- Planning for access reviews
- Creating and configuring access reviews
- Monitoring access review activity
- Manually responding to access review activity

By the end of this chapter, you should be able to configure and manage access reviews, as well as describe the benefits and capabilities of them.

Planning for Access Reviews

The access lifecycle for users should be considered separately for member users and guest users. For member users, the lifecycle is typically tied to a combination of their employment status and the access needs of their department or team as they progress through the employment cycle in the organization. In contrast, guest users are granted access based on external partnerships and collaborations, such as managed services contracts, specific projects, or mergers and acquisitions. These external relationships have defined lifecycles and eventually come to an end, making it essential to implement governance measures to manage these lifecycles effectively.

> **Licensing requirements**
>
> Access reviews are an Entra premium feature and are available as part of Entra ID P2 or through the Entra ID Governance license.

Within Identity Governance, access reviews are used to oversee and manage the lifecycle of access for both member and guest users.

When planning a security strategy, it's important to account for how access to resources and roles will be granted and managed—whether it's through the native group memberships of Teams and sites or through role-assignable groups that are granted administrative privileges. Your access review strategy will likely include a mixture of managing both normal access to sites and teams as well as governance over privileged functions.

Once you have decided what types of resources will be in scope for an access review, you'll also need to make decisions about the frequency of reviews. Your organization may already have established security policies for validating resource assignments; in any case, you should engage with your organization's security team to understand any business requirements prior to implementing a policy.

Since the goal of an access review is determining whether a user still needs access to a resource, you need to decide how to process decisions as well as scenarios where those charged with reviewing fail to complete their reviews in time. You may, for example, decide to remove access for a user if the reviewer fails to respond, which would require the user to re-request access to the resource.

Access reviews should be a key part of an organization's strategy for overseeing the lifecycle of permissions in a tenant.

With planning out of the way, let's get into the practical part of reviewing.

Creating and Configuring Access Reviews

In addition to the access reviews instantiated through an access package configuration (as we covered in *Chapter 14, Planning and Implementing Entitlement Management in Microsoft Entra*), you can also create standalone access reviews to evaluate Microsoft 365 groups, Teams, and applications to ensure your organization's access control requirements are being met.

The configuration settings for an access review will appear familiar, as they are very similar to the review settings configured for an access package.

Configuring an access review requires one of the following roles:

- Global Administrator
- User Administrator (deprecated and will soon no longer work)
- Identity Governance Administrator
- Privileged Role Administrator (for role-assignable group reviews)
- Microsoft 365 group or security group owner (feature in Preview)

To create an access review, use the following steps:

1. Navigate to the Entra admin center (`https://entra.microsoft.com`). Expand **Identity**, select **Identity Governance**, and under the **Access reviews** section, select **Access reviews**.

2. Select **New access review**.

Figure 15.1: Creating a new access review

3. On the **Review type** page, select what type of access review to create—for **Teams + Groups** or **Applications**.

Figure 15.2: Selecting the review type

4. Depending on the type of review selected, you'll be able to select either an application from the configured enterprise application in your tenant or from Microsoft 365 groups and teams.

Figure 15.3: Selecting review options

If you select **Teams + Groups**, you're given options to either select **All Microsoft 365 groups with guest users** or **Select Teams + groups** (which can be any groups you choose in your tenant). There is no option to select all groups in the tenant.

With the **All Microsoft 365 groups with guest users** option, you do have the option to further refine your selection by excluding individual groups as well as filtering on inactive users.

Role-assignable groups

If the group selected is a role-assignable group and has roles assigned to it, all eligible and active users will be included in the review. For more information on **Privileged Identity Management** (**PIM**), see *Chapter 16, Planning and Implementing Privileged Access*.

If you select **Applications**, you can choose from the list of registered enterprise applications and then choose the scope of the access review (either **Guest users only** or **All users** that have access to the application).

In this example, you can choose either option, as the rest of the access review settings are largely the same regardless of the type of access review.

After making your selections, select **Next**.

5. On the **Reviews** page, you can select an option to enable a **Multi-stage review**, which allows you to configure up to three stages of reviewers that can override or continue the approval process.

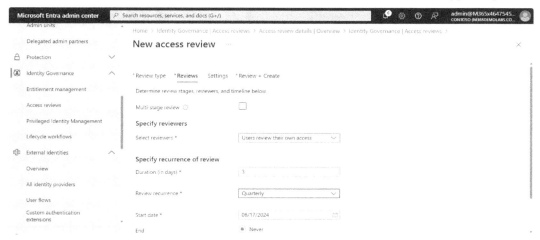

Figure 15.4: Configuring access reviews

Depending on whether you chose **Teams + Groups** or **Applications** as the review type, you will have slightly different options to select for reviewer options.

6. Under **Specify recurrence of review**, you need to select a **Duration (in days)** period, which indicates how long the review period will remain open, a **Review recurrence** (how often the access review will be run) setting, a **Start date** setting, and an **End** date.

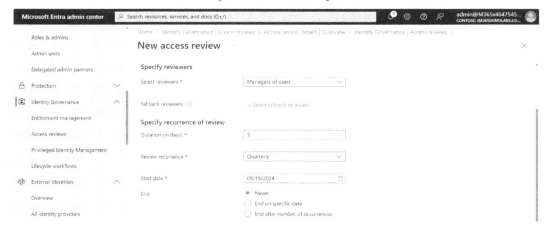

Figure 15.5: Configuring reviewers and recurrence

7. Select **Next**.

8. On the **Settings** page, you can choose how the access review process will handle outcomes, such as auto-applying results to resources. You can also choose options for what to do when reviewers don't follow through on their role in the access review process (such as no change, removing access, approving or extending access, or taking the recommendations that the process recommends for a particular user).

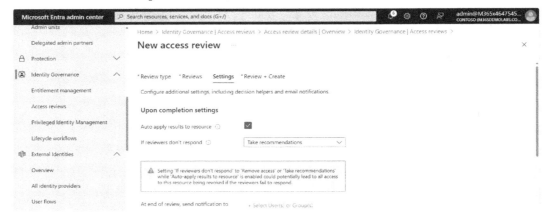

Figure 15.6: Configuring access review settings

Like requests for access packages, access reviews will also generate emails to the assigned reviewers.

9. Scroll to the bottom of the page. Under **Enable reviewer decision helpers**, you can select the **No sign-in within 30 days** checkbox to highlight users who have not signed into the tenant.

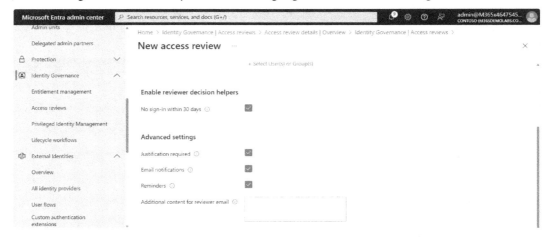

Figure 15.7: Configuring additional options for the access review

Additional options include the following:

- **Justification required**: This requires approvers to provide a reason for their decision. The justification will be recorded in the audit log.

- **Email notifications**: This generates emails for the reviewers when an access review period opens and to the review owner when it is completed.

- **Reminders**: This generates reminder emails to reviewers during the review period for outstanding items.

- **Additional content for reviewer email**: A free-text area to provide additional context for the review process, including comments about a particular project or expanding on the justification.

Select **Next** to continue.

10. On the **Review + Create** page, enter a review name and, optionally, a description. Review the configuration settings and then click **Create**.

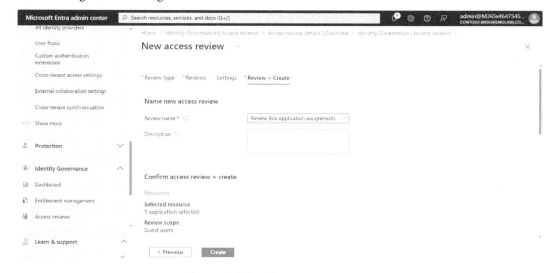

Figure 15.8: Creating an access review

After the access review has been created, you can view the list of configured access reviews on the **Access reviews** page. This page also allows you to update previously configured reviews.

Once you have created and configured your access reviews, you will need to be able to monitor them.

Monitoring Access Review Activity

As part of a governance process, you should periodically monitor the state of your access reviews to ensure they are happening as intended and that those responsible for participating in the review cycle are performing their duties. Monitoring can also help you identify errors in the review cycle, such as if a scenario occurs where an identity to be approved or denied access no longer exists.

The Identity Governance access reviews overview page in the Entra admin center (**Identity Governance | Access reviews | Overview**) can be used to view the current status of access reviews.

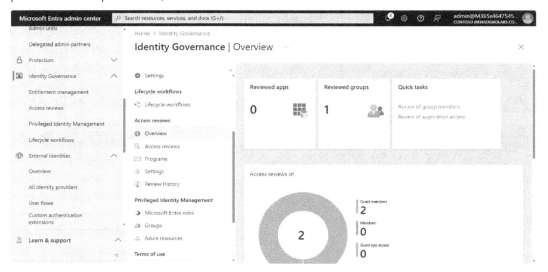

Figure 15.9: Examining the Access reviews dashboard

By selecting the **Access reviews** navigation option, you can see the status of individual access reviews, whether they are upcoming, active, or complete. See *Figure 15.10*.

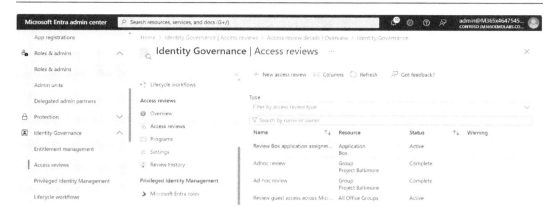

Figure 15.10: Viewing individual access reviews

After activities have started for an access review, you can use the **Results** and **Audit logs** pages of an access review's details to view the actions that have been taken. The results page, as shown in *Figure 15.11*, displays the outcomes of an access review.

Figure 15.11: Reviewing the results of an access review

The results can also be downloaded to a CSV for manipulation in a tool such as Excel or Power BI.

The **Audit logs** entry for an access review shows what actions a reviewer has taken (such as making a single or bulk decision).

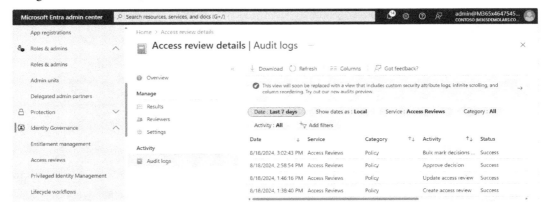

Figure 15.12: Reviewing the audit logs

You can also review the history of all access reviews by running a report. Reports are created on the **Review History** page of Identity Governance, as shown in *Figure 15.13*:

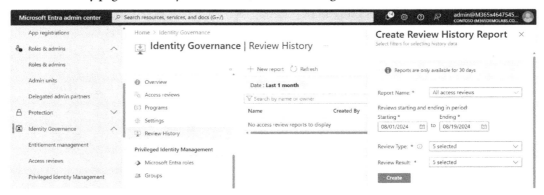

Figure 15.13: Review History

Next, we'll look at responding to access review requests.

Manually Responding to Access Review Activity

When an access review begins, an email similar to the one in *Figure 15.14* is sent to the user or users specified in the access review configuration.

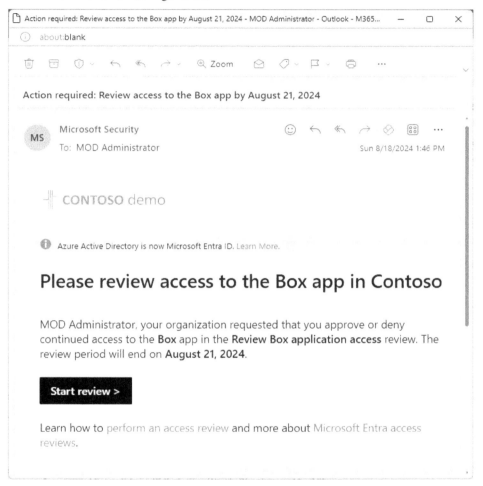

Figure 15.14: An access review email

The reviewer can start the review by clicking on the **Start review** button. Like access package approvals, access reviews are handled through the **My Access** portal.

Figure 15.15: Reviewing the details of an access review

Selecting the **Details** link of a review entry launches a flyout that allows the reviewer to make a decision on the individual item. If the **Enable reviewer decision helpers** option, which is for highlighting users that have been inactive for 30 days, is selected, that information will be present on the flyout as well. See *Figure 15.16*.

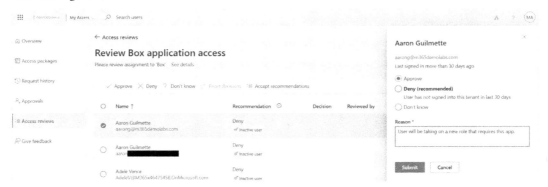

Figure 15.16: Reviewing assignment entry details

After providing justification and selecting an action, clicking **Submit** will save the decision.

You can also bulk-select items to approve or deny. If you have made individual selections, you will be presented with the option to preserve or override the existing decisions.

Figure 15.17: Bulk processing access reviews

Whether you undertake individual or bulk approvals, the results will be available in the Identity Governance reports.

Summary

In this chapter, we covered the standalone access reviews feature, including planning considerations, configuration, and interacting with the access reviews process once it has started.

Access reviews are used to manage the lifecycle of access to applications as well as team and group memberships. Implementing reviews is an integral part of maintaining both a strong security posture as well as compliance in ensuring that users have access to only the resources they need.

In the next chapter, we'll look at the features of PIM.

Exam Readiness Drill – Chapter Review Questions

Apart from mastering key concepts, strong test-taking skills under time pressure are essential for acing your certification exam. That's why developing these abilities early in your learning journey is critical.

Exam readiness drills, using the free online practice resources provided with this book, help you progressively improve your time management and test-taking skills while reinforcing the key concepts you've learned.

HOW TO GET STARTED

- Open the link or scan the QR code at the bottom of this page

- If you have unlocked the practice resources already, log in to your registered account. If you haven't, follow the instructions in *Chapter 19* and come back to this page.

- Once you log in, click the START button to start a quiz

- We recommend attempting a quiz multiple times till you're able to answer most of the questions correctly and well within the time limit.

- You can use the following practice template to help you plan your attempts:

Working On Accuracy		
Attempt	Target	Time Limit
Attempt 1	40% or more	Till the timer runs out
Attempt 2	60% or more	Till the timer runs out
Attempt 3	75% or more	Till the timer runs out
Working On Timing		
Attempt 4	75% or more	1 minute before time limit
Attempt 5	75% or more	2 minutes before time limit
Attempt 6	75% or more	3 minutes before time limit

The above drill is just an example. Design your drills based on your own goals and make the most out of the online quizzes accompanying this book.

First time accessing the online resources? 🔒
You'll need to unlock them through a one-time process. **Head to** *Chapter 19* **for instructions**.

Open Quiz	
https://packt.link/sc300ch15 OR scan this QR code →	

16

Planning and Implementing Privileged Access

Privileged Identity Management (**PIM**) allows organizations to bolster their Zero Trust posture by removing standing privileged access and replacing it with an elevation process. This ensures that the accounts administering resources are granted the necessary permissions and roles when needed and those privileges are revoked when they're no longer required.

The objectives and skills we'll cover in this chapter include the following:

- Planning and managing Microsoft Entra roles in PIM
- Planning and managing Azure resources in PIM
- Planning and configuring groups managed by PIM
- Managing the PIM request and approval process
- Analyzing PIM audit history and reports
- Creating and managing break-glass accounts

By the end of this chapter, you should understand the benefits of privileged identity management, how to configure it, and how to audit the activities related to granting privilege assignments.

Planning and Managing Microsoft Entra Roles

In the previous two chapters, we explored Identity Governance with a focus on user access packages for applications, teams, groups, and SharePoint sites. A critical aspect of identity governance is managing privileged access through administrative user accounts. As organizations continue to add and activate these various administrative roles within their tenants, they inevitably increase the attack surface. Unauthorized access to a compromised account could potentially lead to elevated privileges being exploited.

As identity and access administrators, our responsibility is to protect this layer by implementing the principles of **Zero Trust** and **least privilege** when assigning and managing these administrative accounts. It's essential to develop a clear strategy with defined job tasks for each administrator and role to ensure appropriate role assignments. This strategy should involve meetings with stakeholders (such as security, operations, and service desk management) to discuss the roles each department member needs to perform their job tasks and use that to determine the level of access required. While it may seem convenient to assign everyone to the Global Administrator role, this is not a best practice for securing access to information and resources.

To further secure privileged roles, Microsoft offers PIM as part of its Identity Governance solutions, available with Entra ID Premium P2.

PIM provides **just-in-time** privileged access, meaning users are granted elevated rights only for a limited time after a request workflow rather than being permanently assigned such roles. This reduces the attack surface and the risk of exposing privileged accounts to attacks. PIM includes an approval and justification process for activating privileged roles, along with notifications and an audit trail for these activations.

> **Exam Tip**
>
> PIM supports all Microsoft Entra and Azure roles except the following classic roles: Account Administrator, Service Administrator, and Co-Administrator.

As previously mentioned, PIM requires an Entra ID Premium P2 license. Each user assigned a PIM role must have this license. However, for guest users requiring privileged access through PIM, up to five guests can be assigned PIM roles for every Entra ID Premium P2 license within your tenant.

You can assign multiple groups and users in an assignment. Depending on how your organization is structured, you may choose to implement role-assignable groups, which will contain your users that will be assigned to roles. As part of your management strategy, you should define which roles or resources you want to govern with PIM, as well as the duration of how long access should be granted and any requirements for the request process (such as **multifactor authentication** (**MFA**) or whether requests should be routed for manual approval by another individual).

Next, we'll use assignment strategy information to implement PIM workflows.

Planning and Managing Azure Resources in PIM

Once you have mapped out your administrative strategy (which roles your organization requires, whether assignments will be done directly to the role group or use custom role-assignable groups), it's time to begin configuration.

Adding an Assignment for a Role

In this section, we'll look at creating an assignment for a custom role-assignable group, but the process is the same whether you assign groups or individual users.

To create a PIM assignment, follow these steps:

1. Navigate to the Entra admin center (`https://entra.microsoft.com`). Expand **Identity** and select **Privileged Identity Management**.

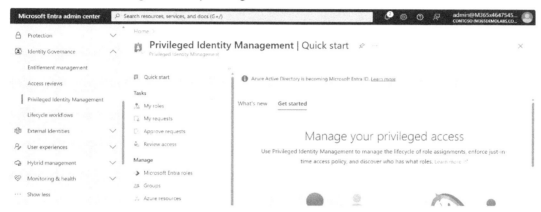

Figure 16.1: Navigating to Privileged Identity Management

2. Under **Manage**, select **Microsoft Entra roles**, and then, under **Manage**, select **Roles**. Click **Add assignments** to create a new PIM role assignment.

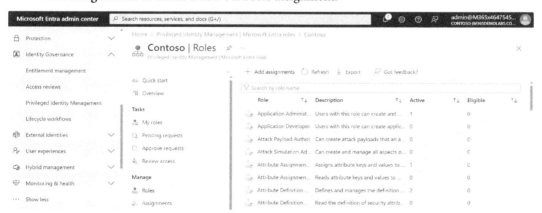

Figure 16.2: Creating a new assignment

3. Under **Select role**, choose the role for which you wish to create an assignment. Under **Scope type**, select the scope (either **Directory** or **Administrative unit**, if the role supports administrative unit scoping).

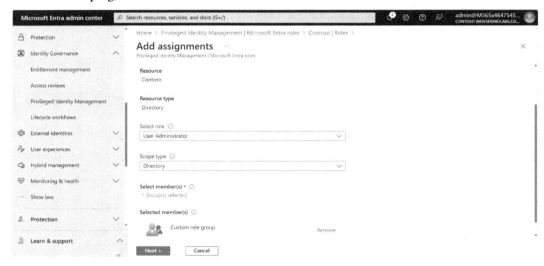

Figure 16.3: Selecting a role and scope

4. Under **Select member(s)**, click **No member selected** and then choose users or groups for assignment. Click **Select** to add them.

Note

In addition to being able to select users or groups, you can also select Enterprise applications for PIM. Since Enterprise apps don't have the ability to interactively assent to MFA challenges, they must be given an active assignment.

---

5. Click **Next**.

6. On the **Settings** tab, choose the required **Assignment type** option:

    - **Eligible** means that users will need to go through the elevation process to gain their rights

    - **Active** means that the role is immediately assigned (just as if the user was added to the role manually)

7. Select whether the assignment is **Permanently eligible** (if the assignment type is set to **Eligible**) or **Permanently assigned** (if the assignment type is set to **Active**).

    If an assignment is not configured to be permanent, choose when the assignment starts and ends. See *Figure 16.4*.

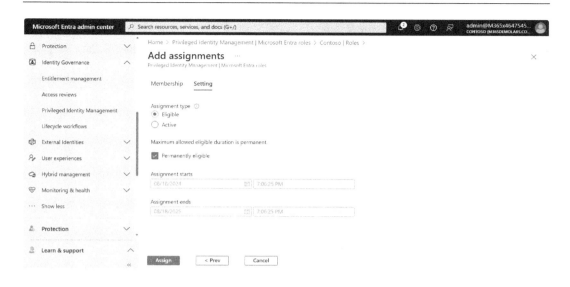

Figure 16.4: Configuring assignment type and duration

> **Best Practices Tip**
>
> To get the most benefit out of PIM (and in alignment with best practices), you should avoid configuring permanent assignments if possible, as it's really no different from just adding the user to the role directly.

8. Click **Assign**.

Next, we'll examine settings for role elevation.

Configuring Settings for Role Elevation

You may need to configure additional approvals, justifications, or other settings for some roles. For example, you may want to configure an additional layer of approval for those elevating to highly privileged roles such as Global Administrator.

If this is the case, you can administer those settings on a per-role basis.

To configure role assignment settings, follow these steps:

1. Navigate to the Entra admin center (`https://entra.microsoft.com`). Expand **Identity | Identity Governance**, and then select **Privileged Identity Management**.
2. Under **Manage**, select **Microsoft Entra roles**.

3. Under **Manage**, select **Settings**.

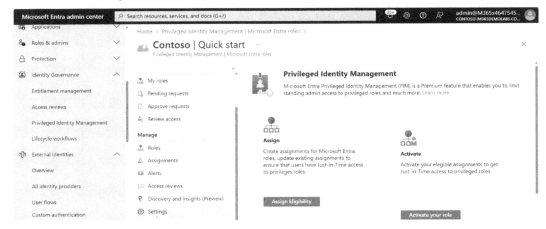

Figure 16.5: Configuring settings for roles

4. Select a role that you wish to modify settings for.

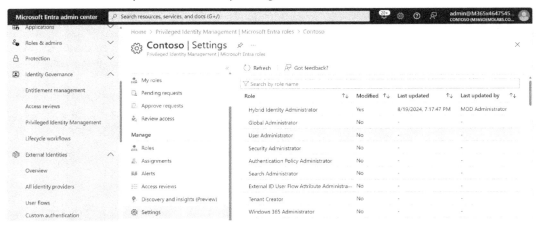

Figure 16.6: Selecting a role to update

5. On the **Role setting details** page, click **Edit**.

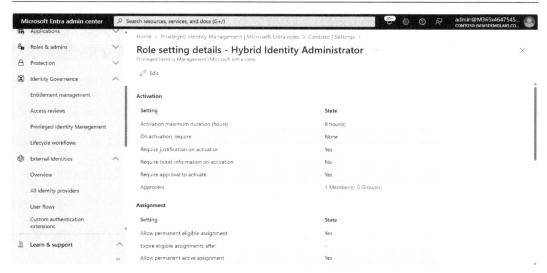

Figure 16.7: Viewing the role setting details

6. Configure the **Activation** settings as necessary, such as the maximum duration, requiring MFA, justification, a ticket, or another approval level, then click **Next: Assignment**.

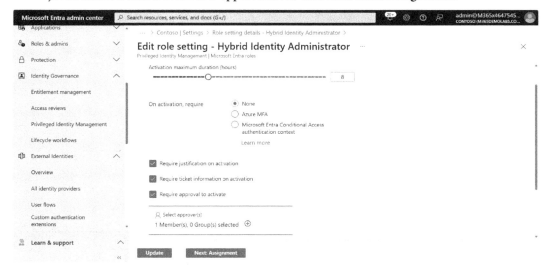

Figure 16.8: Editing the Activation settings

7.   Update the **Assignment** settings and click **Next**.

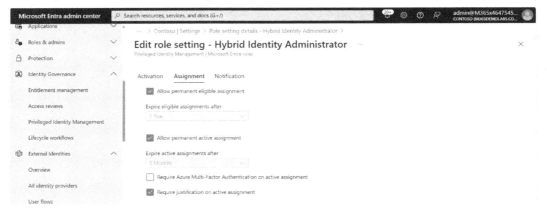

Figure 16.9: Configuring Assignment settings

8.   Edit the **Notification** settings and then click **Update**.

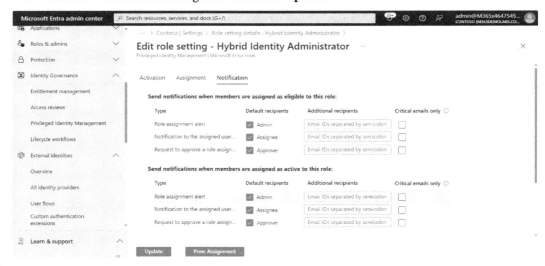

Figure 16.10: Configuring the Notification settings

Repeat the role setting configuration for any other additional roles.

Next, we'll move on to configuring PIM to work with groups.

Planning and Configuring Groups Managed by PIM

In addition to allowing the just-in-time elevation of a user to a particular security role, PIM can also be used to provide just-in-time membership or just-in-time ownership of either a security group or a Microsoft 365 group. These groups can be used to provide access to Microsoft Entra roles or other resources inside the Microsoft 365 environment.

> **Exam Tip**
>
> Only cloud groups with assigned membership can be used with PIM group management. You cannot manage dynamic groups or groups synchronized from the on-premises environment with PIM group management.

You can manage either standard groups or role-assignable groups with PIM group management. From a permissions or role perspective, there are some requirements:

- For role-assignable groups, you must have a Privileged Identity Administrator role or be the owner of the group to add it to PIM.

- For non-role-assignable groups, you need either the Directory Writer, Groups Administrator, Identity Governance Administrator, or User Administrator role (or to be the owner of the group). The PIM rights conferred to the User Administrator role are being deprecated, though they are still valid.

Once a group is brought under PIM, it cannot be removed from PIM—it must be deleted and recreated.

Adding a Group to PIM

To configure groups for PIM, follow these steps:

1. Navigate to the Entra admin center (`https://entra.microsoft.com`). Expand **Identity | Identity Governance**, and then select **Privileged Identity Management**.

2. Under **Manage**, select **Groups**, and then click **Discover groups**.

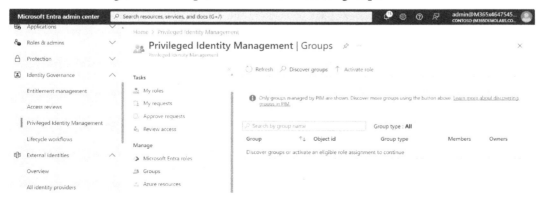

Figure 16.11: Starting a group discovery

3.   Select the groups that you wish to bring under PIM. Notice the groups that are not selectable, such as **Dynamic** groups, **Synced** groups, and **Distribution** groups.

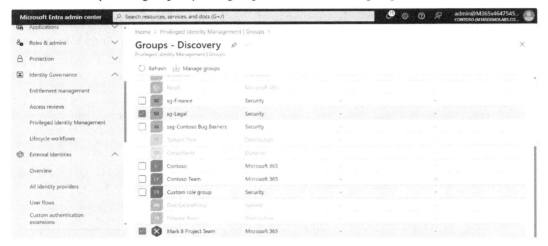

Figure 16.12: Selecting groups to manage

4.   Select **Manage groups**.

5.   In the **Onboard selected groups** window, click **OK**.

Figure 16.13: Confirming groups to bring under management

6.   Close the **Groups - Discovery** page by clicking the **X** button in the window or following the breadcrumb trail back to **Privileged Identity Management**.

7.   Review the groups list to confirm that the groups selected have been added.

Next, we'll look at configuring an assignment for a group.

Adding an Assignment to a Group

After a group has been brought under management for PIM, you can then assign members to be either **Active** or **Eligible** (just like roles). To configure an **Eligible** group member, follow these steps:

1. Navigate to the Entra admin center (`https://entra.microsoft.com`). Expand **Identity | Identity Governance**, and then select **Privileged Identity Management**.

2. Under **Manage**, select **Groups**, and then select a group that you want to configure assignments for. See *Figure 16.14*.

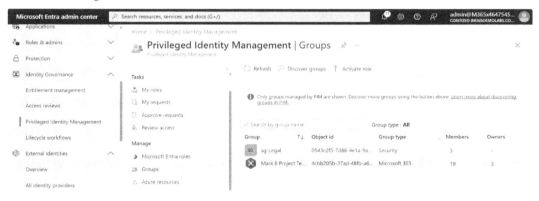

Figure 16.14: Selecting a group

3. Under **Manage**, click **Assignments**.

4. Click **Add assignments**. See *Figure 16.15*.

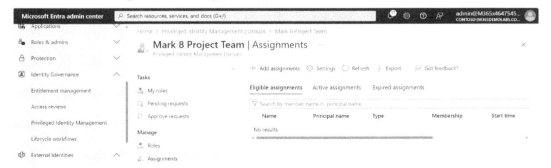

Figure 16.15: Adding an assignment

5.   Under **Select role**, choose the role that you want users who elevate to get, such as **Member** or **Owner**.

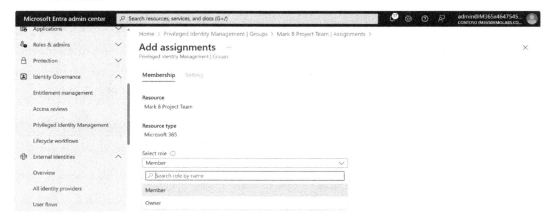

Figure 16.16: Selecting a role assignment

6.   Under **Select member(s)**, click **No member selected**, and then choose the members that you wish to be able to elevate to the selected role for this group.

7.   Once members have been selected, click **Next**.

8.   On the **Setting** tab, choose the required **Assignment type** option, as you did with PIM for roles:

  • **Eligible** means that users will need to go through the elevation process to gain their rights

  • **Active** means that the role is immediately assigned (just as if the user was added to the role manually)

  Choose when the assignment starts and ends. See *Figure 16.17*.

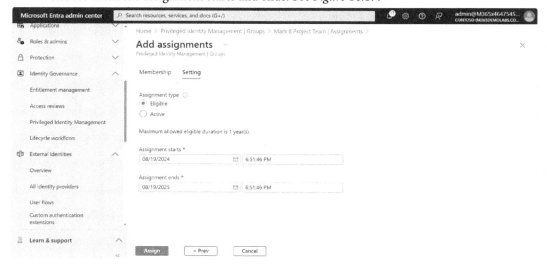

Figure 16.17: Configuring the assignment type

9. Click **Assign**.

You should see the list of eligible members displayed, as shown in *Figure 16.18*.

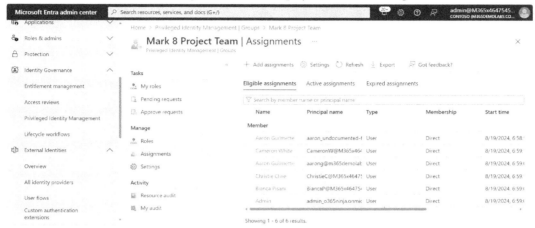

Figure 16.18: Viewing the list of eligible members

Next, we'll look at the process for managing PIM requests and approvals.

Managing the PIM Request and Approval Process

Users who have been assigned to groups or roles can begin their request process by navigating to **Privileged Identity Management**. The easiest location to instruct users to navigate to is `https://aka.ms/pim`, which will display the **My roles** page.

Figure 16.19: The Privileged Identity Management access blade

It can also be accessed through the Entra admin center by navigating to `https://entra.microsoft.com`.

Activating an Assignment

To activate an assignment, follow these steps:

1. Navigate to the **Privileged Identity Management** blade using the PIM short URL: `https://aka.ms/pim`.

2. Expand **Activate** and select **Microsoft Entra roles**.

3. Under the **Eligible assignments** tab, select a role to activate and click **Activate**.

Figure 16.20: Preparing to activate a role

4. Fill out the necessary fields on the **Activate** flyout. If you've modified the required fields by editing the role assignment settings, you'll see those changes here.

Figure 16.21: Requesting the role

If the assignment requires approval, an approval email will be sent. Otherwise, the role will be granted. In the next section, we'll look at the approval process.

Approving an Assignment Request

If an elevation request requires approval, an email will be sent to the approvers configured in the role settings. See the following steps to view and approve a PIM request:

1. Review the PIM email request, as shown in *Figure 16.22*.

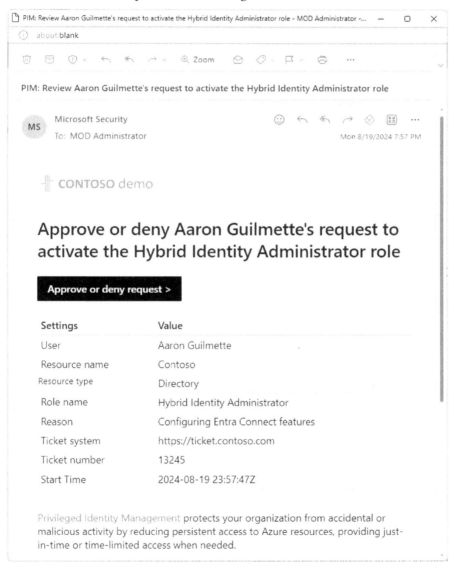

Figure 16.22: Reviewing the PIM request

2. Click **Approve or deny request** to be directed to the PIM blade. Any pending approvals will be listed with the options to approve or deny. See *Figure 16.23*.

Figure 16.23: Viewing the approval requests

3. Click **Approve** or **Deny** to process the request.

4. On the **Approve Request** flyout, enter any justification comments and click **Confirm**.

Figure 16.24: Confirming the request

After the role request has been processed, the requester will be able to see the results in the PIM portal. Next, we'll look at the auditing capabilities of PIM.

Analyzing PIM Audit History and Reports

One of the benefits of utilizing PIM is the ability to audit the use of privileged access. Each use of PIM generates auditable entries that you can review. You can access this history and create your reports from the **Activity** area of Privileged Identity Management.

To view the auditing, follow these steps:

1. Navigate to the Entra admin center (`https://entra.microsoft.com`), expand **Identity Governance**, and select **Privileged Identity Management**.

2. Under **Activity**, select **Resource audit**.

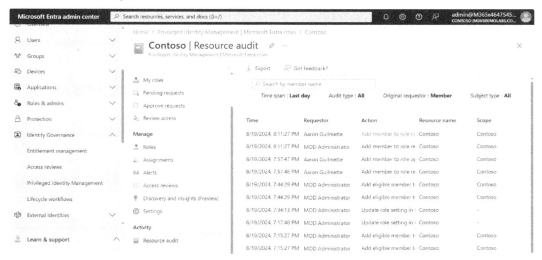

Figure 16.25: Reviewing the audit log

3. To generate a report, select **Export**. The audit events will be compiled into a CSV and made available to download. See *Figure 16.26*.

Figure 16.26: Audit report

The logs are also available in the Microsoft 365 audit logs available in the Microsoft 365 Defender portal (`https://security.microsoft.com/auditlogsearch`) or the Microsoft Purview compliance center (`https://compliance.microsoft.com/auditlogsearch`).

Next, we'll shift gears to the last topic in this exam objective: break-glass accounts.

Creating and Managing Break-Glass Accounts

While it's critical to ensure the security of our identities with features such as MFA, Conditional Access policies, entitlement management, and PIM, it's also crucial to prevent accidental lockout from Entra ID. To safeguard against these scenarios and ensure access during emergencies, you should configure at least two emergency-access accounts, commonly referred to as **break-glass accounts**.

These accounts, which have Global Administrator privileges, allow quick access to resources when other administrator accounts are locked out. They should also be excluded from all Conditional Access policies. The use of these accounts should be strictly limited to emergency situations, and their credentials should be securely stored, such as in a password vault, until needed and then reset after use. A strong authentication method should be enabled, such as a FIDO2 security token.

Break-glass accounts are directly linked to the Entra tenant and should use the initial or tenant domain (`onmicrosoft.com`). They can be used in a number of emergency scenarios, such as when federated identity providers are unavailable or an administrator has misconfigured Conditional Access policies. Other scenarios where these accounts may be necessary include when a Global Administrator loses access to their MFA device, a Global Administrator leaves the company and the account needs to be deleted, or when a storm disrupts cellular services, preventing MFA verification via SMS or text message.

> **Important**
>
> It's important to note that as of July 2024, Microsoft now requires MFA for break-glass accounts. To ensure continued access, you should use an MFA method such as **certificate-based authentication** (**CBA**) or FIDO2 security keys to ensure that a network communication disruption won't prevent you from using the account.

Creating a Break-Glass Account

Break-glass accounts should not be tied to any specific users. Follow these steps to create and secure these accounts:

1. Create a new user account through the Entra admin center (`https://entra.microsoft.com`) or the Microsoft 365 admin center (`https://admin.microsoft.com`), using the initial domain (*tenant.onmicrosoft.com*) as the UPN suffix.

2. Assign the user the **Global Administrator** role. If using PIM, make sure the assignment is set to **Active**.

3. Configure an appropriate strong security method, such as a FIDO2 token.

4. Create a group for the break-glass accounts.

5. Add the break-glass accounts to the group.

6. Exclude the group from each Conditional Access policy.

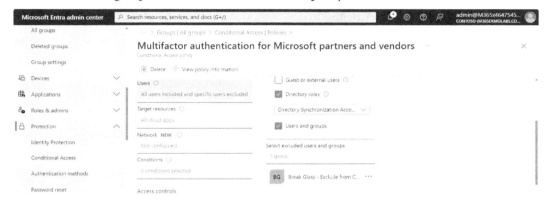

Figure 16.27: Excluding the break-glass group from the Conditional Access policy

Once the accounts have been fully configured, safeguard the credentials in a vault or other safe place.

> **Further Reading**
>
> For more details on emergency-access or break-glass accounts, see `https://learn.microsoft.com/en-us/entra/identity/role-based-access-control/security-emergency-access`.

Monitoring Break-Glass Accounts

Since break-glass accounts should never be used except in cases of emergency, you should configure alerts to notify you when they are used. You can use any system that captures logs for Entra sign-ins.

> **Note**
>
> Log Analytics requires an Azure subscription. If you don't already have an Azure subscription, you will need to activate one.

To do this with Log Analytics and Azure Monitor, follow these steps:

1. Navigate to the Azure portal (`https://portal.azure.com`) and search for **Log Analytics**.

2. Click **Create** to create a new **Log Analytics** workspace.

3. Fill out the details for the **Log Analytics** workspace, including **Resource Group**, **Name**, and **Region**. Select **Review + Create**.

4. Click **Create**.

5. Search for **Microsoft Entra ID**.

6. Under **Monitoring**, select **Sign-in logs** and click **Export Data Settings**.

Figure 16.28: Configuring monitoring settings

7. Click **Add diagnostic setting**.

Figure 16.29: Adding the diagnostic settings

8.   Enter a **Diagnostic Setting name** value.

9.   Under **Logs**, select the categories of logs to send. You'll want at least the **SignInLogs** category, though you may want to choose others for additional monitoring tasks.

10. Under **Destination details**, select **Send to Log Analytics workspace** and select the workspace you created earlier.

Figure 16.30: Configuring the diagnostic logs

11. Click **Save**.

12. In the **Search** box, enter the name of your first break-glass account. Copy the **Object ID** value to a temporary storage location (such as a Notepad window).

Figure 16.31: Copying the object ID

13. Repeat for the other break-glass account.

14. Navigate to **Log Analytics workspaces** and select the new workspace where the identity logs are being sent.

15. Expand **Monitoring** and select **Alerts**.

16. Select **Create | Alert rule**.

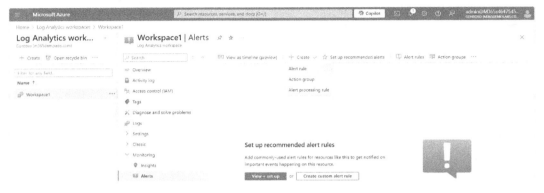

Figure 16.32: Creating a new alert rule

17. On the **Condition** tab, under **Signal name**, select **Custom log search**.

Figure 16.33: Configuring the signal name

18. In the **Search query** text area, enter the query:

```
SigninLogs
| project UserId
| where UserId == "3122d1e4-fc6c-4d48-b6ac-8f284a31abf6" or
UserId == "7e10c929-845d-4acb-a6e3-2cbd335a50c2"
```

Replace the object GUID values with the IDs for your break-glass accounts. See *Figure 16.34*.

Figure 16.34: Configuring a search query

19. Under **Alert logic**, select **Greater than** for the **Operator** field. Enter **0** for the **Threshold** value, and then select a **Frequency of evaluation** value that meets your organization's risk and cost needs (as this will generate an Azure cost to run the query).

Figure 16.35: Configuring the alert logic

20. Click **Next**.

Figure 16.36: Configuring the action

21. Configure the alert actions, such as generating an email to a monitored mailbox. Click **Next**.

22. Edit the **Severity** level to meet your organization's requirements. Add an alert rule name and then click **Review + create**.

Figure 16.37: Configuring the alert rule details

23. Review the settings and click **Create**.

Keep an eye on the monitor to ensure your break-glass account hasn't been compromised.

Summary

In this chapter, you learned about the security features and benefits of PIM, such as removing standing access to privileged roles and implementing a just-in-time access workflow. You also learned how to review the audit logs for PIM and how to create and manage break-glass accounts for emergency situations.

In the next chapter, we'll start exploring how to monitor identity.

Exam Readiness Drill – Chapter Review Questions

Apart from mastering key concepts, strong test-taking skills under time pressure are essential for acing your certification exam. That's why developing these abilities early in your learning journey is critical.

Exam readiness drills, using the free online practice resources provided with this book, help you progressively improve your time management and test-taking skills while reinforcing the key concepts you've learned.

HOW TO GET STARTED

- Open the link or scan the QR code at the bottom of this page

- If you have unlocked the practice resources already, log in to your registered account. If you haven't, follow the instructions in *Chapter 19* and come back to this page.

- Once you log in, click the START button to start a quiz

- We recommend attempting a quiz multiple times till you're able to answer most of the questions correctly and well within the time limit.

- You can use the following practice template to help you plan your attempts:

Working On Accuracy		
Attempt	Target	Time Limit
Attempt 1	40% or more	Till the timer runs out
Attempt 2	60% or more	Till the timer runs out
Attempt 3	75% or more	Till the timer runs out
Working On Timing		
Attempt 4	75% or more	1 minute before time limit
Attempt 5	75% or more	2 minutes before time limit
Attempt 6	75% or more	3 minutes before time limit

The above drill is just an example. Design your drills based on your own goals and make the most out of the online quizzes accompanying this book.

First time accessing the online resources? 🔓
You'll need to unlock them through a one-time process. **Head to** *Chapter 19* **for instructions**.

Open Quiz `https://packt.link/sc300ch16` OR scan this QR code →	

17

Monitoring Identity Activity Using Logs, Workbooks, and Reports

A robust **identity and access management (IAM)** strategy in Azure requires constant monitoring. This chapter will discuss designing a comprehensive strategy for monitoring your Azure environment to guarantee optimal security and observance of compliance regulations. We'll dive into the intricacies of reviewing and analyzing various log types, including sign-in logs that track user login attempts, audit logs that record administrative actions, and provisioning logs that detail user account creation and modification activities. Analyzing these logs will give you valuable insights into user activity patterns and potential security concerns.

We'll also explore configuring diagnostic settings to streamline log collection and analysis. This ensures you have the right data readily available for review. To further empower your analysis, you'll learn how to leverage **Kusto Query Language (KQL)** queries within Log Analytics. KQL allows you to extract specific data from logs and generate insightful reports, enabling you to identify trends and anomalies that might indicate suspicious activity.

We will also gain proficiency in utilizing Entra ID workbooks and reports. These prebuilt tools provide visual representations of identity activity trends, making it easier to identify potential security risks. Finally, we'll explore the **identity secure score** metric, a valuable tool that helps you assess and improve your organization's security posture. By understanding your Identity Secure Score, you can identify areas for improvement and implement targeted security measures to strengthen your organization's defenses.

The objectives and skills we'll cover in this chapter include the following:

- Designing a strategy for monitoring Microsoft Entra
- Reviewing and analyzing sign-in, audit, and provisioning logs
- Configuring Diagnostic Settings

- Monitoring Microsoft Entra by using KQL queries in Log Analytics
- Using logs, workbooks, and reports to monitor identity activity
- Monitoring and improving the security posture by using the Identity Secure Score

By the end of this chapter, you'll be equipped with the knowledge and skills to become an expert in monitoring identity activity within Microsoft Entra, ensuring a more secure and compliant environment for your users.

Designing a Strategy for Monitoring Microsoft Entra

As you've already seen throughout this book, Entra ID secures access to your organization's most critical asset: data. As an identity and access administrator, your role includes both managing the identity components (provisioning and deprovisioning accounts, and configuring conditional access policies) and monitoring the system for potential risks or unauthorized access attempts.

Microsoft provides a variety of tools that can help you achieve this task, from things such as sign-in logs and audit logs to comprehensive threat hunting. To create an effective monitoring strategy for Microsoft Entra, utilize the platform's built-in monitoring tools and integrate them with Azure Monitor logs. You can set up alerts for critical events such as failed login attempts, suspicious activities, and access modifications. You can further enhance your monitoring capabilities by using a **security information and event management (SIEM)** tool such as Microsoft Sentinel for advanced threat analysis and detection.

As you design your strategy, you may find it useful to capture and analyze data related to the following:

- **User activity**: Track logins, unusual behavior, and location-based anomalies
- **Privileged access**: Monitor administrative actions and high-risk access changes
- **Application access**: Ensure secure usage and prevent unauthorized access
- **Identity governance**: Oversee provisioning, deprovisioning, and compliance adherence

Additionally, you should consider establishing robust log retention policies and customizing alert thresholds to align with your organization's security needs and regulatory requirements.

Using Entra ID workbooks and logs to monitor user activities, sign-in patterns, and potential security threats helps maintain a secure environment while regular reviews and audits ensure compliance with security policies and help identify anomalies.

Proactive Monitoring and Alerts

Move beyond reactive security by implementing a continuous monitoring strategy. Set up alerts to notify you of any unusual activity related to user identities and access attempts. The following is a step-by-step guide on how to set up alerts:

1. Sign in to the Azure portal (`https://portal.azure.com`).

2. In the Azure portal, search for and select **Monitor**.

3. Select **Alerts**. Click **Create** and then select **Alert rule**.

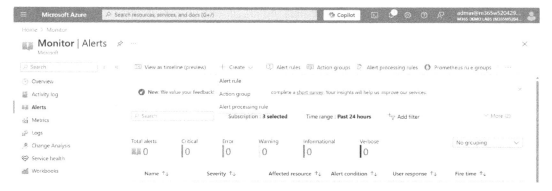

Figure 17.1: Alerts page in Azure Monitor

4. Choose the resources you want to monitor and click **Apply**. This could be a managed identity, a specific application, or another relevant resource that's available in the subscription. In this example, we'll choose a managed identity.

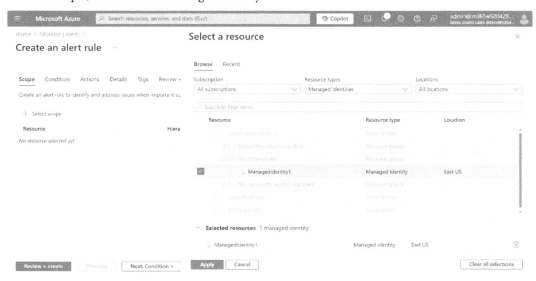

Figure 17.2: Selecting a resource

5. Click **Next**.

6. On the **Condition** tab, select a signal name. The signal is the property that you want to monitor that will trigger the alert.

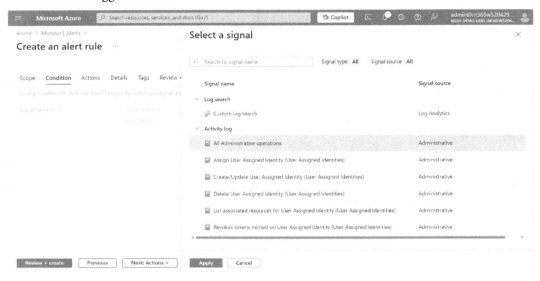

Figure 17.3: Selecting a condition or signal

7. Depending on the type of signal you choose, you may be presented with additional filtering options such as a time span or status. Select the options appropriate for your alert rule and click **Next**.

8. On the **Actions** tab, select the actions to take. You can typically choose to generate an email or app notification in this section (though you can create additional actions later).

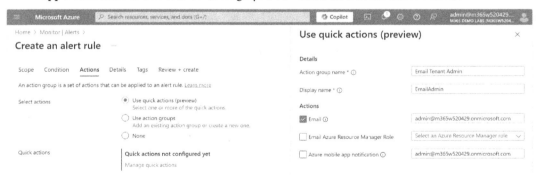

Figure 17.4: Configuring actions

9. Click **Next**.

10. On the **Details** tab, choose a resource group to store the alert rule in. Additionally, you'll need to configure an alert rule name. You can also choose additional *property:value* pairs to include in the email message payload.

11. Click **Review + create**.

After an alert has been created, you can view the alert and configure further automation actions, such as Azure CLI or PowerShell commands, Azure Automation tasks, or additional action groups that can launch other automation, as shown in *Figure 17.5*:

Figure 17.5: Creating an action group with additional automation actions

Azure Monitor alerts are a simple but powerful tool to help stay on top of what's happening in your environment. As you're designing your strategy, you may want to consider investigating various monitors and alerts for components across your Azure environment.

In the following sections, we'll look at some ideas for the types of things to incorporate into your overall monitoring strategy.

User Accounts

Monitoring user accounts is essential for maintaining security and compliance within your organization. Track the creation of new user accounts to ensure they are legitimate and comply with your organization's policies. This helps prevent unauthorized access by ensuring that only authorized personnel can create accounts. Additionally, analyze usage patterns to detect deviations from the norm, such as unusual login times or access to resources with which the user typically does not interact.

Pay close attention to anomalies such as multiple failed sign-in attempts, which could indicate a brute-force attack, and suspicious privilege escalation attempts, where a user tries to gain higher access levels without proper authorization.

Applications and Services

Keeping an eye on applications and services is crucial for identifying unauthorized access and maintaining control over your IT environment. Monitor access logs for critical applications and services to understand who is accessing what and when. This provides valuable insights into potential unauthorized access and helps you take corrective action promptly.

Additionally, identify any unauthorized application registrations to prevent shadow IT, where users register and use applications without the knowledge or approval of the IT department. This practice can expose the organization to security risks and compliance issues.

Devices

Keep track of device registrations to ensure that only authorized devices access your network. You will also need to monitor the usage of registered devices to detect any suspicious behavior, such as multiple devices accessing a single account. This could indicate that credentials have been compromised and used on unauthorized devices, posing a significant security risk.

Infrastructure

Monitor the health of Entra ID Connect to ensure that synchronization between your on-premises AD and Microsoft Entra is functioning correctly. This helps maintain the integrity and consistency of your identity data. Regularly check synchronization logs to identify any issues that might prevent data from being accurately synced between your on-premises and cloud environments.

You should monitor any changes to federation settings, as these can impact how users authenticate and access resources. Unauthorized changes could compromise your security posture. Finally, review audit logs for any administrative actions to ensure that all administrative changes are authorized and comply with your organization's policies, providing a trail for accountability and compliance.

Leveraging Powerful Tools

Azure Monitor is a comprehensive tool that collects telemetry data from various Azure services, including Entra ID. This data encompasses metrics, logs, and traces, providing a holistic view of your environment's performance and health. With Azure Monitor, you can set up custom alerts based on specific conditions, such as unusual sign-in attempts or changes in user roles. These alerts help you stay proactive by notifying you of potential issues before they escalate. Additionally, Azure Monitor integrates with Log Analytics, allowing you to use KQL for advanced querying and analysis. This enables you to create detailed reports and dashboards, helping you gain deeper insights into your data and identify trends or anomalies.

SIEM integration enhances your incident response capabilities by integrating Entra ID logs with your existing SIEM solution, such as Azure Sentinel or Splunk. SIEM solutions aggregate and analyze log data from multiple sources, providing a centralized view of security events. By configuring alert rules and workflows within your SIEM, you can automate the detection and response to security incidents. This includes setting up alerts for specific events, such as failed login attempts or unauthorized access, and defining workflows for incident investigation and remediation. SIEM integration ensures comprehensive security monitoring across your entire IT environment, helping to correlate events from different sources and providing a more accurate and complete picture of potential threats.

Microsoft Defender for Identity (**MDI**), formerly known as **Azure Advanced Threat Protection** (**ATP**), monitors domain activities for suspicious behavior. This includes tracking user activities, analyzing authentication patterns, and identifying potential security threats. The tool is designed to detect suspicious behaviors such as lateral movement, where an attacker moves within the network to gain access to additional resources, or privilege escalation attempts, where an attacker tries to gain higher access levels. When suspicious activities are detected, MDI triggers alerts, allowing security teams to investigate further. These alerts provide detailed information about the detected threat, helping you understand the context and take appropriate action to mitigate the risk.

In the next section, we'll analyze crucial data sources such as sign-in, audit, and provisioning logs to gain deeper insights into your identity activity.

Reviewing and Analyzing Sign-In, Audit, and Provisioning Logs

The SC-300 exam emphasizes your ability to analyze sign-in, audit, and provisioning logs effectively. Sign-in logs provide a detailed record of user attempts to access your Azure and Entra ID environments. These logs reveal valuable information such as the user identity, application accessed, sign-in location, date and time, and success or failure of the attempt. You can filter these logs based on specific criteria by leveraging the Entra ID or Entra ID admin center.

> **Note**
>
> Different licensing levels are required to filter and view logs in Microsoft Entra ID. Basic filtering capabilities are available with a Microsoft Entra ID P1 license, which allows you to access and filter audit logs. However, for more advanced features such as **Privileged Identity Management** (**PIM**) and access reviews, a Microsoft Entra ID P2 license is necessary. These advanced licenses provide enhanced security and management features, ensuring organizations maintain robust identity and access management practices. Understanding these licensing requirements is crucial for leveraging the full potential of Entra ID's filtering capabilities.

Maintaining logs is important for monitoring and securing your Azure and Entra ID environments. The system automatically generates and stores logs, capturing detailed information about various activities. To ensure logs are kept effectively, you should enable diagnostic settings in the Azure portal to capture all relevant logs, including sign-in logs, audit logs, and provisioning logs. Additionally, setting appropriate retention policies is essential to determine how long logs are stored. This ensures that you can access historical data for analysis and compliance purposes, allowing you to review past activities and identify any long-term trends or recurring issues.

Logs capture various information from activities within your Azure and Entra ID environments. Sign-in logs provide a detailed record of user attempts to access your environments, including user identity, application accessed, sign-in location, date and time, and the success or failure of the attempt. Audit logs record changes applied to your tenant, such as user and group management activities and updates to resources. Provisioning logs capture events related to the provisioning of user accounts, detailing the steps taken to provision users and any errors encountered. This comprehensive logging ensures that all critical activities are documented, providing a clear audit trail for security and compliance purposes.

Logs are stored in various locations depending on your configuration. They can be streamed to Azure Monitor, where they can be stored and analyzed using Log Analytics. This allows for advanced querying and visualization of log data. Logs can be stored in an Azure Storage account for long-term retention and archival purposes, ensuring that historical data is preserved for future reference. Logs can also be sent to Event Hubs for real-time streaming to other systems, such as SIEM tools, enabling immediate analysis and response. Logs can be accessed directly within the Microsoft Entra admin center for immediate review and analysis, providing a convenient way to monitor and manage log data.

You can view and filter logs directly in the Microsoft Entra admin center, allowing you to access relevant information and perform basic analysis quickly. For more advanced analysis, Azure Monitor and Log Analytics offer powerful tools for creating custom queries using KQL, enabling you to generate detailed reports and dashboards. Workbook templates in Azure Monitor can be utilized to visualize log data and gain insights into user activities and security events. Additionally, logs can be accessed programmatically through the Microsoft Graph API, allowing integration with other tools and automated analysis. Logs can also be downloaded as CSV or JSON files for offline analysis or integration with third-party tools, providing flexibility in managing and analyzing log data.

You can detect potential security threats by focusing on anomalies such as failed login attempts from unusual locations, repeated logins from a single IP address within a short timeframe, or sign-ins outside of regular work hours. These anomalies might indicate brute-force attacks, compromised credentials, or unauthorized access attempts, allowing you to take proactive measures to mitigate these risks.

When users report login issues, analyzing sign-in logs can help pinpoint the root cause of the problem. By examining these logs, you can identify errors related to **multi-factor authentication** (**MFA**) failures, application permission issues, or violations of Conditional Access policies. This detailed analysis enables you to address users' specific issues, ensuring a smoother and more secure login experience.

Monitoring the adoption of MFA is a critical aspect of maintaining security. By reviewing sign-in logs, you can track which users have registered their MFA devices and identify those who still need to do so. This information helps you prioritize user education and enforce MFA policies more effectively. Ensuring widespread adoption of MFA is a significant step in enhancing overall security by adding an extra layer of protection against unauthorized access.

Audit logs chronicle all administrative actions performed within your Azure and Entra ID environment. These logs capture details such as the administrator who acted, the specific action taken (for example, user creation, permission assignment, or group modification), the target resource, and the date and time. Scrutinizing these logs is important for maintaining organizational accountability and ensuring administrators adhere to established security protocols. By tracking administrator activity through audit logs, you can monitor their actions and verify compliance with security policies. This tracking helps identify potential policy violations or unauthorized configuration changes, allowing you to address these issues promptly and maintain a secure environment.

Audit logs become invaluable when investigating security incidents. In the event of a security breach, these logs provide a detailed, chronological record of all activities, which can help identify the source of the incident and the actions taken by the attacker. This information is important for understanding how the breach occurred and taking swift remediation action to prevent further compromise. By analyzing the audit logs, you can organize the sequence of events and develop a comprehensive response strategy.

Audit logs also play a critical role during compliance audits. They prove your organization's adherence to security regulations and standards. When undergoing a compliance audit, these logs demonstrate your commitment to monitoring user activity and maintaining a secure identity landscape. They show that your organization is proactive in ensuring security and compliance, which can help build trust with stakeholders and regulatory bodies.

Provisioning logs track user account creation, modification, and deletion within your Azure and Entra ID environment. These logs capture details such as the user being provisioned, the source system (for example, HR system or identity provider), the type of provisioning action (create, update, or delete), and the date and time of the action.

Ensuring user lifecycle management by provisioning logs is vital in verifying that user accounts are created, updated, and disabled promptly according to your organization's lifecycle management policies. This ensures no orphaned accounts, which can pose a significant security risk by providing unauthorized access to sensitive information.

These logs help pinpoint the root cause of problems encountered during user provisioning. For instance, you can identify errors related to user attribute mapping, which might occur if there is a mismatch between the attributes defined in your directory and those required by the application. Additionally, provisioning logs can reveal connector configuration problems, such as incorrect settings, connectivity issues, and permission limitations that might prevent successful user provisioning.

Auditing identity provider activity is another critical aspect of provisioning logs. If your organization leverages an identity provider for user provisioning, these logs serve as a comprehensive audit trail of their activities. This audit trail helps ensure the identity provider provides users with your organization's security policies. By reviewing these logs, you can verify that the identity provider is adhering to the agreed-upon standards and promptly addressing any deviations or issues.

By effectively understanding and utilizing provisioning logs, you can enhance your organization's security posture, streamline user management processes, and ensure compliance with internal policies and external regulations.

Remember, the effectiveness of log analysis hinges on a well-defined strategy. Consider integrating your logs with Azure Sentinel or other SIEM tools for advanced threat detection and correlation. Furthermore, leverage KQL to write custom queries for in-depth log analysis and create workbooks within the Entra ID admin center for comprehensive reporting and visualization of security trends.

The next section will cover configuring diagnostic settings specifically tailored for your Azure tenant. This process ensures precise log collection and streamlined analysis, contributing to effective monitoring and security management.

Configuring Diagnostic Settings

Diagnostic settings allow you to route Entra ID logs to various destinations for analysis and long-term storage. This empowers you to gain deeper insights into user sign-in activity, directory service access attempts, and administrative actions within your tenant. This information proves invaluable for troubleshooting security incidents, identifying suspicious behavior, and ensuring compliance with organizational policies.

Here are some examples of destinations where you can route Entra ID logs using diagnostic settings:

- **Log Analytics workspace**: Routing logs to a Log Analytics workspace allows you to use Azure Monitor to analyze and visualize the data. This is useful for creating custom dashboards, setting up alerts, and running complex queries to gain insights into user activities and security events.

- **Azure Storage account**: Storing logs in an Azure Storage account is ideal for long-term retention and archival purposes. This can help you meet compliance requirements by keeping a historical record of all activities. You can also access these logs later for detailed forensic analysis.

- **Event Hub**: Sending logs to an Event Hub enables real-time data streaming to other systems, such as SIEM tools (for example, Splunk or Azure Sentinel). This is beneficial for real-time monitoring and immediate response to security incidents.

- **Azure Monitor**: Integrating logs with Azure Monitor allows you to leverage its full monitoring and alerting capabilities. You can set up automated responses to specific events, ensuring that any suspicious activities are promptly addressed.

- **Third-party tools**: You can route logs to third-party tools for specialized analysis and reporting. For example, you might use tools such as Splunk, Sumo Logic, or QRadar to gain deeper insights and integrate with other security data sources.

In this example, we'll configure Entra ID to send logging data to a Log Analytics workspace:

1. Navigate to the Azure portal (`https://portal.azure.com`) and sign in with your credentials.

2. Search for and select **Microsoft Entra ID** from the available services.

3. In the Entra ID blade, navigate to **Monitoring** and select **Diagnostic settings**, as shown in *Figure 17.6*.

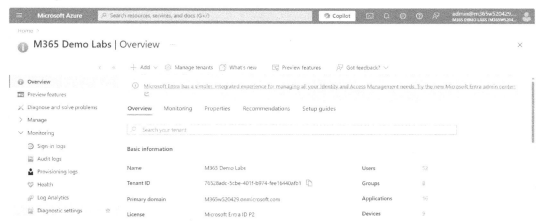

Figure 17.6: Diagnostic settings

4.   Click **Add diagnostic setting** to select the log categories you want to capture.

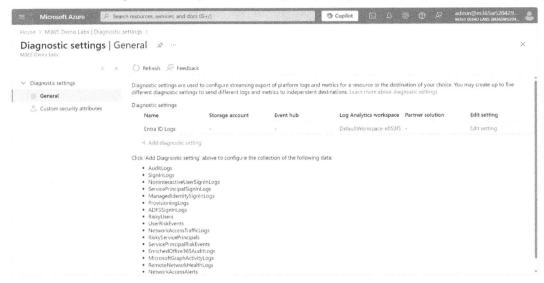

Figure 17.7: Adding diagnostic settings

5.   Select the categories of settings that you want to capture, such as **AuditLogs** and **SignInLogs**. In this example, all settings have been selected:

- **SignInLogs**: Captures details of user sign-in attempts

- **AuditLogs**: Records changes applied to your tenant, such as user and group management activities

- **ProvisioningLogs**: Captures events related to the provisioning of user accounts

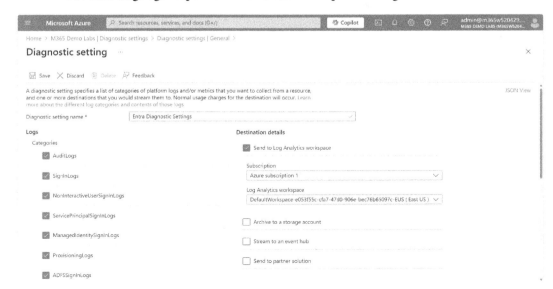

Figure 17.8: Configuring logging categories to send to a Log Analytics workspace

6. Choose where you want to send the logs. In this example, choose **Send to Log Analytics workspace**. You can also select one or more of the following destinations:

    - **Log Analytics workspace**: This is for advanced querying and analysis using KQL. If you choose Log Analytics, select the workspace you want to send the logs to.

    - **Azure Storage account**: For long-term retention and archival purposes. If you choose Azure Storage, select the storage account and configure the retention period.

    - **Event hub**: This is for real-time streaming to other systems, such as SIEM tools. If you choose event hub, select the event hub namespace and the specific event hub to send the logs to.

    - **Partner solution**: This could be a third-party solution available from the Azure marketplace or a custom solution that you or another organization have developed.

7. After configuring the destinations and log categories, click **Save** to apply the diagnostic settings.

Ensure logs are captured and sent to the selected destinations. You can check the Log Analytics workspace, Azure Storage account, or event hub for incoming log data.

Next, we'll take a deeper dive into some of the commonly used logging categories for Entra ID diagnostic settings.

Choosing Log Categories

Entra ID offers a variety of log categories that can be streamed to your chosen destination. These categories include sign-ins, directory service access requests, administrative actions, and PIM activities. It's essential to choose the categories most relevant to your monitoring needs. The SC-300 exam might assess your ability to identify the appropriate log categories for specific scenarios, such as investigating suspicious sign-in attempts.

Use the following table to identify the appropriate log categories for different scenarios:

Category	Scenario	Example
Sign-ins	Investigating suspicious sign-in attempts	This category includes logs of all user sign-in activities. By analyzing these logs, you can detect anomalies such as failed login attempts from unusual locations, repeated logins from a single IP address, or sign-ins outside of regular work hours. These insights help identify potential security threats like brute-force attacks or compromised credentials.

Category	Scenario	Example
Directory service access requests	Monitoring access to directory services	Logs in this category capture requests made to directory services, such as querying user information or accessing group memberships. This is useful for tracking who is accessing sensitive directory data and ensuring that only authorized users have access.
Administrative actions	Ensuring accountability and compliance	This category includes actions performed by administrators, such as creating or deleting user accounts, changing permissions, or modifying group memberships. Monitoring these logs helps ensure administrators adhere to security protocols and allows you to identify unauthorized configuration changes.
PIM activities	Managing and auditing privileged access	PIM logs capture activities related to managing privileged roles, such as role assignments, activations, and approvals. These logs are essential for auditing privileged accounts and ensuring elevated permissions are granted and used appropriately.
Conditional Access	Troubleshooting access issues	Logs in this category provide information about Conditional Access policies and their enforcement. They help you understand why a user was granted or denied access based on the conditions set in your policies. This is particularly useful for troubleshooting login issues related to conditional access.

Table 17.1: Diagnostic settings log categories

It is important to understand diagnostic settings and their configuration for effective Entra ID monitoring and auditing. By strategically routing logs to appropriate destinations, you can gain valuable insights into user activity, identify potential security threats, and maintain a compliant and secure Entra ID environment.

In the next section, we'll troubleshoot the diagnostic settings for Entra ID.

Troubleshooting Diagnostic Settings for Entra ID

In order to effectively use monitoring tools such as Azure Monitor, you need to ensure that the appropriate log sources are configured and validate that data is flowing to the correct destinations.

Validating Diagnostic Settings Configuration

If logs are not appearing in your destination, use the following steps to validate your configuration:

1. Navigate to the Azure portal (`https://portal.azure.com`) and sign in.

2. Search for and select **Entra ID**.

3. Expand **Monitoring** and select **Diagnostic settings**.

4. Select **Edit setting** next to the diagnostic setting entry to validate.

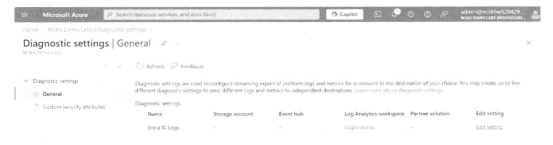

Figure 17.9: Editing diagnostic settings

5. Ensure the necessary log categories are correctly configured. If any required log categories are missing, update the diagnostic settings to include them.

6. Under **Destination details**, ensure the correct destinations are set:

   - For Log Analytics, ensure that the correct Azure subscription and Log Analytics workspace are selected

   - For Storage accounts, ensure that the correct Azure subscription and Storage account are selected

   - For Event Hubs, ensure that the correct Azure subscription, event hub namespace, event hub name, and event hub policy name are selected

7. Click **Save**.

For all destinations, ensure that the identity you are using to configure the diagnostics setting has the appropriate permissions to the resource (such as *Owner* or *Contributor* to subscriptions, Log Analytics workspaces, or Storage accounts). If the destination is a Storage account, you'll need to ensure that you have *Contributor* permissions to use it.

Checking Log Destinations

You can also check the destinations to make sure they are receiving data.

Log Analytics Workspace

To validate data in a workspace, follow these steps:

1. Navigate to your Log Analytics workspace in the Azure portal.

2. Select **Logs**.

3. Use KQL to query the logs and ensure data is ingested correctly. For example, run a basic query to check for recent sign-in attempts:

```
SigninLogs
| where TimeGenerated > ago(1d)
| summarize count() by ResultType
```

Figure 17.10: Running a sample KQL query to check for log data

You can also look for any errors or anomalies in the workspace's activity log that might indicate issues with log ingestion. You can also view recent changes under **Diagnose and solve problems** to see whether a recent change may have impacted your configuration, as shown in *Figure 17.11*:

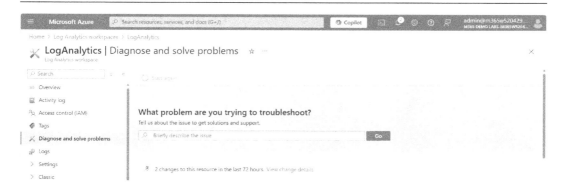

Figure 17.11: Reviewing the Diagnose and solve problems page

Next, let's look at validating a storage account.

Storage Account

Older Entra log data can be exported to a Storage account. Data is stored as blobs in the Storage account and can be viewed using a tool such as Azure Storage Explorer.

Once authenticated to Storage Explorer, follow these steps to locate archived log data:

1. Expand the subscription containing the Storage account.

2. Expand **Storage Accounts**.

3. Select the Storage account containing the archived log data.

4. Expand **Blob Containers** and select the container for the log category you want to view.

Logs are stored in a folder structure corresponding to year/month/day/hour/minute, as shown in *Figure 17.12*:

Figure 17.12: Viewing blob storage in Azure Storage Explorer

The log data itself is stored in a JSON file. You can download it and view it using applications such as Notepad or Visual Studio Code, as shown in *Figure 17.13*:

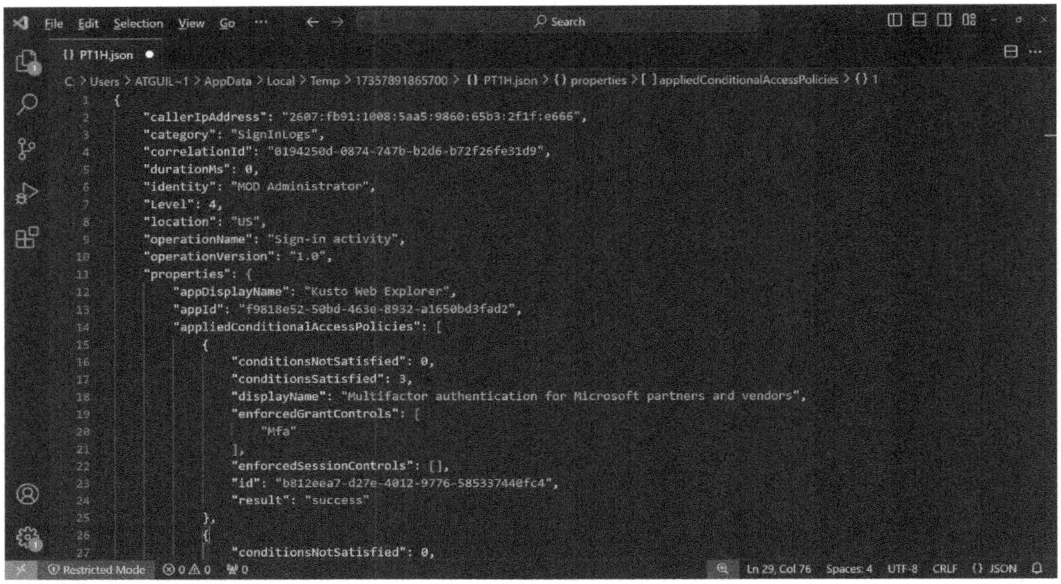

Figure 17.13: Viewing JSON Entra log data

You can use tools such as Visual Studio Code or PowerShell to further search and process the output.

Event Hub

You can check to see that data is flowing into an event hub by navigating to the event hub in the Azure portal and viewing the data on the **Overview** tab.

Figure 17.14: Viewing content in an event hub

The next section will explore utilizing KQL queries within Log Analytics. These queries monitor Microsoft Entra, allowing administrators to extract specific data from logs and generate insightful reports.

Monitoring Microsoft Entra by Using KQL Queries in Log Analytics

Proficiency in KQL queries within Log Analytics is essential for robustly monitoring your Microsoft Entra environment. By utilizing KQL, you gain the ability to analyze extensive Entra ID logs collected via diagnostic settings. These insights are derived from user activity patterns, security posture, and potential threats, enabling proactive security measures.

Understanding Log Schemas

To effectively analyze Entra ID logs within Log Analytics, it is crucial to familiarize yourself with the schema of these logs. This involves understanding the structure and meaning of different log fields. For instance, the **ResultType** field indicates whether a sign-in attempt was successful or failed, providing immediate insight into the outcome of authentication attempts. The **UserPrincipalName** field identifies the user involved in the log entry, essential for tracking individual user activities. The **Location** field shows the origin of the sign-in attempt, helping to identify potentially suspicious access from unusual locations.

> **Further Reading**
>
> For a deeper dive into the schema of logs, see `https://learn.microsoft.com/en-us/entra/identity/monitoring-health/concept-activity-log-schemas`. The Microsoft Learn Azure Monitor documentation also has detailed schema values for sign-in logs (`https://learn.microsoft.com/en-us/azure/azure-monitor/reference/tables/signinlogs`) and activity logs (`https://learn.microsoft.com/en-us/azure/azure-monitor/reference/tables/microsoftgraphactivitylogs`).

The SC-300 exam may present you with log data samples and require you to identify specific fields for analysis, so practicing with sample logs and understanding the significance of each field is vital.

Understanding KQL Syntax

Grasping the fundamentals of KQL is essential for querying and analyzing log data effectively. Start by learning basic KQL operators such as `where`, `project`, and `summarize`, which help you filter and manipulate data. Additionally, familiarize yourself with common functions such as `count`, `avg`, and `percentiles`, which are useful for performing calculations on your data. Time filters are also important, as they allow you to narrow down your queries to specific periods, making it easier to analyze recent activities or historical trends. Practice writing basic queries to filter logs based on specific criteria, such as finding all failed sign-ins within the last hour or identifying users signing in from unusual locations. This foundational knowledge will be crucial for the SC-300 exam.

Once you are comfortable with basic KQL syntax, it is important to deepen your understanding by exploring advanced KQL features. Learn how to use joins to correlate data from different tables, which is useful for combining related information from multiple sources. Practice aggregations to group and summarize data, allowing you to generate comprehensive reports and insights. Experiment with visualizations to present data in charts and graphs, making it easier to interpret and communicate your findings.

The SC-300 exam may assess your ability to combine these techniques to create complex queries for specific scenarios. For example, you might need to join the **Sign-ins** and **Risky Users** tables to identify high-risk users with recent failed sign-in attempts. Developing proficiency in these advanced techniques will enhance your analytical capabilities and prepare you for the exam.

> **Further Reading**
>
> For a deeper exploration of KQL, we recommend reviewing the Microsoft Learn training on KQL: https://learn.microsoft.com/en-us/training/modules/write-first-query-kusto-query-language/.

The following query examples can give you an idea of how KQL can be used to discover and investigate log data.

This quick query searches the `SigninLogs` table in Azure Monitor for all sign-in-related events:

```
SigninLogs | project UserDisplayName,
Identity,UserPrincipalName,  AppDisplayName, AppId,
ResourceDisplayName
```

The following figure shows the results:

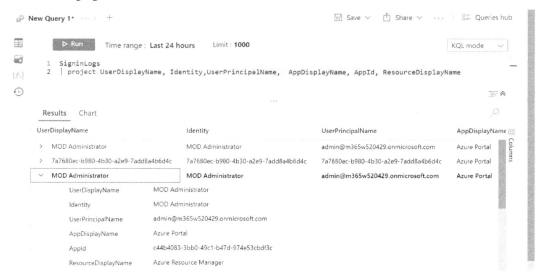

Figure 17.15: Running a KQL query to view all sign-in data

This next query allows you to display the reasons behind failed sign-ins (such as bad passwords):

```
SigninLogs | where ResultType != 0 | summarize Count=count() by
ResultDescription, ResultType | sort by Count desc nulls last
```

Figure 17.16 displays the outcome.

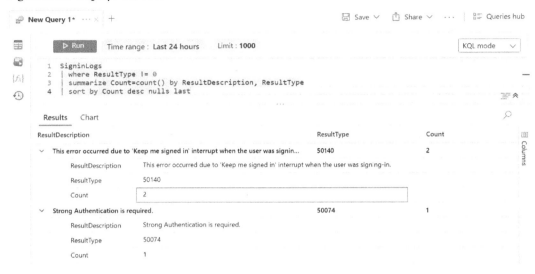

Figure 17.16: Searching log entries for failed logins

> **Further Reading**
>
> You can find more examples of common KQL queries here: `https://learn.microsoft.com/en-us/azure/azure-monitor/reference/queries/signinlogs`.

Focusing on KQL queries relevant to common monitoring scenarios is essential for the SC-300 exam. One key use case is identifying suspicious sign-in attempts. Craft queries to filter for failed sign-ins, focusing on specific locations, user identities, or applications to detect potential security threats. Another important use case is monitoring privileged user activity.

By understanding KQL fundamentals and applying them to relevant Entra monitoring scenarios, you'll gain a significant advantage in the SC-300 exam. Remember, practice is vital! Utilize Microsoft documentation and online resources to test your KQL skills and build confidence in writing effective queries for comprehensive Entra monitoring.

Next, we will explore monitoring identity activity through logs, workbooks, and reports.

Using Logs, Workbooks, and Reports to Monitor Identity Activity

Analyzing Entra ID workbooks and reporting allows you to monitor user activity effectively. By analyzing workbook data, you can track user sign-ins, access patterns, and other activities, providing insight into how users interact with your systems. This capability is essential for detecting anomalies and identifying potential security threats early.

Workbooks can highlight suspicious sign-in attempts, such as those from unusual locations or devices, which might indicate compromised accounts. A clear view of potential risks enables you to take proactive measures to mitigate them, such as enforcing stricter access controls or initiating security investigations.

Another critical aspect is ensuring compliance with organizational policies and regulatory requirements. Workbooks can help verify that user activities adhere to these policies. They also provide detailed logs of user activities, which are essential for audits and compliance checks.

Translating raw data into visual reports and narratives allows you to make data-driven decisions. For example, if you notice a spike in sign-in attempts from a foreign country where your organization doesn't operate, analyzing the workbook data can help determine whether these attempts are legitimate or indicate a potential security breach. This insight enables you to respond appropriately.

Entra ID workbooks provide a visual canvas within Log Analytics for creating interactive reports that combine KQL queries, charts, and text narratives. You can build workbooks to analyze specific security concerns, such as tracking suspicious sign-in attempts across different locations over time.

Microsoft provides prebuilt workbooks for common monitoring scenarios. To identify the appropriate prebuilt workbook for a specific situation, you need to understand the purpose and data presented by each workbook. For instance, the *Risky Users* workbook is ideal if you need to investigate risky user activity. This workbook helps you monitor and analyze users flagged as risky based on their sign-in behavior and other activities. It includes details on risky sign-ins, user risk levels, and remediation actions taken.

You can see a list of prebuilt workbooks by navigating to the Entra admin center (`https://entra.microsoft.com`), expanding **Monitoring & health**, and then selecting **Workbooks**. See *Figure 17.17*.

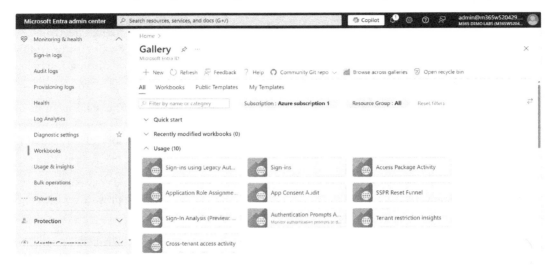

Figure 17.17: Entra monitoring workbooks

Next, we'll look at using one of the prebuilt workbooks.

Using a Prebuilt Workbook

When monitoring sign-in activity, the **Sign-In Analysis** workbook provides valuable insights into user sign-in patterns, helping you identify unusual or suspicious sign-in attempts. This workbook shows sign-in success and failure rates, locations, and device information, making it easier to spot anomalies.

To examine a workbook, follow these steps:

1. Navigate to the Entra admin center (`https://entra.microsoft.com`).
2. Expand **Identity**, expand **Monitoring & health**, and then select **Workbooks**.
3. Under **Usage**, select **Sign-In Analysis**.

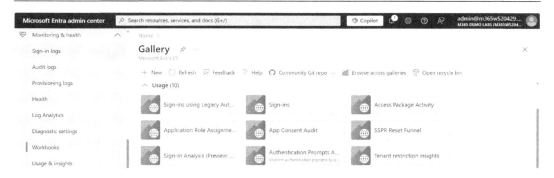

Figure 17.18: Selecting a workbook

4. Review the contents of the workbook.

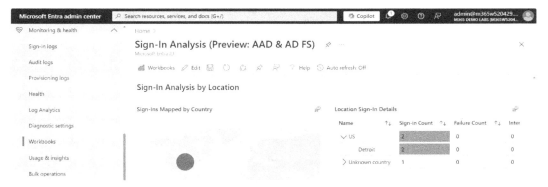

Figure 17.19: Reviewing the workbook

Workbooks can be used to quickly visualize patterns and data throughout your Microsoft 365 environment.

Customizing a Workbook

You can easily customize a workbook or create your own. To customize prebuilt workbooks, follow these steps:

1. Navigate to the Entra admin center (https://entra.microsoft.com).

2. Expand **Identity**, expand **Monitoring & health**, and then select **Workbooks**.

3. Under **Usage**, select a workbook such as **Sign-In Analysis**. Click **Edit**.

4.   Select an item to edit by clicking the button labeled **Edit**.

Figure 17.20: Editing a workbook item

5.   Edit the properties of the query, such as a time range, visualization, or size. When finished, scroll to the bottom of the item you're editing and click **Done Editing**.

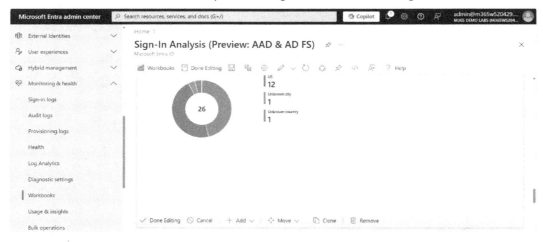

Figure 17.21: Saving changes to a query

At the top of the pane, click **Done Editing** to save changes to your workbook.

Workbooks can be shared with other IAM administrators for collaboration and improved security posture.

You can transform raw log data into valuable insights that strengthen your Entra environment by effectively utilizing Entra ID workbooks and reports. Practice building and customizing workbooks using KQL queries and familiarize yourself with prebuilt offerings. This hands-on approach will equip you to excel in the SC-300 exam and confidently navigate real-world Entra monitoring scenarios.

The next topic we will explore is how to monitor and improve Azure's security posture using the Identity Secure Score.

Monitoring and Improving the Security posture by using the Identity Secure Score

The built-in tools within Entra provide a holistic view of your Entra ID configuration, highlighting potential security weaknesses and recommending remediation steps. Understanding your Identity Secure Score is essential for demonstrating your expertise.

To access the Identity Secure Score dashboard, follow these steps:

1. Sign in to the Microsoft Entra admin center (`https://entra.microsoft.com`) with at least a *Global Reader* role.

2. Expand **Protection** and select **Identity Secure Score**.

Figure 17.22: Identity Secure Score

The Identity Secure Score is calculated automatically. It monitors your Microsoft Entra ID configuration and updates your score as changes are made. This automated process ensures that your score reflects the current state of your security posture, highlighting areas that need attention and providing recommended actions to improve your overall security. The score is calculated based on various security best practices categorized into six core areas:

- **Identity governance**: This assesses the strength of your MFA configuration, privileged access controls, and password policies

- **Endpoint security**: This evaluates the health of devices accessing your Entra ID environment, focusing on device registration and compliance

- **Threat protection**: This analyzes your configuration to detect and prevent suspicious activity, including Conditional Access policies and Entra ID Protection

- **Automation and orchestration**: This assesses your use of automation tools for security tasks such as privileged access management and identity lifecycle management

- **Security awareness and training**: This evaluates your organization's commitment to user security awareness training

- **Baseline configuration**: This measures your adherence to Microsoft's recommended security settings for Entra ID

Identity Secure Score assigns a numerical value (0–100) representing your overall security posture. A higher score indicates a more secure configuration.

Each category within the score provides detailed recommendations for improvement. These recommendations can include enabling specific security features, configuring Conditional Access policies, or deploying MFA. You can select an improvement action to display detailed information, including the item's impact on the score and steps on how to remediate the item.

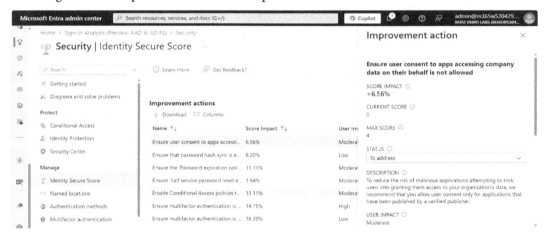

Figure 17.23: Reviewing an improvement action

Identity Secure Score is not a one-time assessment. It continuously monitors your Entra ID configuration and updates your score as changes are made.

You can demonstrate a proactive approach to security within your Entra ID environment by actively monitoring and improving your Identity Secure Score. Familiarize yourself with the different score categories, understand the meaning of recommendations, and practice interpreting sample scores to succeed in the SC-300 exam and maintain a robust security posture in your organization.

Summary

In this chapter, you learned the fundamentals of designing a robust monitoring strategy for your Azure environment, ensuring optimal security and compliance with regulations. You explored various log types, such as sign-in (user login attempts), audit (administrative actions), and provisioning (user account management), to gain valuable insights into user activity and identify potential security concerns.

You configured diagnostic settings to streamline log collection and analysis, ensuring the correct data is readily available for review. You learned how to leverage KQL queries within Log Analytics to empower your analysis further. KQL allows you to extract specific data from logs and generate insightful reports, enabling you to identify trends and anomalies that might indicate suspicious activity.

Additionally, you gained proficiency in utilizing Entra ID workbooks and reports. These prebuilt tools visually represent identity activity trends, simplifying the identification of potential security risks. Finally, you explored the Identity Secure Score metric, a valuable tool that helps you assess and improve your organization's security posture. By understanding your Identity Secure Score, you can identify areas for improvement and implement targeted security measures to strengthen your organization's defenses.

In the next chapter, you will focus on planning and implementing Microsoft Entra Permissions Management. This chapter will guide you through understanding and managing permissions, ensuring your organization adheres to the principle of least privilege and maintains a secure identity environment.

Exam Readiness Drill – Chapter Review Questions

Apart from mastering key concepts, strong test-taking skills under time pressure are essential for acing your certification exam. That's why developing these abilities early in your learning journey is critical.

Exam readiness drills, using the free online practice resources provided with this book, help you progressively improve your time management and test-taking skills while reinforcing the key concepts you've learned.

HOW TO GET STARTED

- Open the link or scan the QR code at the bottom of this page

- If you have unlocked the practice resources already, log in to your registered account. If you haven't, follow the instructions in *Chapter 19* and come back to this page.

- Once you log in, click the START button to start a quiz

- We recommend attempting a quiz multiple times till you're able to answer most of the questions correctly and well within the time limit.

- You can use the following practice template to help you plan your attempts:

Working On Accuracy		
Attempt	Target	Time Limit
Attempt 1	40% or more	Till the timer runs out
Attempt 2	60% or more	Till the timer runs out
Attempt 3	75% or more	Till the timer runs out
Working On Timing		
Attempt 4	75% or more	1 minute before time limit
Attempt 5	75% or more	2 minutes before time limit
Attempt 6	75% or more	3 minutes before time limit

The above drill is just an example. Design your drills based on your own goals and make the most out of the online quizzes accompanying this book.

First time accessing the online resources? 🔒

You'll need to unlock them through a one-time process. **Head to *Chapter 19* for instructions**.

Open Quiz	
https://packt.link/sc300ch17	
OR scan this QR code →	

18

Planning and Implementing Microsoft Entra Permissions Management

Managing access permissions has become a critical component of maintaining security and compliance. However, with the increasing complexity of multi-cloud environments, managing access permissions has become a very complex task. Microsoft **Entra Permissions Management** (**EPM**) is a **cloud infrastructure entitlement management** (**CIEM**) solution designed to provide comprehensive visibility into permissions assigned to all identities across multiple cloud platforms. Microsoft EPM helps organizations effectively secure and manage cloud permissions by detecting, automatically adjusting, and continuously monitoring permissions to ensure the principle of least privilege is consistently applied across all major cloud providers. By doing this, organizations can reduce the attack surface and prevent data breaches. The solution also provides a unified view of access policies across multiple cloud platforms, making it easier to manage and enforce security policies.

The objectives and skills we'll cover in this chapter include the following:

- Onboarding Azure subscriptions to Permissions Management

- Evaluating and remediating risks relating to Azure identities, resources, and tasks

- Evaluating and remediating risks relating to Azure's highly privileged roles

- Evaluating and remediating risks relating to the **Permission Creep Index** (**PCI**)

- Configuring activity alerts and triggers for Azure subscriptions

By the end of this chapter, you should understand where EPM fits into a secure multi-cloud environment. You will also understand how to get started and begin using the tool to benefit from its insights and alerts.

Onboarding Azure Subscriptions to Permissions Management

As organizations continue to adopt Microsoft Azure as their cloud platform of choice, managing access permissions across multiple subscriptions can become a daunting task. Onboarding Azure subscriptions to Permissions Management is an essential step in securing your cloud environment and ensuring that only authorized users have access to sensitive resources. By onboarding Azure subscriptions to Permissions Management, you can gain a unified view of access policies across your entire Azure environment, making it easier to manage and enforce security policies.

Onboarding Azure subscriptions to Permissions Management is important for several reasons. It will improve your security by giving you visibility into access permissions across multiple Azure subscriptions, allowing you to identify over-privileged users, unused permissions, and excessive access to sensitive resources.

It also reduces compliance risks. Onboarding Azure subscriptions to Permissions Management helps ensure that you are meeting regulatory requirements and industry standards for cloud security and compliance.

Finally, by automating access management processes, you can reduce the administrative burden of managing access permissions across multiple Azure subscriptions.

Next, we will walk through the step-by-step process of onboarding Azure subscriptions to Permissions Management, including setting up the necessary configuration, connecting to Azure subscriptions, and verifying successful onboarding.

In order to successfully onboard a subscription, you will need to assign the reader role to the intended subscription so Entra Permissions Management can read the objects. You can use either PowerShell or the Azure CLI to add the necessary role. The syntax for each command is as follows.

Use the following in PowerShell:

```
New-AzRoleAssignment -ApplicationId b46c3ac5-9da6-418f-
a849-0a07a10b3c6c -RoleDefinitionName "Reader" -Scope "/
subscriptions/<subscriptionID>"
```

Use the following in the Azure command line:

```
az role assignment create --assignee b46c3ac5-9da6-418f-a849-
0a07a10b3c6c --role "Reader" --scope /subscriptions/<subscriptionID>
```

Once you've granted permissions for the subscriptions you want to onboard, it's time to configure Permissions Management. Use the following steps to get started:

1. Navigate to the Microsoft Entra admin portal (`https://entra.microsoft.com`).

2. Click on **Permissions Management**.

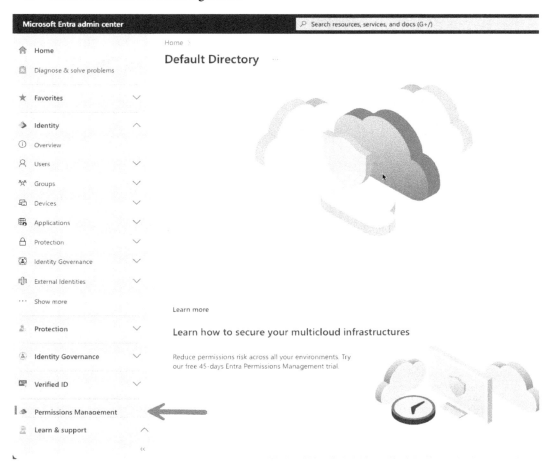

Figure 18.1: Selecting Permissions Management in the Entra admin center

3. You will need to subscribe to EPM, but you can get a free trial as well, which will fully cover performing the tasks in this chapter. Once this is done, please click **Launch portal**.

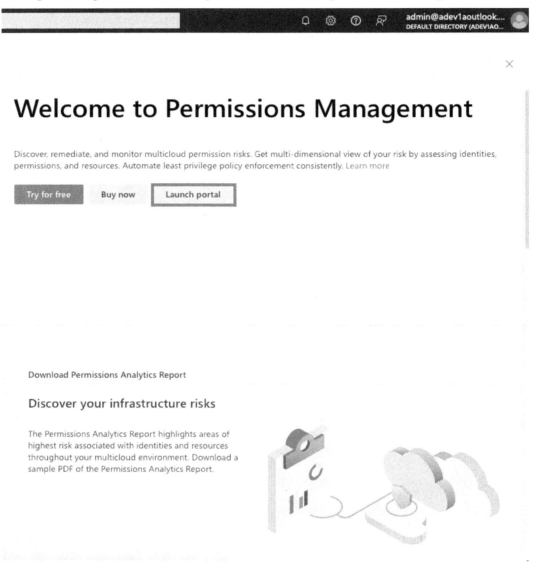

Figure 18.2: Launch the EPM portal

4. Once you launch the portal for the first time, you will be directed to the **Data Collectors** page. On this page, you can choose against which cloud service provider the collector will run. You can choose between **AWS** for Amazon Web Services, **Azure** for Microsoft Azure, or **GCP** for Google Cloud Platform. In our case, we will onboard an Azure subscription. Once you configure a data collector, the tool will begin gathering information from your onboarded subscriptions.

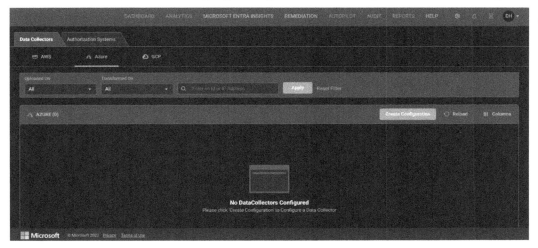

Figure 18.3: Data Collector configuration

5. Click **Create Configuration**.

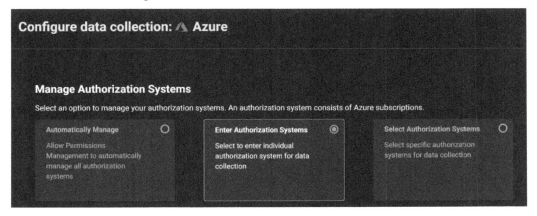

Figure 18.4: Selecting the onboarding method

You have three options when deciding how data collection is going to be handled:

- **Automatically Manage**: Permissions Management automatically discovers, onboards, and monitors all current and future subscriptions.

- **Enter Authorization Systems**: Manually enter individual subscriptions for Permissions Management to discover, onboard, and monitor. You can enter up to 100 subscriptions per data collector.

- **Select Authorization Systems**: Permissions Management automatically discovers all current subscriptions. Once discovered, you select which subscriptions to onboard and monitor.

In this example, we're going to onboard a single subscription to EPM. Choose **Enter Authorization Systems** to continue.

6. Review the configuration. Scroll to the bottom of the page. Select **Verify Now & Save** to start gathering data for the newly onboarded subscription.

Figure 18.5: Confirming a new data collector

Now that you have successfully onboarded your first subscription, we will begin looking at how you can monitor this subscription and use the tools in Entra to identify and remediate risks.

Evaluating and Remediating Risks Relating to Azure Identities, Resources, and Tasks

A subtle yet pervasive threat often evades detection: *permissions creep*. This phenomenon occurs when users accumulate excessive permissions over time, posing significant security risks to organizations. The PCI serves as a vital metric for assessing this vulnerability.

Once you have onboarded a subscription and the tool has begun to collect data, you will be able to monitor the status of your environments using the **DASHBOARD** tab. On this tab, you can view a heat map of your PCI.

The PCI analyzes how permissions change with time, especially those resources that become more privileged, a common occurrence as applications are added and changed. The higher the PCI score, the more risk is being posed by potentially excessive permissions. We will look more closely at the PCI in the last section of this chapter. For now, it is only necessary to understand what it is. Later, you will learn how to use it to remediate common issues related to the PCI risk score.

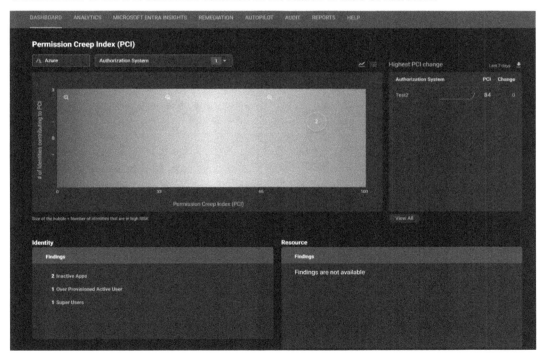

Figure 18.6: The DASHBOARD tab

As you can see, the dashboard also highlights the resource contributing the most to the PCI. Additionally, you can drill down into each finding to gather more information and plan for remediation.

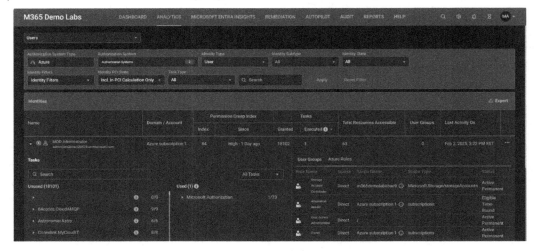

Figure 18.7: The Analytics tab

The analytics tab shows trends and historical changes that have occurred over time. Let's break down this screen and discuss its constituent parts in detail.

Figure 18.8: Filtering the Analytics data

The top pane in this window gives you the ability to filter exactly what you need to see. Here are the fields and their meanings:

- **Authorization System Type**: This will generally be the cloud provider.

- **Authorization System**: For each authorization system (cloud provider), choose which instance you want to filter on. For Azure, this would generally be the subscription or subscriptions that you added. For AWS, it could be the specific account you added.

- **Identity Type**: This value enables you to select the entity that you wish to filter to see permissions for. For Azure, this would be user, managed identity, or application.

- **Identity Subtype**: You can specify **All**, **ED (Enterprise Directory)**, **Local**, or **Cross Account**.

- **Identity State**: You can specify **All**, **Active**, or **Inactive** based on what user or identity type you want to drill down into.

- **Identity Filters**: You can specify **All** or only **High-risk** identities.

- **Identity PCI State**: In this box, you can specify identities included in the PCI calculation.

- **Task Type**: In this box, you can specify all tasks or only high-risk tasks.

Once you specify the filter criteria, you can click **Apply** to execute the filter. Additionally, you can reset the filter to start over.

The next section on this screen lists the data collector and any resources in scope for this query.

Figure 18.9: Filter results

If you expand the data collector, you will see the list of used and unused tasks or rules that have been used to calculate the current risk score.

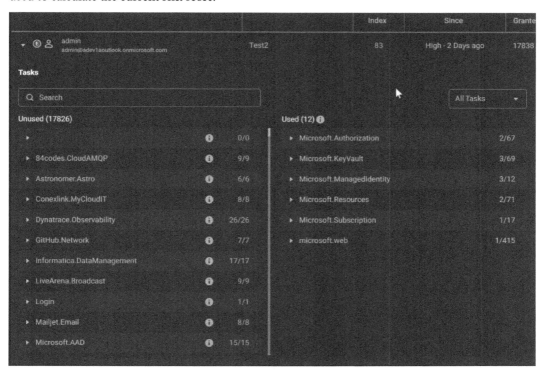

Figure 18.10: Used and unused rules

On this screen, you can review the rules that have and have not been used for this scope. The left side, **Unused**, lists the rules that have not been applied. The right side, **Used**, shows those rules that were used in the current scope. If you drill down into any of the rules listed on the **Used** side, you can see what rule is contributing to PCI.

Figure 18.11: Expanding the data

At this point, you have learned how to onboard a subscription to EPM and identify risks related to identities, applications, and resources. We will now look at risks specifically related to Azure's highly privileged roles.

Evaluating and Remediating Risks Relating to Azure's Highly Privileged Roles

Microsoft EPM performs several checks against deployed resources and will highlight misconfigured and overly permissive roles. One of the most concerning threat vectors is the misuse of highly privileged or administrative roles. These should be tracked and remediated quickly to avoid unnecessary exposure. Some examples of highly privileged roles are as follows:

- **Global Administrator**: Manages all aspects of an organization's Entra ID environment, including user and group management, security settings, and subscription administration

- **Subscription Owner**: Exercises control over a specific Azure subscription, encompassing resource creation, access management, and billing configurations

- **Azure Security Center Contributor**: Holds the authority to configure and manage security policies and threat detection and response across subscriptions

- **Contributor/Owner Roles for Critical Resources**: Users with elevated permissions on sensitive resources such as key vaults, storage accounts, or **Azure Kubernetes Service (AKS)** clusters

While this is not an exhaustive list of all highly privileged roles, it is an example of some highly privileged roles that can be abused. You should always consider the principle of least privilege when deploying roles. This means simply granting the minimum permissions necessary to perform the required activities. It is not uncommon for these roles to be used where they are not needed, either for troubleshooting or because the required privileges are not known and therefore have been applied incorrectly. EPM will factor highly privileged roles into the PCI and the risk score will increase accordingly.

Here's an example of what you will see when this happens:

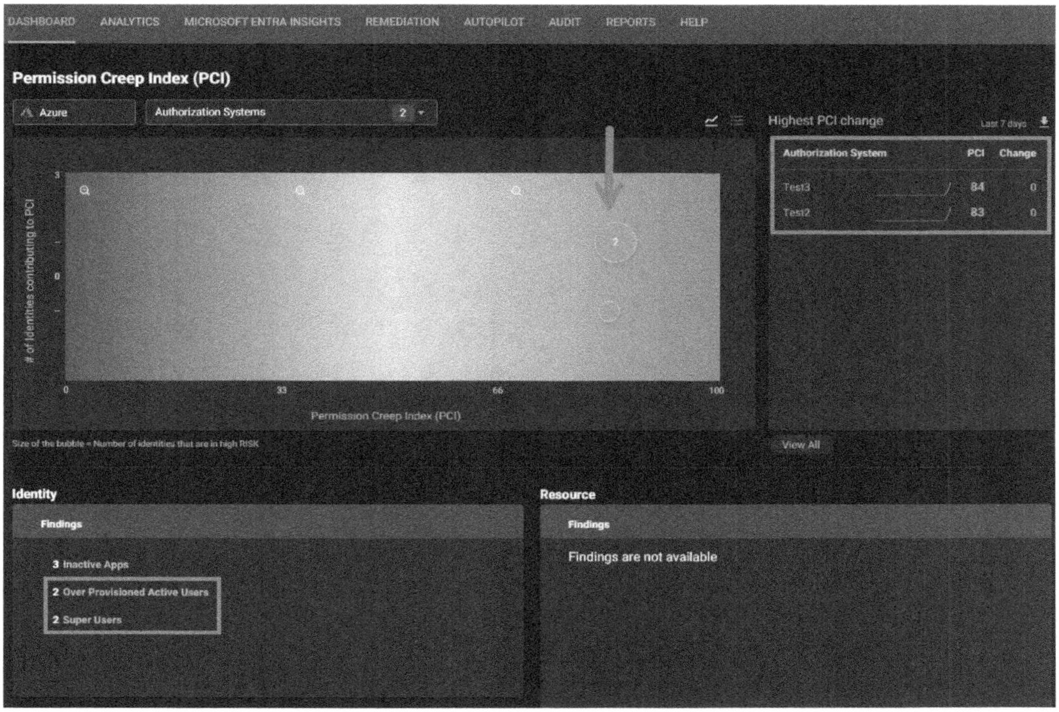

Figure 18.12: EPM dashboard

Figure 18.12 depicts a dashboard (labeled **DASHBOARD**) that provides you with an overview of your current risk status. By reviewing the heat map, you can see bubbles depicting the risky permissions. In the box to the right, you will see the authorization system; in this case, it shows two different subscriptions, listing the overall PCI score and, more importantly, the change in the past seven days. You will want to investigate all recent increases to the PCI immediately, but you should also verify all high-risk subscriptions listed. In the **Findings** box, you can see that your issues include overprovisioned active users and super users.

You should investigate each finding, evaluating the following items:

- Assess whether users possess more privileges than necessary for their roles
- Identify highly privileged accounts without MFA enabled, increasing vulnerability to phishing attacks
- Determine whether sufficient logging and alerting mechanisms are in place to detect suspicious activity
- Evaluate whether a single user holds multiple high-risk roles, amplifying the potential impact of a compromised account

There are several ways to gather more information on the user or role that is generating the risk. The **ANALYTICS** tab, shown in *Figure 18.13*, lists users with highly privileged roles.

Figure 18.13: Identities on the ANALYTICS tab

Based on the evaluation outcomes, implement the following remediation strategies:

- Privilege reduction and **just-in-time (JIT)** access:

  - Revoke unnecessary permissions using Azure's built-in role definitions or custom roles

  - Implement JIT access using privileged identity management for highly privileged roles, ensuring users only possess elevated permissions when required

- MFA enforcement and Conditional Access:

  - Mandate MFA for all highly privileged accounts

  - Configure Conditional Access policies to restrict sign-in attempts from untrusted locations or devices

- Enhanced monitoring and auditing:

  - Enable Azure Security Center's threat protection and advanced threat analytics

  - Set up custom alerts for suspicious activity on highly privileged accounts using Azure Monitor

- Role segregation and user education:

  - Enforce role segregation by assigning users to a single high-risk role whenever possible

  - Provide regular security awareness training for users with highly privileged roles, emphasizing the importance of secure practices

You can view and add roles related to the Microsoft Entra admin center and Entra Permissions Management on the **REMEDIATION** tab.

Figure 18.14: The role remediation tab

Next, we'll see some of Entra's tools for identifying and resolving risks related to permissions.

Evaluating and Remediating Risks Relating to the Permission Creep Index (PCI)

In this section, we will learn how to evaluate and remediate risks associated with the PCI, providing actionable guidance on mitigating "permissions creep" and ensuring least-privilege access in Azure environments.

As discussed in previous sections, the PCI measures the percentage increase in permissions assigned to users or service principals compared to a baseline. A higher PCI indicates a greater accumulation of unnecessary privileges, amplifying the attack surface.

Evaluating Risks

The PCI allows you to assess the risks associated with over-permissioning in the Azure environment. Evaluating risks involves the following steps:

1. Establish a baseline of assigned permissions for users and service principals.
2. Periodically reassess permissions and calculate the PCI to detect deviations from the baseline.
3. Set risk-based thresholds (for example, 10% or 20%) to trigger remediation efforts.

With EPM, these steps are performed automatically and continually updated. The results are readily available in the dashboard and allow you to drill down into the high-risk items to gather more information. The availability of this information enables you to determine the best course of action.

When evaluating risks, focus on the following factors:

- Identify users or service principals with significantly elevated permissions over time
- Determine whether users retain permissions after role changes or project completion
- Assess the frequency and effectiveness of access reviews in identifying and removing excessive permissions
- Evaluate whether users are granted only the necessary permissions for their tasks

While it's important to minimize privilege accumulation, it's also important to balance the role assignments with factors such as the size of the technical administrative staff and the inherent risk based on the organization's industry or datasets. Keep in mind that different types of organizations have different appetites for risk, depending on the type of data being stored and the risks if that data becomes compromised.

Remediating Risks

Based on the evaluation outcomes, you can implement a variety of remediation strategies:

- Privilege reduction and right-sizing:

  - Revoke unnecessary permissions using Azure's built-in role definitions or custom roles
  - Implement JIT access for elevated privileges, ensuring users only possess necessary permissions when required

- Role rotation, revocation, and access reviews:

  - Establish a role rotation policy to regularly review and update user permissions
  - Configure the automated revocation of permissions upon role changes or project completion
  - Schedule regular access reviews (for example, quarterly) to identify and remove excessive permissions

- Least privilege enforcement:

  - Implement **attribute-based access control** (**ABAC**) to grant permissions based on user attributes and resource tags
  - Utilize Azure's built-in roles and custom roles to ensure users have only the necessary permissions

The information provided by Permissions Management gives insight into permissions creep and allows you to make informed decisions about managing permissions-based risks across environments.

Next, we'll look at configuring alerts to warn of permissions-based risks.

Configuring Activity Alerts and Triggers for Azure Subscriptions

The PCI in EPM provides a snapshot of our permissions landscape. However, once we've identified potential vulnerabilities, it's crucial to address two key questions:

- How can we stay informed when issues arise related to our permissions management?

- What strategies can we employ to prevent permissions creep and safeguard our environment?

Alerts offer a powerful solution, enabling the creation of detections and notifications tailored to our specific needs within EPM. There are four primary alert types, each designed to tackle unique aspects of our environment:

- **Operational insights alerts**: These alerts focus on real-time events unfolding within your architecture. By monitoring logged activities that could impact your permissions posture, you can generate targeted alerts based on actual environmental occurrences.

- **Rule-based anomaly detection alerts**: More analytically inclined, these alerts hinge on identifying deviations from the norm. Rule-based alerts are high-precision alerts that identify potential risks based on particular actions occurring over a specified period of time—such as the first time an identity accesses a resource during the time interval or the number of times an identity invokes a specific task during the time frame.

- **Statistical anomaly alerts**: Building further upon Entra's AI and ML capabilities, these alerts analyze historical log data to discern normalized behavior for each identity and cloud service provider. By evaluating the emerging model against this baseline, they can notify you of deviations from the expected normative specifications.

- **Permissions analytics alerts**: Tied directly to the permissions analytics report, these are pre-configured alerts centered around specific categories or types of notifications, providing immediate, actionable insights into your Permissions Management ecosystem.

You can navigate to the alerts configuration page by clicking the alert icon, depicted as a bell, at the top of the page.

Figure 18.15: Accessing alerts

When you enter the **Alerts** area, you will see any alerts that have been triggered and you can configure new alert triggers.

Figure 18.16: The Alerts area

Let's create an activity alert to understand the process better:

1. Navigate to the Entra admin center (`https://entra.microsoft.com`).
2. Select **Permissions Management** and then click **Launch Portal**.
3. Click the alert bell in the top navigation bar.
4. Click **Create Alert Trigger**.

Figure 18.17: Viewing the Alerts tab

5. In the **Create Alert Trigger** dialog box, enter an alert name.

Figure 18.18: Specifying an alert name

6. In the **Authorization System Type** row, click the empty field to select an authorization system type (**Azure**, **GCP**, or **AWS**).

Figure 18.19: Selecting an authorization system type

7. Under **Authorization System**, select an onboarded subscription.

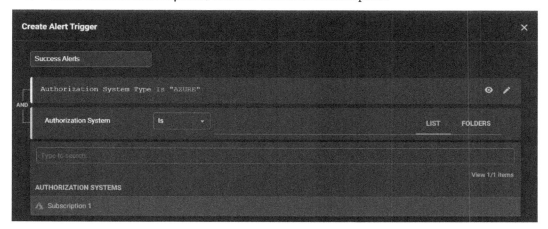

Figure 18.20: Selecting an authorization system

8. Select the type of alert or activity you wish to base the alert on, such as a resource name, resource type, state, task name, or username.

Figure 18.21: Selecting a type of alert

9. Depending on the type of alert selected, you may be prompted to choose from a dropdown or will need to enter your own data. See *Figure 18.22*.

Figure 18.22: Configuring the type values

10. If desired, click **Add** to add another condition to alert on and repeat the steps. Click **Save** when finished.

Once you have configured one or more alert triggers (either activity, rule-based anomaly, or statistical anomaly), you will see alerts in the corresponding **Alerts** activity window as they are generated.

Figure 18.23: Viewing generated alerts

You can look into each area to learn what types of alerts have been generated, including information on what identity or service principal generated the identity, as well as the types of tasks performed or resources acted upon.

Summary

In this chapter, you learned about onboarding subscriptions into Entra Permissions Management. Permissions Management allows you to gather role and permission-based information across your onboarded subscriptions—whether they're Azure, GCP, or AWS.

You then examined how to evaluate and remediate risks related to resources and highly privileged roles. You also learned how to use the PCI to monitor identity and permission-based risks over time.

Finally, you learned about the different types of alerts available (activity, rule-based anomaly, and statistical anomaly) and how to configure them.

And with that, you've reached the end of the book! To complete your preparation, be sure to review the end-of-chapter questions, practice exams, flashcards, and other materials available online at https://www.packtpub.com. Good luck!

Exam Readiness Drill – Chapter Review Questions

Apart from mastering key concepts, strong test-taking skills under time pressure are essential for acing your certification exam. That's why developing these abilities early in your learning journey is critical.

Exam readiness drills, using the free online practice resources provided with this book, help you progressively improve your time management and test-taking skills while reinforcing the key concepts you've learned.

HOW TO GET STARTED

- Open the link or scan the QR code at the bottom of this page

- If you have unlocked the practice resources already, log in to your registered account. If you haven't, follow the instructions in *Chapter 19* and come back to this page.

- Once you log in, click the START button to start a quiz

- We recommend attempting a quiz multiple times till you're able to answer most of the questions correctly and well within the time limit.

- You can use the following practice template to help you plan your attempts:

Attempt	Target	Time Limit
Working On Accuracy		
Attempt 1	40% or more	Till the timer runs out
Attempt 2	60% or more	Till the timer runs out
Attempt 3	75% or more	Till the timer runs out
Working On Timing		
Attempt 4	75% or more	1 minute before time limit
Attempt 5	75% or more	2 minutes before time limit
Attempt 6	75% or more	3 minutes before time limit

The above drill is just an example. Design your drills based on your own goals and make the most out of the online quizzes accompanying this book.

First time accessing the online resources? 🔒

You'll need to unlock them through a one-time process. **Head to** *Chapter 19* **for instructions**.

Open Quiz	
https://packt.link/sc300ch18	
OR scan this QR code →	

19

Accessing the Online Practice Resources

Your copy of *Microsoft Identity and Access Administrator SC-300 Exam Guide, Second Edition* comes with free online practice resources. Use these to hone your exam readiness even further by attempting practice questions on the companion website. The website is user-friendly and can be accessed from mobile, desktop, and tablet devices. It also includes interactive timers for an exam-like experience.

How to Access These Materials

Here's how you can start accessing these resources depending on your source of purchase.

Purchased from Packt Store (packtpub.com)

If you've bought the book from the Packt store (`packtpub.com`) eBook or Print, head to `https://packt.link/sc300unlock`. There, log in using the same Packt account you created or used to purchase the book.

Packt+ Subscription

If you're a *Packt+ subscriber*, you can head over to the same link (`https://packt.link/sc300practice`), log in with your `Packt ID`, and start using the resources. You will have access to them as long as your subscription is active.

If you face any issues accessing your free resources, contact us at `customercare@packt.com`.

Purchased from Amazon and Other Sources

If you've purchased from sources other than the ones mentioned above (like *Amazon*), you'll need to unlock the resources first by entering your unique sign-up code provided in this section. **Unlocking takes less than 10 minutes, can be done from any device, and needs to be done only once**. Follow these five easy steps to complete the process:

STEP 1

Open the link `https://packt.link/sc300unlock` OR scan the following **QR code** (*Figure 19.1*):

Figure 19.1: QR code for the page that lets you unlock this book's free online content

Either of those links will lead to the following page as shown in *Figure 19.2*:

Figure 19.2: Unlock page for the online practice resources

STEP 2

If you already have a Packt account, select the option **Yes, I have an existing Packt account**. If not, select the option **No, I don't have a Packt account**.

If you don't have a Packt account, you'll be prompted to create a new account on the next page. It's free and only takes a minute to create.

Click **Proceed** after selecting one of those options.

STEP 3

After you've created your account or logged in to an existing one, you'll be directed to the following page as shown in *Figure 19.3*.

Make a note of your unique unlock code:

```
LPD1398
```

Type in or copy this code into the text box labeled **Enter Unique Code**:

UNLOCK YOUR PRACTICE RESOURCES

You're about to unlock the free online content that came with your book. Make sure you have your book with you before you start, so that you can access the resources in minutes.

Microsoft Identity and Access Administrator: SC-300 Exam Guide Second Edition

Book   ISBN: 9781836200390

Aaron Guilmette   •   James Hardiman   •   Doug Haven and 1 more   •   Mar 2025   •   821 pages

ENTER YOUR PURCHASE DETAILS

Enter Unique Code *

E.g 123456789   ⑦ Where To Find This?

☐   Check this box to receive emails from us about new features and promotions on our other certification books. You can opt out anytime.

REQUEST ACCESS

Figure 19.3: Enter your unique sign-up code to unlock the resources

> **Troubleshooting tip**
>
> After creating an account, if your connection drops off or you accidentally close the page, you can reopen the page shown in *Figure 19.2* and select **Yes, I have an existing account**. Then, sign in with the account you had created before you closed the page. You'll be redirected to the screen shown in *Figure 19.3*.

STEP 4

> **Note**
>
> You may choose to opt into emails regarding feature updates and offers on our other certification books. We don't spam, and it's easy to opt out at any time.

Click **Request Access**.

STEP 5

If the code you entered is correct, you'll see a button that says **OPEN PRACTICE RESOURCES**, as shown in *Figure 19.4*:

Figure 19.4: Page that shows up after a successful unlock

Click the **OPEN PRACTICE RESOURCES** link to start using your free online content. You'll be redirected to the Dashboard shown in *Figure 19.5*:

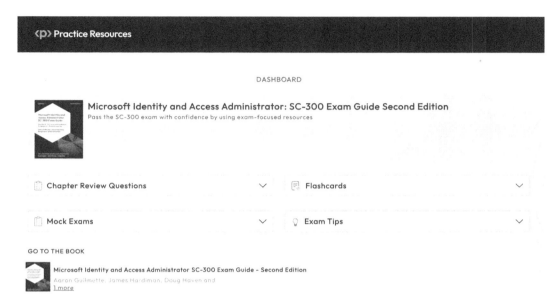

Figure 19.5: Dashboard page for SC-300 practice resources

Bookmark this link

Now that you've unlocked the resources, you can come back to them anytime by visiting `https://packt.link/sc300practice` or scanning the following QR code provided in *Figure 19.6*:

Figure 19.6: QR code to bookmark practice resources website

Troubleshooting Tips

If you're facing issues unlocking, here are three things you can do:

- Double-check your unique code. All unique codes in our books are case-sensitive and your code needs to match exactly as it is shown in *STEP 3*.

- If that doesn't work, use the **Report Issue** button located at the top-right corner of the page.

- If you're not able to open the unlock page at all, write to `customercare@packt.com` and mention the name of the book.

Share Feedback

If you find any issues with the platform, the book, or any of the practice materials, you can click the **Share Feedback** button from any page and reach out to us. If you have any suggestions for improvement, you can share those as well.

Back to the Book

To make switching between the book and practice resources easy, we've added a link that takes you back to the book (*Figure 19.7*). Click it to open your book in Packt's online reader. Your reading position is synced so you can jump right back to where you left off when you last opened the book.

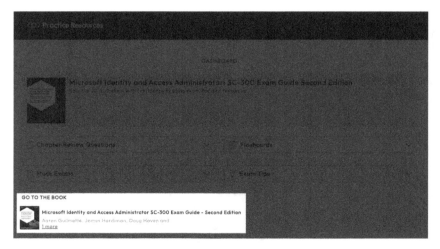

Figure 19.7: Dashboard page for SC-300 practice resources

> **Note**
> Certain elements of the website might change over time and thus may end up looking different from how they are represented in the screenshots of this book.

Index

www.packtpub.com

Subscribe to our online digital library for full access to over 7,000 books and videos, as well as industry leading tools to help you plan your personal development and advance your career. For more information, please visit our website.

Why subscribe?

- Spend less time learning and more time coding with practical eBooks and Videos from over 4,000 industry professionals

- Improve your learning with Skill Plans built especially for you

- Get a free eBook or video every month

- Fully searchable for easy access to vital information

- Copy and paste, print, and bookmark content

At www.packtpub.com, you can also read a collection of free technical articles, sign up for a range of free newsletters, and receive exclusive discounts and offers on Packt books and eBooks.

Other Books You May Enjoy

If you enjoyed this book, you may be interested in these other books by Packt:

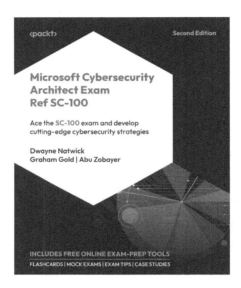

Microsoft Cybersecurity Architect Exam Ref SC-100

Dwayne Natwick, Graham Gold and Abu Zobayer

ISBN: 978-1-83620-851-8

- Design a zero-trust strategy and architecture
- Evaluate GRC technical and security operation strategies
- Apply encryption standards for data protection
- Utilize Microsoft Defender tools to assess and enhance security posture
- Translate business goals into actionable security requirements
- Assess and mitigate security risks using industry benchmarks and threat intelligence
- Optimize security operations using SIEM and SOAR technologies
- Securely manage secrets, keys, and certificates in cloud environments

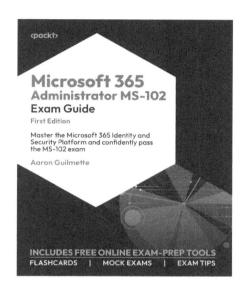

Microsoft 365 Administrator MS-102 Exam Guide

Aaron Guilmette

ISBN: 978-1-83508-396-3

- Implement and manage Microsoft 365 tenants
- Administer users, groups, and contacts in Entra ID
- Configure and manage roles across Microsoft 365 services
- Troubleshoot identity synchronization issues
- Deploy modern authentication methods to enhance security
- Analyze and respond to security incidents using Microsoft 365 Defender
- Implement retention policies and sensitivity labels
- Establish data loss prevention for enhanced information protection

Share Your Thoughts

Now you've finished *Microsoft Identity and Access Administrator SC-300 Exam Guide, Second Edition*, we'd love to hear your thoughts! Scan the QR code below to go straight to the Amazon review page for this book and share your feedback or leave a review on the site that you purchased it from.

https://packt.link/r/1836200390

Your review is important to us and the tech community and will help us make sure we're delivering excellent quality content.

Download a Free PDF Copy of This Book

Thanks for purchasing this book!

Do you like to read on the go but are unable to carry your print books everywhere?

Is your eBook purchase not compatible with the device of your choice?

Don't worry, now with every Packt book you get a DRM-free PDF version of that book at no cost.

Read anywhere, any place, on any device. Search, copy, and paste code from your favorite technical books directly into your application.

The perks don't stop there, you can get exclusive access to discounts, newsletters, and great free content in your inbox daily.

Follow these simple steps to get the benefits:

1. Scan the QR code or visit the link below:

https://packt.link/free-ebook/9781836200390

2. Submit your proof of purchase.

3. That's it! We'll send your free PDF and other benefits to your email directly.

www.ingramcontent.com/pod-product-compliance
Lightning Source LLC
LaVergne TN
LVHW080109070326

832902LV00015B/2493